MEI

Foundations of Advanced Mathematics

DIANA COWEY
DAVE FAULKNER
ROGER PORKESS
DAVID SNELL

Hodder Murray
A MEMBER OF THE HODDER HEADLINE GROUP

Acknowledgements

The authors and publishers would like to thank the following companies, institutions and individuals who have given permission to reproduce copyright material. The publishers will be happy to make suitable arrangements with any copyright holder whom it has not been possible to contact.

Mike Agliolo/Science Photo Library page 167 bottom left
Camelot/The National Lottery page 128
Earth Satellite Corporation/Science Photo Library page 95
Adrienne Hart-Davis/Science Photo Library page 138 bottom right
Peter Menzel/Science Photo Library page 138 middle left
Pekka Parvianen/Science Photo Library page 167 top left
Tanya Piejus pages 40, 138 top left, centre middle
Mauro Fermarielto/Science Photo Library page 138 calipers
R.D. Battersby/Tografox page 138 fan of gauges
Ed Young/Science Photo Library page 138 sextant
Tek Image/Science Photo Library page 138 speedometer
David Vaughan/Science Photo Library page 138 bottom left
Jonathan Watts/Science Photo Library page 167 right

Orders: please contact Bookpoint, 130 Milton Park, Abingdon, Oxon OX14 4SB.
Telephone: (44) 01235 827720, Fax: (44) 01235 400454.
Lines are open from 9.00–5.00, Monday to Saturday, with a 24 hour message answering service.

British Library Cataloguing in Publication Data
A catalogue entry for this title is available from The British Library

Foundations of advanced mathematics. – (MEI)

ISBN: 978 0 340 86926 0

First Published 1997
Second edition 2003
Impression number 10 9 8 7 6 5 4 3
Year 2007

Typeset by Pantek Arts Ltd, Maidstone, Kent.
Printed in Great Britain for Hodder Murray, a division of Hodder Education, an Hachette Livre UK Company, 338 Euston Road, London NW1 3BH by Martins the Printers, Berwick upon Tweed.

Introduction

This book has been written to support the MEI intermediate Free Standing Mathematics Qualification, Foundations of Advanced Mathematics. The course serves serveral purposes.

- It provides access to AS and A Levels in Mathematics for those who are not ready to start those courses.
- It provides access to Higher Education courses where a reasonable level of numeracy and Mathematics is expected.
- It is a worthwhile course in its own right, and as support for other subjects (e.g. Sciences).
- It provides the basis for an Advanced VCE unit.

If you would like to take AS or A Level Mathematics but feel that you are not quite ready to embark on that level of work, you will find that this book fills just the gaps that are worrying you. All the explanations and worked examples appear on the left hand pages; the right hand pages are reserved for exercises and activities. This means that there is a lot for you to do. By the time you have successfully worked your way through the book, you will be well placed to continue Mathematics at a higher level.

If you are using this book for a Mathematics unit in a vocational course, you will find it particularly helpful that many of the questions are set in contexts from the world of work. You will see examples of how Mathematics is used in the area that you are studying and in others that are closely related to it. Remember that to use Mathematics you need more than just a collection of techniques: you also need to know when to use them.

This is the second edition of this book. There have been some changes to the content, in line with the demands of the specification, and the data in a number of the questions have been brought up to date.

The authors would like to thank all those who have helped with this book: particularly Mike Jones, who worked through the draft typescript with meticulous attention to detail, and Karen Eccles who, on many pages, had the difficult task of trying to squeeze the text into the available space.

Mathematics in Education and Industry is a curriculum development body which aims to promote the links between Education and Industry in Mathematics and to produce relevant examination and teaching syllabuses and support material. Since its foundation in the 1960s, MEI has provided syllabuses for GCSE (or O Level), Additional Mathematics and A Level. Foundations of Advanced Mathematics is part of the present suite.

For more information about MEI syllabuses and materials, write to MEI Office, Albion House, Market Place, Westbury BA13 3DE or visit the MEI website www.mei.org.uk.

Contents

Chapter 

Calculations

Introduction

Whatever your interests or line of work, it is useful to be able to do quick calculations by hand or in your head. Even when you are using a calculator, you need to be very clear about what you are asking it to do, and what sort of answer you expect.

The *positive integers* are 1,2, 3, … .

Much of the work at the start of this chapter will be familiar to you. Use these pages to make sure you are confident on the basics of factors, multiples, fractions and decimals before proceeding further.

Factors, multiples and primes

Carl has bought 12 square flagstones and plans to use all of them to lay a rectangular patio. The diagram below illustrates different rectangles he can make using all the flagstones.

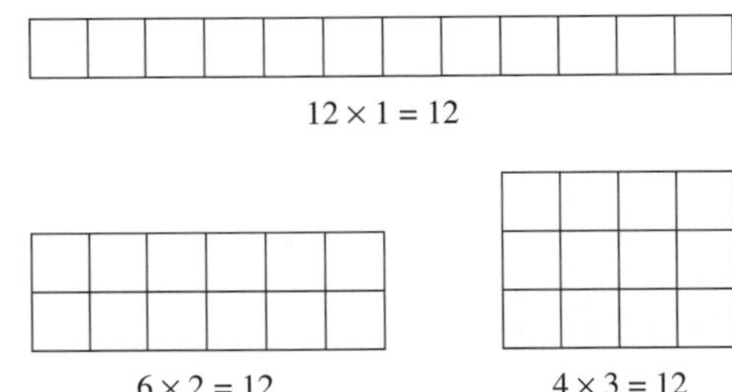
12 × 1 = 12
6 × 2 = 12
4 × 3 = 12

When you write $4 \times 3 = 12$

- 4 and 3 are *factors* of 12. (Other factors of 12 are 1, 2, 6 and 12.)
- 12 is a *multiple* of 4 and 3. (It is also a multiple of 1, 2, 6 and 12.)

Carl finds that the supplier has given him 13 flagstones. In this case, the only possible arrangement is 13 × 1. This is because 13 is a *prime number*; the only factors of 13 are 1 and 13 itself.

The concepts of factors and multiples can be extended to two (or more) numbers.

The *highest common factor* (HCF) of 12 and 18 can be found as follows:

6 is the highest common factor.

The factors of 12 are 1, 2, 3, 4, 6, 12.
The factors of 18 are 1 2, 3, 6, 9, 18.

The HCF of 12 and 18 is 6.

The *lowest common multiple* (LCM) of 12 and 18 can be found as follows:

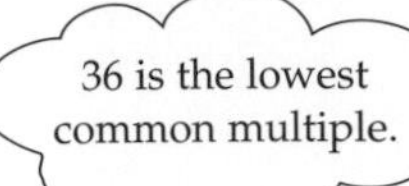
36 is the lowest common multiple.

The multiples of 12 are 12, 24, 36, 48 …
The multiples of 18 are 18, 36, 54, 72 …

The LCM of 12 and 18 is 36.

Exercise

1. Find all the factors of the following numbers.
   (i) 14 (ii) 24 (iii) 20 (iv) 56 (v) 99
2. Write down the first 5 multiples of the following numbers.
   (i) 2 (ii) 5 (iii) 7 (iv) 11 (v) 17
3. Which of the following statements are true and which are false?
   (i) 26 is a multiple of 2
   (ii) 2 is a factor of 8
   (iii) 8 is a multiple of 16
   (iv) 216 is a multiple of 3
   (v) 9 is a multiple of 81
   (vi) 11 is a factor of 99
4. Find the highest common factor of
   (i) 9 and 15 (ii) 20 and 30 (iii) 18 and 24
   (iv) 36 and 54 (v) 14 and 42 (vi) 7 and 10
   (vii) 12, 24 and 30 (viii) 50, 75 and 100.
5. Find the lowest common multiple of
   (i) 2 and 5 (ii) 4 and 6 (iii) 10 and 15
   (iv) 18 and 27 (v) 30 and 40 (vi) 4 and 8
   (vii) 2, 3 and 4 (viii) 6, 8 and 9.
6. Write out all the prime numbers between 20 and 30.
7. (i) Which of the following are prime numbers?
   12, 15, 26, 27, 29, 31, 33, 37, 39
   (ii) Are there any even prime numbers?

Investigations

1. The *consecutive sum* of the numbers 1, 2 and 3 is 6:
   $$1 + 2 + 3 = 6.$$
   The consecutive sum of the numbers 2, 3 and 4 is 9:
   $$2 + 3 + 4 = 9.$$
   Work out the sums of some other sets of three consecutive numbers. What do you notice?

   Now work out the sums of some sets of four consecutive numbers, to see if there are any patterns, then try five consecutive numbers. Can you make any general statements about consecutive sums?

2. A *perfect number* is one that is half the sum of all its factors including 1.

   The number 8 is not perfect. Its factors are 1, 2, 4 and 8, and $1 + 2 + 4 + 8 = 15$: 8 is not half of 15. There is one perfect number between 4 and 10, and another between 25 and 30. What are they?

   The number 496 is a perfect number. What are its factors?

Activity

Copy and complete the following stock-taking sheet. The unit size indicates how many there are in a box or packet. The supplier will not split a box.

The maximum must not be exceeded, but you are to bring the stock as close as possible to its maximum level.

Item	Balance of stock	Maximum	Minimum	Unit size	Number of units to be ordered
A4 files	28	40	5	5	
Erasers	51	200	50	10	
Print cartridges	12	20	5	3	
HB pencils	21	100	10	20	
Box files	17	80	10	4	
30 cm rulers	137	200	30	50	
Staple guns	5	16	5	2	

Fractions

Equivalent fractions

(a)

This fraction is $\frac{2}{4}$.

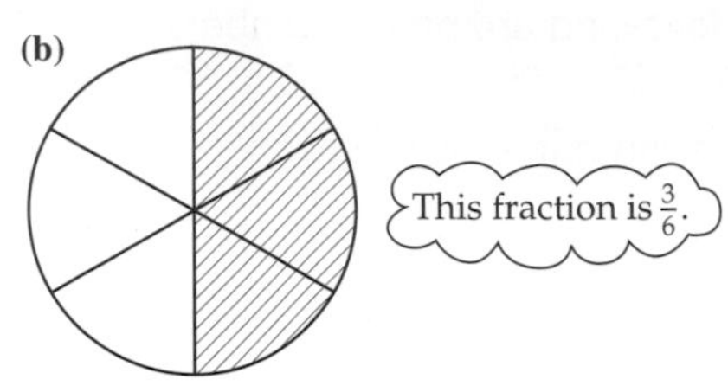
(b)
This fraction is $\frac{3}{6}$.

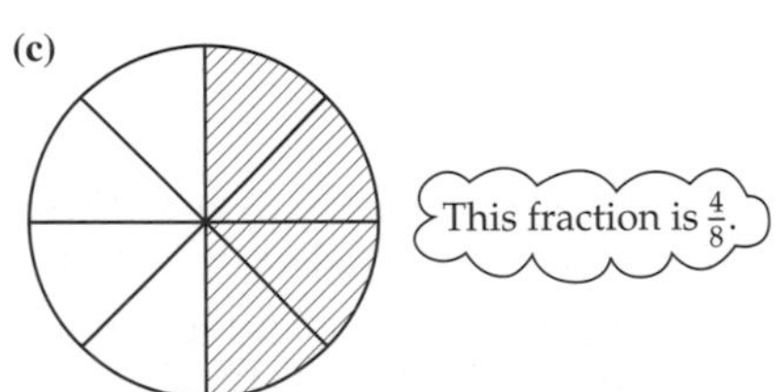
(c)
This fraction is $\frac{4}{8}$.

These three pizzas have been cut into pieces. In each case if you ate the shaded part you would have eaten half a pizza.

You can write $\frac{1}{2} = \frac{2}{4} = \frac{3}{6} = \frac{4}{8}$. These are all equivalent fractions.

You can write out a set of equivalent fractions for any fraction, for example:

$$\frac{3}{5} = \frac{6}{10} = \frac{9}{15} = \frac{12}{20} = \ldots$$

Writing a fraction in its simplest form (or lowest terms)

Look at the fraction $\frac{36}{48}$. You can divide the top and the bottom by 12:

$$\frac{36}{48} = \frac{3}{4}$$

Dividing the top and the bottom by the same number is called *cancelling*. The fraction $\frac{3}{4}$ cannot be cancelled any further, since the top and bottom have no common factors. We say that $\frac{3}{4}$ is the *simplest form* of the fraction $\frac{36}{48}$.

Top-heavy fractions

The fractions $\frac{3}{2}$, $\frac{5}{4}$ and $\frac{13}{5}$ are called *top-heavy* (or *improper*) fractions.

They have the top (*numerator*) larger than the bottom (*denominator*).

Top-heavy fractions can be written as *mixed numbers*, that is as a mixture of whole numbers and fractions. The fractions above can be written as $1\frac{1}{2}$, $1\frac{1}{4}$ and $2\frac{3}{5}$.

Example

(i) Write $\frac{37}{5}$ as a mixed number.

(ii) Write $2\frac{5}{6}$ as a top-heavy fraction.

Solution

(i) To write $\frac{37}{5}$ as a mixed number, divide the top by the bottom:

$$37 \div 5 = 7, \text{ remainder } 2.$$

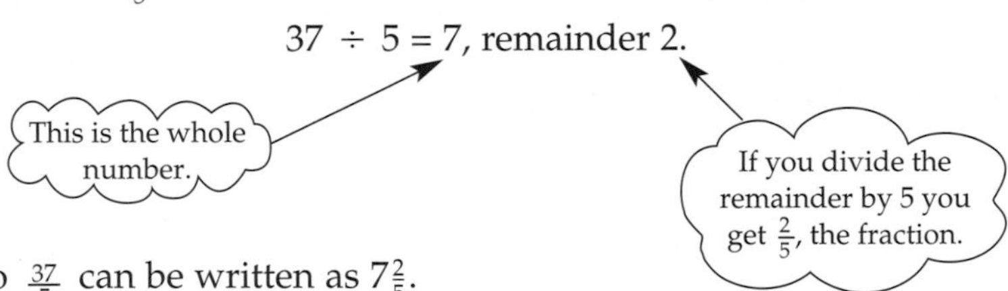
This is the whole number.
If you divide the remainder by 5 you get $\frac{2}{5}$, the fraction.

So $\frac{37}{5}$ can be written as $7\frac{2}{5}$.

Here 2 is written as $\frac{12}{6}$.

(ii) To write $2\frac{5}{6}$ as a top-heavy fraction, you write $2\frac{5}{6} = \frac{12}{6} + \frac{5}{6} = \frac{17}{6}$.

Reciprocals

The reciprocal of 2 is $1 \div 2 = \frac{1}{2}$.

The reciprocal of 3 is $1 \div 3 = \frac{1}{3}$.

Exercise

1. Copy and complete the following:
   (i) $\frac{1}{2} = \frac{}{4} = \frac{}{6} = \frac{}{8} = \frac{}{10} = \frac{}{12}$
   (ii) $\frac{1}{4} = \frac{}{8} = \frac{}{12} = \frac{}{16} = \frac{}{20} = -$
   (iii) $\frac{3}{5} = \frac{6}{} = \frac{9}{} = - = - = -$
   (iv) $\frac{5}{12} = - = - = - = - = -$

2. Write the following fractions in their simplest form.
   (i) $\frac{4}{8}$ (ii) $\frac{5}{20}$ (iii) $\frac{6}{24}$ (iv) $\frac{9}{36}$ (v) $\frac{23}{46}$
   (vi) $\frac{15}{20}$ (vii) $\frac{16}{24}$ (viii) $\frac{18}{27}$ (ix) $\frac{16}{36}$ (x) $\frac{24}{36}$

3. Would you rather win $\frac{2}{3}$ or $\frac{3}{5}$ of a lottery prize? Why?

4. Football League Club A with a ground capacity of 55 000 allocates 5500 seats to away supporters. Football League Club B with a ground capacity of 20 000 allocates 2 000 seats to away supporters. Which of the clubs allocates the larger fraction of seats to away supporters?

5. What fraction lies half way between
   (i) $\frac{2}{5}$ and $\frac{3}{5}$? (ii) $\frac{3}{7}$ and $\frac{4}{7}$?
   (iii) $\frac{2}{3}$ and $\frac{3}{4}$? (iv) $\frac{1}{2}$ and $\frac{3}{4}$?
   (v) $\frac{1}{3}$ and $\frac{1}{4}$?

6. Arrange the following fractions in order of size, smallest first.
   (i) $\frac{1}{3}, \frac{1}{4}, \frac{1}{2}$ (ii) $\frac{3}{10}, \frac{1}{4}, \frac{1}{2}$

7. A paper-making plant takes its pulp from 3 sources: $\frac{1}{4}$ of it is new pulp, $\frac{5}{9}$ is 'post-consumer' pulp (from recycled waste paper), and $\frac{7}{36}$ is 'pre-consumer' pulp (made from the waste created during the production process). Which source provides the biggest part of the pulp?

8. Change the following top-heavy fractions to mixed numbers.
   (i) $\frac{4}{3}$ (ii) $\frac{5}{2}$ (iii) $\frac{10}{7}$ (iv) $\frac{7}{3}$
   (v) $\frac{10}{4}$ (vi) $\frac{20}{6}$ (vii) $\frac{26}{8}$ (viii) $\frac{56}{3}$

9. Change the following mixed numbers to top-heavy fractions.
   (i) $2\frac{1}{4}$ (ii) $3\frac{5}{6}$ (iii) $9\frac{1}{4}$ (iv) $10\frac{1}{10}$ (v) $1\frac{1}{11}$

10. An interior decorator mixes 10 litres of a blue paint with 50 litres of a pink paint.
    (i) What fraction of the resulting mauve paint is made up of blue paint?
    (ii) Write this fraction in its simplest form.

    The decorator finds that she does not have enough paint to finish the job. She needs another 20 litres of mauve paint. Her supplier has plenty of the pink paint, but only 2 litres of the blue paint left in stock.
    (iii) Is this sufficient to finish the job?

11. Write down the reciprocal of
    (i) 4 (ii) 5 (iii) 10.

Activities

1. Suppose the fraction $\frac{1}{2}$ can be represented on a graph by the point (2, 1) as shown below. The equivalent form $\frac{2}{4}$ can be represented by the point (4, 2), $\frac{3}{6}$ by (6, 3) and so on.

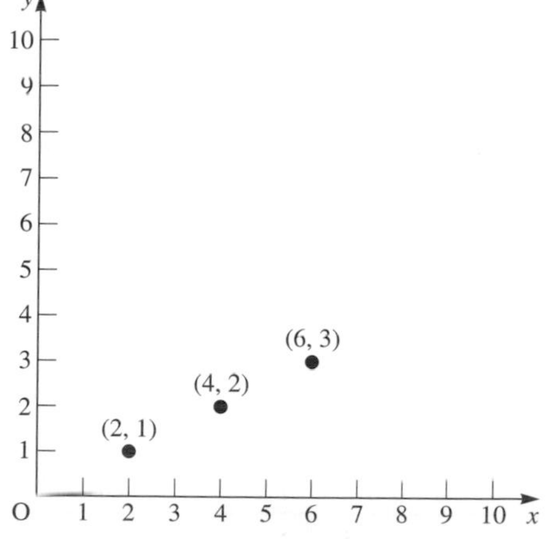

   (i) Draw x and y axes on paper with a 1 cm square grid so that they are both numbered from 0 to 10. Plot the three points given above on your graph, and notice the pattern they produce.
   (ii) Use your graph to find 2 more fractions which are equivalent to $\frac{1}{2}$.
   (iii) Experiment with other simple fractions. Find three other sets of equivalent fractions from your graph paper.

2. Design a poster for a mathematics classroom to illustrate the idea of equivalent fractions.

Fractions: review of basic calculations

You must be careful when adding and subtracting fractions. The first two examples show what you should do.

Example

Add $\frac{3}{4}$ to $\frac{1}{10}$.

Solution

$$\frac{3}{4} + \frac{1}{10} = \frac{15}{20} + \frac{2}{20}$$

$$= \frac{17}{20}$$

First you write down equivalent fractions with the same bottom line. This is called a *common denominator:* in this case 20 is the *lowest* common denominator.

Now you can add the top lines to get the answer.

Example

Subtract $\frac{3}{5}$ from $\frac{2}{3}$.

Solution

$$\frac{2}{3} - \frac{3}{5} = \frac{10}{15} - \frac{9}{15}$$

$$= \frac{1}{15}$$

This time the lowest common denominator is 15.

Multiplying fractions is much easier than adding or subtracting them, since you do not need to find a common denominator first.

Example

Multiply $\frac{3}{4}$ by $\frac{4}{5}$.

Solution

$$\frac{3}{4} \times \frac{4}{5} = \frac{12}{20}$$

$3 \times 4 = 12$

$4 \times 5 = 20$

This can be cancelled to give $\frac{3}{5}$.

Notice that you could have cancelled before multiplying $\frac{3}{\cancel{4}_1} \times \frac{\cancel{4}^1}{5} = \frac{3}{5}$

Example

Find $\frac{3}{5}$ of $2\frac{1}{2}$.

Solution

$$_1\frac{3}{\cancel{5}} \times {}^1\frac{\cancel{5}}{2} = \frac{3}{2}$$

$$= 1\frac{1}{2}$$

the word 'of' is replaced by $\times$ and $2\frac{1}{2}$ is written as $\frac{5}{2}$.

In this case we have multiplied before cancelling.

The next example shows how to divide by a fraction.

Example

Find $\frac{5}{12} \div \frac{10}{21}$.

Solution

$$\frac{5}{12} \times \frac{21}{10} = \frac{105}{120} = \frac{7}{8}$$

The second fraction is turned upside down and the $\div$ becomes $\times$.

In this case we have multiplied before cancelling.

Exercise

1. Write the answers to the following calculations in their simplest form.
   (i) $\frac{5}{8} + \frac{2}{8}$ (ii) $\frac{3}{10} + \frac{2}{10}$
   (iii) $\frac{3}{8} + \frac{1}{4}$ (iv) $\frac{1}{10} + \frac{2}{3}$
   (v) $\frac{7}{8} + \frac{1}{3}$ (vi) $\frac{1}{5} + \frac{3}{8}$
   (vii) $2\frac{3}{8} + 1\frac{1}{4}$ (viii) $5\frac{7}{8} + 2\frac{1}{4}$
2. A quality control inspector found that $\frac{1}{15}$ of a batch of CDs were oversized and $\frac{1}{10}$ were undersized. What fraction were defective, and what fraction were good?
3. Write the answers to the following calculations in their simplest form.
   (i) $\frac{5}{8} - \frac{2}{8}$ (ii) $\frac{7}{10} - \frac{3}{10}$
   (iii) $\frac{5}{16} - \frac{1}{4}$ (iv) $\frac{7}{8} - \frac{2}{3}$
   (v) $2\frac{3}{4} - 1\frac{1}{2}$ (vi) $3\frac{1}{4} - 1\frac{1}{2}$
   (vii) $2\frac{1}{10} - 1\frac{1}{2}$ (viii) $2\frac{2}{3} - 1\frac{1}{4}$
4. Joe's roll of 13-amp wire has $10\frac{1}{4}$ m left on it. He lets Sara have $3\frac{1}{2}$ m of it. He then realises he needs $6\frac{1}{10}$ m of the same wire for another use. Has he enough left on the roll?
5. Write the answers to the following calculations in their simplest form.
   (i) $\frac{1}{2} \times \frac{1}{2}$ (ii) $\frac{1}{4} \times \frac{1}{2}$
   (iii) $\frac{5}{8} \times \frac{3}{10}$ (iv) $\frac{1}{10} \times \frac{1}{10}$
   (v) $1\frac{1}{4} \times \frac{3}{5}$ (vi) $\frac{1}{3} \times 2\frac{4}{7}$
   (vii) $2\frac{1}{2} \times 3\frac{1}{5}$ (viii) $2\frac{6}{7} \times 5\frac{5}{6}$
6. What is the floor area of a rectangular office, $6\frac{1}{2}$ m long and $3\frac{1}{4}$ m wide?
7. Write the answers to the following calculations in their simplest form.
   (i) $\frac{3}{4} \div \frac{1}{2}$ (ii) $\frac{1}{3} \div \frac{3}{10}$
   (iii) $1\frac{3}{4} \div \frac{1}{10}$ (iv) $1\frac{1}{2} \div \frac{1}{8}$
   (v) $3\frac{1}{2} \div \frac{3}{4}$ (vi) $\frac{1}{4} \div 2$
   (vii) $3\frac{3}{4} \div 2\frac{1}{4}$ (viii) $1\frac{1}{2} \div 2\frac{2}{5}$
8. In a general hospital, $\frac{1}{2}$ of the employees are doctors and nurses, $\frac{2}{5}$ are ancillary/support staff and the rest are administrative. What fraction are administrative?
9. Work out the reciprocal of
   (i) $1\frac{1}{2}$ (ii) $2\frac{3}{4}$.
10. On one day in June, 2000 people went through the turnstiles of a leisure park. Of these, $\frac{4}{5}$ were children.
    How many adults visited the park on that day?
11. A painting is sold for £18 000.The artist receives $\frac{5}{6}$ of this and the gallery exhibiting her work receives the rest. How much money does the gallery receive?
12. On each bounce, a ball rises to $\frac{4}{5}$ of its previous height. It is initially dropped from a height of 250 cm.
    (i) How high will the ball rise after its third bounce?
    (ii) How far below its original starting height will it be then?
    (iii) How many bounces will there be before it fails to reach half of its original height?
13. Ron is making doors by nailing $\frac{1}{4}$ inch hardboard onto a wooden frame $\frac{5}{8}$ inch thick. He is using $\frac{1}{2}$ inch nails.

Stage 1

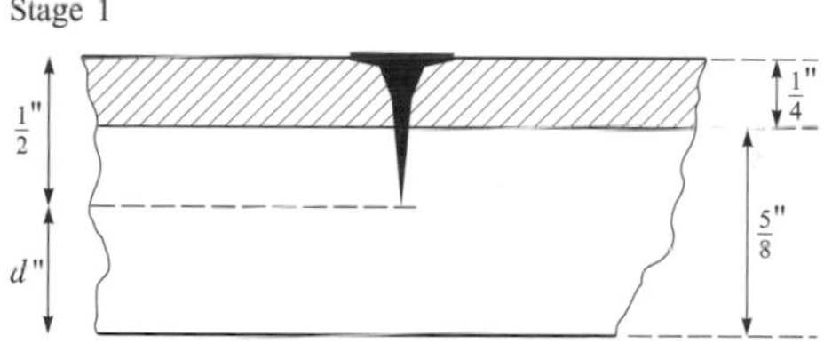

Stage 2

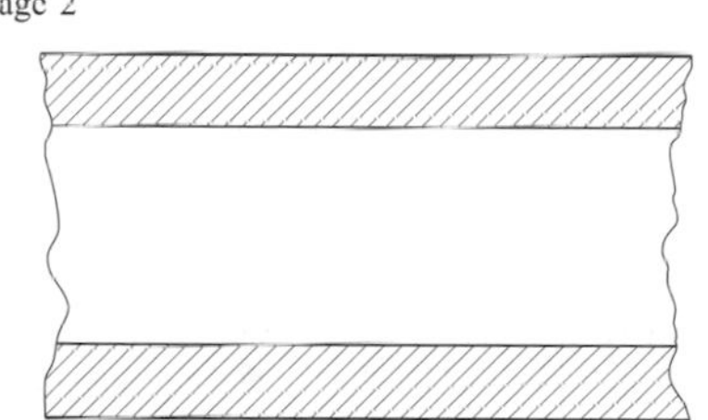

Not drawn to scale

Which one of these four statements is false?
A At stage 1, the door is $\frac{7}{8}$ inch thick.
B The distance d is less than $\frac{5}{16}$ inch.
C A stack of 10 doors at stage 1 will be 9 inches high to the nearest inch.
D When Ron adds the layer of hardboard on the other side of the frame (stage 2) the thickness of the door is $\frac{9}{7}$ times its thickness at stage 1.

Activity

Many calculators have the facility to do fraction calculations. Check whether yours does.
To make sure you know how to calculate with fractions on your calculator, repeat questions 1, 3, 5 and 7. You should get the same answers by calculator as you did by hand!

Decimals: review of basic calculations

Although calculations involving decimals are easy to do on a calculator, you also need to be able to do them by hand, and to have a sense of the size of the numbers you are dealing with. The examples on this page cover the basic procedures.

Addition and subtraction are straightforward provided you set out your calculation with the decimal points in the same vertical line.

Example

Work out (i) $1.1 + 0.994$, (ii) $1.1 - 0.994$.

Solution

(i) Addition (ii) Subtraction

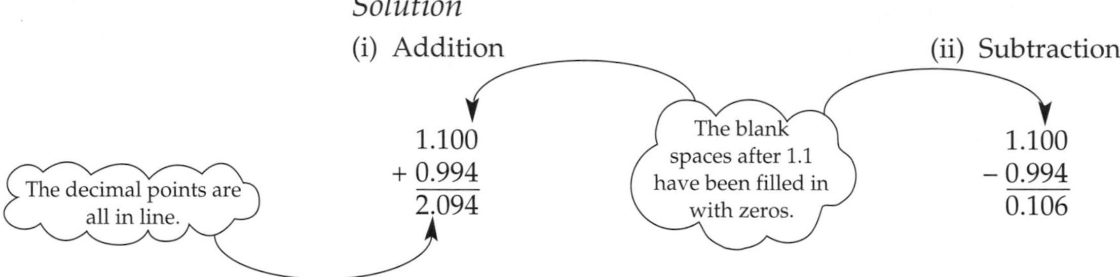

The blank spaces after 1.1 have been filled in with zeros.

You can multiply decimals by converting them into fractions. However it is more usual to use the rules illustrated in the next example. The two methods are actually the same and you might like to ask yourself why.

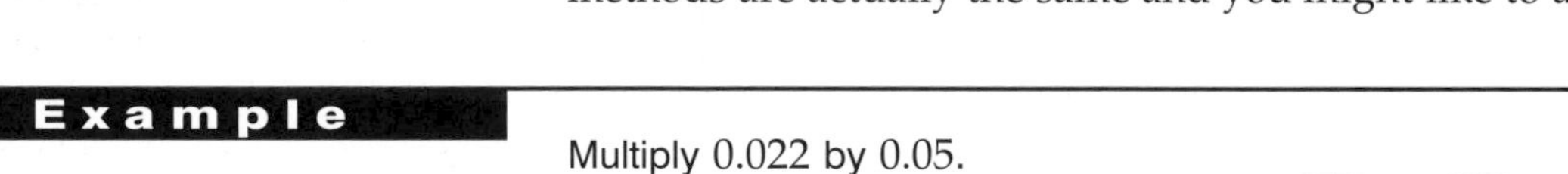

Example

Multiply 0.022 by 0.05.

Solution

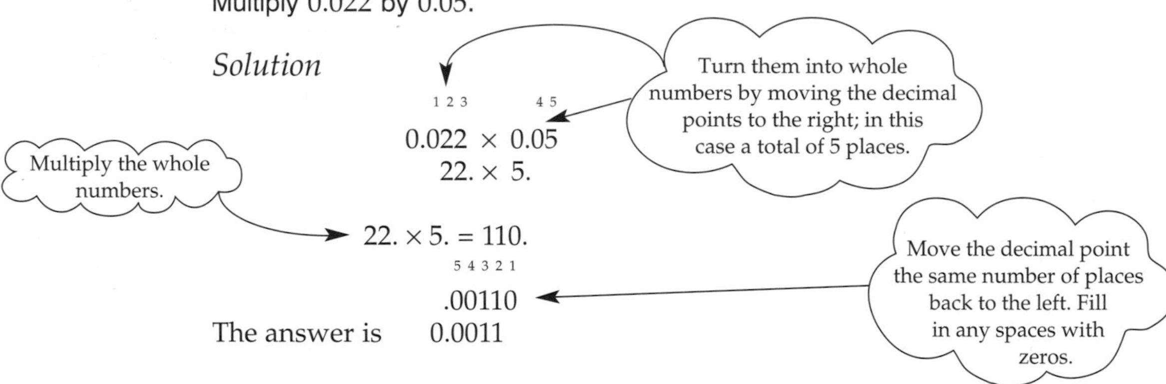

1 2 3 4 5
0.022 × 0.05
22. × 5.
Turn them into whole numbers by moving the decimal points to the right; in this case a total of 5 places.
Multiply the whole numbers.
22. × 5. = 110.
5 4 3 2 1
.00110
Move the decimal point the same number of places back to the left. Fill in any spaces with zeros.

The answer is 0.0011

The method for dividing by a decimal is shown in the next example.

Example

Divide 15.42 by 0.3.

Solution

The calculation can be written as $\dfrac{15.42}{0.3}$,

which is the same as $\dfrac{154.2}{3}$

Move the decimal point to the right to make the bottom line a whole number. Move the point on the top line the same number of spaces to the right.

Now divide to get the answer: 51.4

Exercise

Use pencil and paper methods and show your workings. You may use a calculator to check your answers.

1. Work out
   (i) $12.34 + 234.62 + 1.69$
   (ii) $34.6 + 23 + 3 + 0.06$
   (iii) $16.32 - 9.45$
   (iv) $16 - 1.34$

2. What is the cost of production of a gold ballpen if the top costs £1.46, the body £3.24, the refill 39p and the labour 64p per pen?

3. The diagram below shows the measurements of a living-room wall with a fireplace.
   (i) How wide is the fireplace?
   (ii) How wide is the doorway?

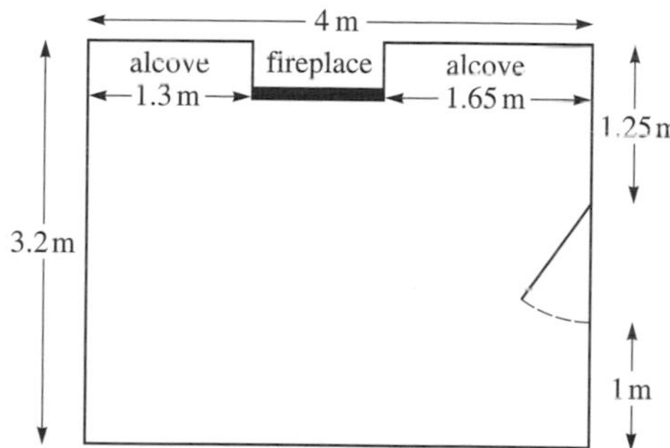

4. Work out
   (i) 14.32×5
   (ii) 14.36×13
   (iii) 0.56×9
   (iv) 2.3×1.6
   (v) 0.59×0.345

5. Shelly wants to decorate her living-room. She has written down the price of all the materials involved, and how much of each she needs. Work out how much it will cost in total.
   8.5 m of carpet @ £12.55 a metre
   4.5 litres of gloss paint @ £4.75 a litre
   12 rolls of wallpaper @ £6.20 a roll.

6. Work out
   (i) $34.5 \div 5$
   (ii) $0.144 \div 1.2$
   (iii) $1.26 \div 0.009$

7. A road traffic engineer has to place 6 lamp-posts along one side of a stretch of road which is 432.5 m long. The lamp-posts must be equally spaced, and there must be one at each end of the stretch of road. What distance should he leave between the lamp-posts?

8. The diagram shows the top view of Mike and Debra's cars in their garage. The total length of the garage is 5.2 m.

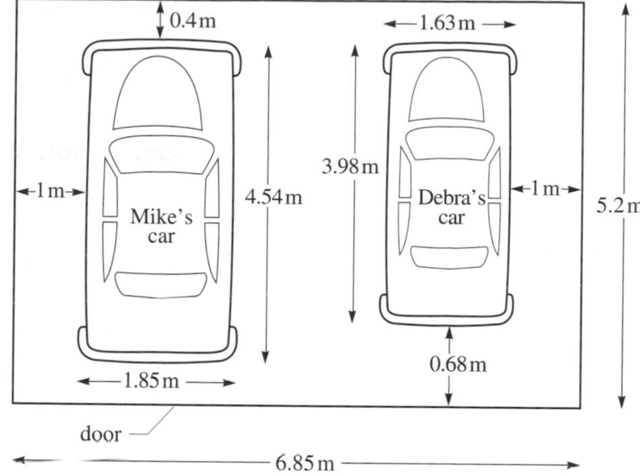

   (i) Find the distance between Mike's car and the door.
   (II) Find the distance between Debra's car and the back wall.
   (iii) How big is the gap between the cars?

   Mike thinks Debra is encroaching on his parking space, and Debra thinks Mike is encroaching on hers. Eventually they agree to draw a line down the centre of the garage. They are both to park in the centre of their own half.
   (iv) How far will each car now be from its nearest wall?
   (v) What is the gap between the cars under the new arrangement?

9. Two taxi companies, A and B, provide these flow diagrams so that their passengers can calculate their fares if they know how far they are going to travel.

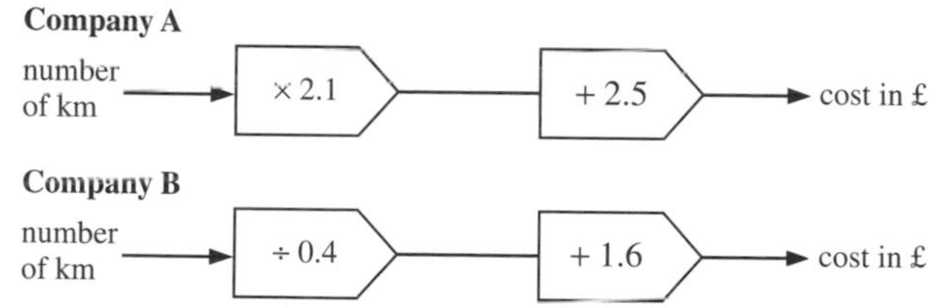

   (i) Which company is cheaper if you want to travel 2.8 km?
   (ii) Which company is cheaper if you want to travel 16.4 km?
   (iii) Find the distance at which both companies charge the same fare.

Equivalent forms

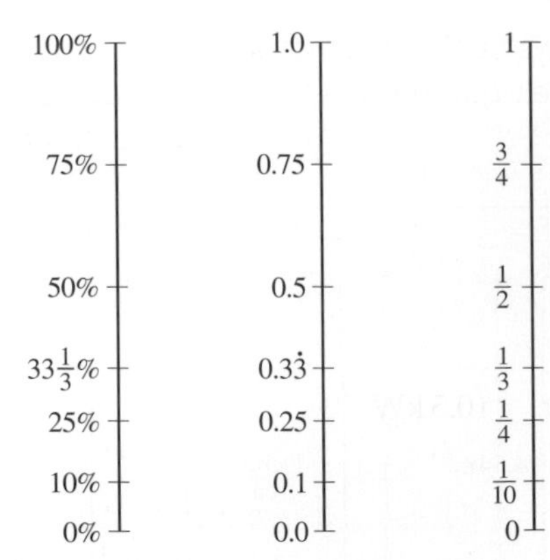

You probably already know that 50% and 0.5 are both equivalent to $\frac{1}{2}$: for example $\frac{1}{2}$ a litre of cola is the same as 0.5 litres, and 50% of the population is the same as $\frac{1}{2}$ the population. There may be other fractions for which you know straight away the percentage and decimal equivalents. Some of the most commonly-used fractions and their equivalents are shown in the diagram on the left.

You need to be able to convert between fractions, decimals and percentages, whatever the numbers involved. The diagram can help you to check your answers, as well as giving some clues about how to do the conversions. Here are some examples.

Example

Write 25% as a decimal.

Solution

$25 \div 100 = 0.25$

Divide the percentage by 100 to find the decimal equivalent.

Example

Write 0.345 as a percentage.

Solution

$0.345 \times 100\% = 34.5\%$

Multiply the decimal by 100 to find a percentage.

Example

Write $\frac{3}{5}$ as a decimal.

Solution

$3 \div 5 = 0.6$

Divide the top number of a fraction by the bottom number to find a decimal.

Example

Write $\frac{3}{5}$ as a percentage.

Solution

$\frac{3}{5} = \frac{3}{5} \times 100\% = 60\%$

Multiply the fraction by 100 to get a percentage.

Example

Write 36% as a fraction.

Solution

$36\% = \frac{36}{100} = \frac{9}{25}$

Remember that 'per cent' means 'out of 100'.

Example

Write 0.125 as a fraction.

Solution

$0.125 = \frac{125}{1000} = \frac{1}{8}$

Write as a fraction and then cancel down if possible.

Exercise

1. Copy and complete the following table.

Fraction	Decimal	Percentage
$\frac{1}{4}$		
	0.2	
		40
$\frac{5}{8}$		
	0.075	
		10
	0.24	
$\frac{1}{3}$		
	0.6	
$2\frac{1}{2}$		
		45
		12

2. You are offered the option of a wage rise of 10% of your annual salary, and no bonus at the end of the financial year, or a bonus of $\frac{1}{8}$ of your annual salary and no wage rise.
What are the advantages and disadvantages of each option?

3. The power used by an electrical appliance is measured in watts (W). Write the following appliances in order of increasing power consumption. (1 kW = 1000 W.)
Storage heater: $2\frac{1}{2}$ kW,
Infra-red lamp: 0.25 kW
Vacuum cleaner: $\frac{3}{4}$ kW
Electric kettle: 750 W
Electric cooker: 10.5 kW
Light bulb: 60 W

4. Three people who live on steep hills describe the gradients of the roads outside their houses as follows.
(a) 1 in 9 (b) 12% (c) 0.1
Assuming that they are all telling the truth, who lives on the steepest hill?

5. A team of scientists is studying a mould. The proportion of the available area which the mould occupies is recorded at midday each day. All their measurements are accurate but they write the figure differently as follows:
Tuesday (Cindy) 0.54
Wednesday (Ali) $\frac{11}{20}$
Thursday (Kwame) 53%

On which of the three days was the mould the largest?

Investigation

$\frac{1}{7} \equiv 0.142857142857142\ldots$

This is a *recurring decimal*. It can be written as $0.\dot{1}4285\dot{7}$, and can be represented by the path shown in the diagram.

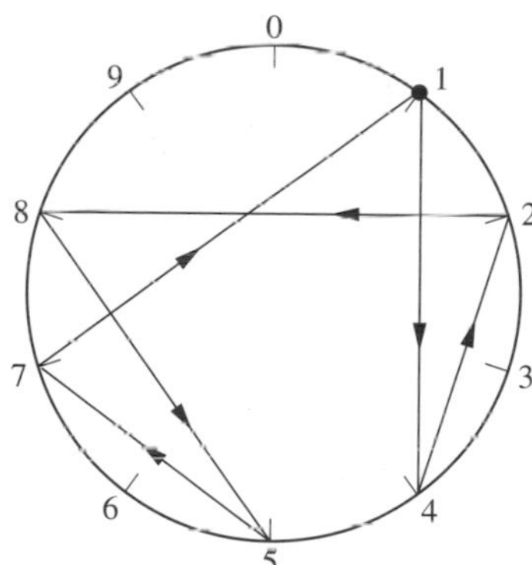

Draw some other circles like this, with dots every 36°, and see what patterns are produced by other fractions which have recurring decimals as equivalents. (Alternatively you could use a plastic bowl with a clothes peg attached every 36°, and use string to show the pattern.)

You might like to start with $\frac{3}{7}$ and $\frac{5}{7}$, then try $\frac{1}{11}$, $\frac{3}{11}$ etc.

Percentages

You will often meet situations in which quantities change. The next example shows how such change can be written as a percentage. Notice that the percentage is always based on the original amount.

Example

One year a manufacturer sold walking boots to the value of £112 432. The following year this increased to £119 262. Express the increase as a percentage.

Solution

$$\text{Percentage increase} = \frac{\text{actual increase}}{\text{original amount}} \times 100$$

$$= \frac{(119\,262 - 112\,432)}{112\,432} \times 100$$

$$= 6.07\% \text{ (to 2 decimal places).}$$

It is a common mistake to use the new amount rather than the original amount on the bottom line.

The next two examples involve a quantity being changed by a given percentage.

Example

What is the price of a £33 coat in a sale where all prices are reduced by 20%?

Solution

The original price of £33 is taken as 100%:

$$100\% = £33.00$$
$$\text{so} \quad 1\% = £\,0.33$$
$$\text{and} \quad 20\% = £\,6.60.$$

Marketed price = £33.00 and reduction = £ 6.60 so

$$\text{actual price} = £33.00 - £6.60 = £26.40.$$

An alternative method is to say that after a reduction of 20% the actual price is 80%:

100% = £33.00
1% = £ 0.33
80% = £26.40

NOTE

You would often set out the calculation of 20% of £33.00 as

$$\tfrac{20}{100} \times £33.00 = £6.60.$$

Example

VAT is charged at $17\frac{1}{2}$%. The total cost of a car service is £70.50 inclusive of VAT. How much VAT has been added?

Solution

The cost excluding VAT is 100%. $117\frac{1}{2}\% = £70.50$

$1\% = £\,0.60$

The VAT is $17\frac{1}{2}\% = £10.50$

The total cost is £60 + £10.50 = £70.50.

Divide both sides by $117\frac{1}{2}$.

Multiply both sides by $17\frac{1}{2}$.

Exercise

1. A car auctioneer receives 10% commission from the seller on each car that he sells. How much does he receive for selling the following cars?
   (i) A Peugeot 205 at £4800
   (ii) A Nissan Cherry 1.3 at £600
   (iii) A Ford Sierra 2.8i at £1775

2. Calculate the price of each of the following when VAT at 17.5% is added.
   (i) A pair of jeans priced at £28 excl. VAT.
   (ii) A colour TV priced at £314 excl. VAT.
   (iii) A telephone bill at £87.14 excl. VAT.
   (iv) A car exhaust priced at £36.12 excl. VAT.
   (v) A pair of shoes priced at £44.44 excl. VAT.

3. Hamish went into a hardware store and asked the price of 100 m of chicken mesh. The shop assistant told him £64 plus VAT. Neither Hamish nor the shopkeeper had a calculator to hand. Luckily Hamish remembered that VAT was set at 17.5% rather than 18%, because it made calculations easier! He demonstrated his method as follows.

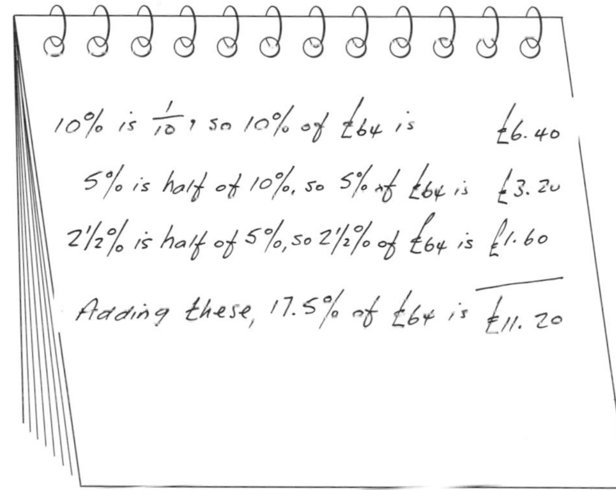

10% is 1/10, so 10% of £64 is £6.40
5% is half of 10%, so 5% of £64 is £3.20
2½% is half of 5%, so 2½% of £64 is £1.60
Adding these, 17.5% of £64 is £11.20

   Use this method to check your answers to question **2.**

4. Calculate the percentage increase or decrease in the following situations.
   (i) A baby weighs 3.3 kg at birth and 3.8 kg two weeks later.
   (ii) John's annual salary goes up from £10 495 to £10 812.
   (iii) A metal bar of length 1 m is cut down to 88 cm.
   (iv) A village shop buys 25 tubes of toothpaste for £20, and sells them for 90p each.
   (v) Slimmer of the year Laura Oakes who originally weighed 17 stones has lost 6.5 stones.

5. The following is an extract from the Terms and Conditions section of an antique auctioneer's guide.

   The purchaser of any lot shall pay the hammer price plus a premium (together with VAT on the premium at the appropriate rate and on the hammer price at the appropriate rate if applicable on that lot). Except for specialised sales of coins, medals, stamps and wine, the premium shall be calculated at 15% on the first £30 000 of hammer price of each lot plus 10% of that hammer price in excess of £30 000.

   Assuming VAT is payable at 17.5% on the items below, calculate the total price the purchaser pays for each one.

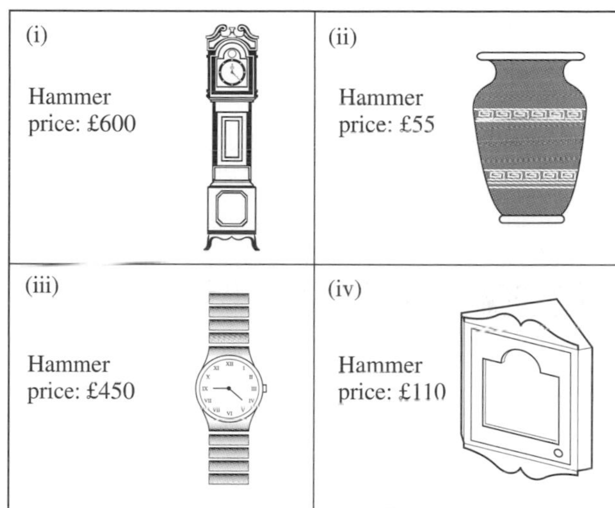

6. The following telephone bills include VAT at 17.5%.
   (i) Georgia's bill is £76.14.
   How much is the bill before VAT is added?
   (ii) Harry's bill is £104.81.
   How much VAT is included in the bill?
   (iii) Rakhee's bill is £56.83.
   How much is the bill before VAT is added?
   (iv) Christopher's bill is £82.77.
   How much VAT is included in the bill?

7. In a sale everything is reduced by 30%.
   (i) Explain why the sale price is 70% of the original price.
   (ii) The sale price of a pair of shoes is £14. Show that their original price was £20.
   (iii) Find the original price of a coat which was £38.50 in the sale.
   (iv) Debbie saved £9 on a skirt by buying it in the sale. How much did she spend?

Positive and negative numbers

Adding and Subtracting. Look at Sarah's bank balance below.

Transaction	Calculation	Balance
Initial Balance		£20
She pays in her wages of £230	20 + 230= +250	£250
She buys a television for £265	+250 – 265 = –15	–£15
She pays her electricity bill of £65	–15 – 65 = –80	–£80
She pays in her next wages of £230	–80 + 230 = +150	£150

Sometimes it is helpful to use a *number line* as shown below. Subtraction is moving to the left, addition is moving to the right. Addition of a negative number is the same as subtraction.

negative ← –5 –4 –3 –2 –1 0 +1 +2 +3 +4 +5 → positive

Some people add positive and negative numbers by using the two rules given below. Use the number line to see for yourself why these rules work.

1. *Same signs:*

If the numbers are both positive or both negative, add them and give the answer the same sign as the original numbers:

$$2 + 3 = (+2) + (+3) = +5; \quad -2 - 3 = (-2) + (-3) = -5$$

2. *Different signs:*

If one number is positive and one negative, subtract the smaller one from the larger and give the answer the same sign as the larger:

$$2 - 3 = (+2) + (-3) = -1; \quad -2 + 3 = (-2) + (+3) = +1$$

In this case the sign of the larger number is –

Multiplying and Dividing. The rules for getting the signs right are illustrated below for multiplication; those for division are the same.

$$(+12) \times (+2) = +24 \qquad + \times + \rightarrow +$$
$$(-12) \times (-2) = +24 \qquad - \times - \rightarrow +$$
$$(+12) \times (-2) = -24 \qquad + \times - \rightarrow -$$
$$(-12) \times (+2) = -24 \qquad - \times + \rightarrow -$$

Sequence of operations

What is the value of $2 + 3 \times 4$?

Do you work it out as $2 + 3 = 5$ and then $5 \times 4 = 20$

or $3 \times 4 = 12$ and $2 + 12 = 14$?

Ideally, brackets are used to indicate which part is worked out first. However, if there are no brackets the rule is that division and multiplication are done before addition and subtraction.

Brackets
Indices
Division
Multiplication
Addition
Subtraction

So $2 + 3 \times 4 = 2 + 12 = 14$ is correct.

Multiplication is done before addition.

Exercise

1. Draw number lines to illustrate the following additions.
   (i) $(+4) + (+2) = 6$
   (ii) $(-3) + (+4) = (+1)$
   (iii) $(-4) + (-2) = (-6)$

2. State whether my bank account is 'in the red' or 'in the black', and by how much, after each of the following transactions:
   (i) balance £14, pay in £15;
   (ii) balance £13, withdraw £12.50;
   (iii) overdrawn by £4, withdraw £3.50;
   (iv) balance £158.57, withdraw £160;
   (v) overdrawn by £15.60, pay in £23.89;
   (vi) overdrawn by £39, pay in £29.89.

3. Write down the size of the temperature change in each of the following cases, and say whether it is a rise or a fall.

	Initial temp. (°C)	*Final temp.* (°C)
(i)	27	15
(ii)	17	18
(iii)	–4	15
(iv)	–1	–10

4. Write down the answers to the following.
   (i) $(+7) + (+6)$ (ii) $(+7) + (-4)$
   (iii) $(-7) + (-9.8)$ (iv) $(-1.8) + (+10)$
   (v) $(-6) - (+4)$ (vi) $(-2) - (-6)$
   (vii) $(+4) - (-4.5)$ (viii) $(-7) - (+3)$.

5. Write down the answers to the following.
   (i) $(+3) \times (-2)$ (ii) $(-2) \times (-2)$
   (iii) $(-2) \times 1.5$ (iv) $(-0.7) \times (-6)$
   (v) $(-8) \div 2$ (vi) $(-16) \div 4$
   (vii) $100 \div (-2)$ (viii) $(-4.2) \div (-5.6)$

6. Write down the answers to the following:
   (i) $(3 + 4) \times 2$ (ii) $3 + (4 \times 2)$
   (iii) $3 + 4 \times 2$ (iv) $2 \times 7 - 3$
   (v) $2 \times (7 - 3)$ (vi) $20 \div 2 + 8$
   (vii) $20 \div (2 + 8)$ (viii) $12 - 3 \times 2 + 1$

7. A credit limit is the amount by which a bank allows someone to go into debt. I have reached my credit limit of £500, and persuaded my bank manager to double my credit limit.
   (i) What will my bank balance be if I make full use of the new credit limit?
   (ii) How would you write down your calculation using positive and negative numbers (as in **5**)?

8. The pit-head of a coal mine is 232 metres above sea level. The deepest part of the mine is 651 metres below ground level.
   (i) How far below sea level is the deepest part?

   The lift from the deepest part gets stuck 405 m above its lowest point.
   (ii) How far above or below sea level is the lift?

9. The gauge on an oil tank is marked so that when it reads 0 the tank is not actually empty (but should be refilled).

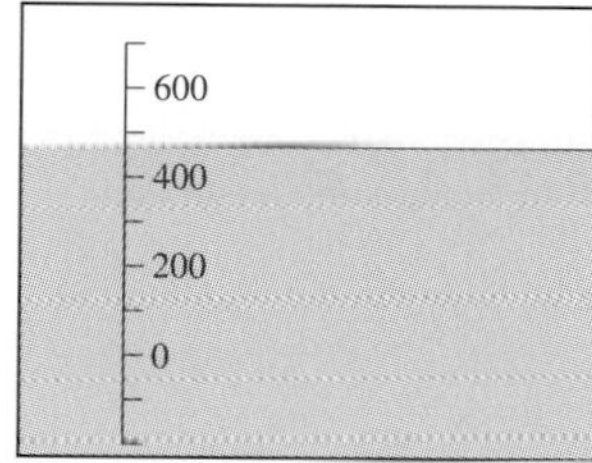

   On December 1st the gauge reads 320 litres. On January 30th it reads –70 litres.
   (i) What is the average rate of oil consumption in litres per day?

   On January 30th more oil is put in. It is intended that this should last until May 10th when the gauge should read zero. The same rate of consumption is assumed.
   (ii) To what level is the tank filled? (Assume that it is not a leap year.)

Activity

If you have a $\boxed{\pm}$ key on your calculator, you can use it to do all of the questions in the exercise above. However, you need to be careful about the order in which you press the keys.

(i) Make sure you can do the calculations correctly.
(ii) Write a short, clear set of instructions so that someone borrowing your calculator would be able to do calculations of this sort.

Investigation

Look at the pattern of numbers below.

1	= +1
1 – 2	= –1
1 – 2 + 3	= +2
1 – 2 + 3 – 4	= –2

Continue the pattern for several more rows. Write down anything you notice about your answers

Now try this pattern:

2
2 – 4
2 – 4 + 6
2 – 4 + 6 – 8

Investigate other patterns with alternating + and – signs.

Powers or indices

Writing $3 \times 3 \times 3 \times 3 \times 3$ takes time and is hard to read. A short way of writing it is 3^5. You read this as 'three to the *power* five'. The 5 is called the *index* (plural indices) and 3^5 is called *index form*.

Notice that 3^5 is **not** the same as 3×5: $3^5 = 243$, but $3 \times 5 = 15$.

Negative powers

Look at this pattern of powers of 4.

In the first column the power of 4 is decreasing by 1 each time.

$$4^3 = 4 \times 4 \times 4 = 64$$
$$4^2 = 4 \times 4 = 16$$
$$4^1 = 4 = 4$$

In the last column each number is the previous number divided by 4.

You would expect the next 3 lines to look like this.

$$4^0 = 1 = 1$$
$$4^{-1} = 1 \div 4 = \tfrac{1}{4} \quad \text{or} \quad 0.25$$
$$4^{-2} = 1 \div (4 \times 4) = \tfrac{1}{16} \quad \text{or} \quad 0.0625.$$

These three lines are very important because they allow you to give a meaning to zero and negative powers.

Example

(i) Write 10^{-2} as a fraction.

(ii) What is the value of 5^0?

Solution

(i) $10^{-2} = \dfrac{1}{10^2} = \dfrac{1}{10 \times 10} = \dfrac{1}{100}$.

(ii) Any number to the power zero has a value of 1.

Sometimes you will need to work with numbers in index form. In particular, powers of 10 occur when using SI units (page 20) and with standard form (page 22).

Example

Work out the following, giving your answers in index form.

(i) $10^5 \times 10^2$ (ii) $10^6 \div 10^2$ (iii) $(10^3)^2$

Solution

(i) $10^5 \times 10^2 = (10 \times 10 \times 10 \times 10 \times 10) \times (10 \times 10) = 10^7$

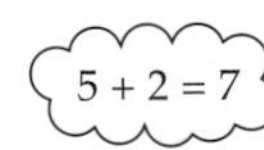

(ii) $10^6 \div 10^2 = \dfrac{10 \times 10 \times 10 \times 10 \times 10 \times 10}{10 \times 10} = 10^4$

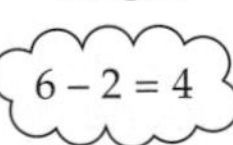

(i) $(10^3)^2 = (10 \times 10 \times 10) \times (10 \times 10 \times 10) = 10^6$

Exercise

1. Copy and complete the table below and describe the pattern of numbers in the right-hand column.

2^5	$2 \times 2 \times 2 \times 2 \times 2$	
2^4		
2^3	$2 \times 2 \times 2$	8
2^2		4
2^1		
2^0		
2^{-1}	$\frac{1}{2}$	
2^{-2}		
2^{-3}	$\frac{1}{2\times2\times2}$	$\frac{1}{8}$
2^{-4}		

2. Work out the values of the following, without using a calculator.
   (i) 3^2 (ii) 4^3 (iii) 3^4 (iv) 5^3 (v) 10^4

3. Work out the values of the following, without using a calculator. Leave your answers as fractions.
   (i) 2^{-2} (ii) 3^{-4} (iii) 5^{-2} (iv) 10^{-3} (v) 4^{-1}

4. In each of these, write the first number as a power of the second number.
   (i) 16; 2 (ii) 27; 3
   (iii) 100; 10 (iv) $\frac{1}{100}$; 10
   (v) $\frac{1}{16}$; 2 (vi) $\frac{1}{49}$; 7
   (vii) 16; 4 (viii) 1; 5

5. Work out the values of the following, giving your answers in index form.
   (i) $10^6 \times 10^4$ (ii) $10^5 \div 10^2$
   (iii) $10^{11} \times 10^8$ (iv) $10^{21} \div 10^{12}$
   (v) $(10^4)^2$ (vi) $10^{13} \div 10$
   (vii) $(10 \times 10^7) \div 10^4$ (viii) $10^{-2} \times 10^6$
   (ix) $(10^{-3})^2$ (x) $10^3 \div 10^{-3}$
   (xi) $(10^{-4} \times 10) \div 10^6$ (xii) $10^{11} \times 10^{-25}$

6. (i) Show that 600 can be written as
   $$2^3 \times 3 \times 5^2.$$
   This is called the *prime factorisation* of 600.
   (ii) Work out the values of
   (a) $2^3 \times 5$ (d) $3^2 \times 5^2$
   (b) $2 \times 3^2 \times 7$ (e) $2^6 \times 5^6$.
   (c) $2^5 \times 5^2 \times 11$

7. Write the following numbers as the products of the powers of prime numbers (i.e. their prime factorisations).
   (i) 60 (iv) 4000
   (ii) 196 (v) 78125
   (iii) 1200

8. (i) Show that $\frac{9}{16}$ can be writtem as $2^{-4} \times 3^2$.
   (ii) Write the following as fractions.
   (a) $2^{-2} \times 3$ (d) $2^{-1} \times 3^2 \times 5^{-1}$
   (b) 2×3^{-2} (e) $2^3 \times 5^{-3}$
   (c) $2^{-2} \times 3^{-1} \times 5$

9. Write the following fractions as the products of the powers of prime numbers.
   (i) $\frac{3}{10}$ (iv) $\frac{25}{36}$
   (ii) $\frac{4}{9}$ (v) $\frac{8}{27}$
   (iii) $\frac{9}{11}$

10. On January 1st each year I hide some money under my mattress. The first time I did this was on January 1st 1991, when I hid £1. On January 1st 1992 I hid twice that amount, i.e. £2, so that I had a total of £3 under the mattress. The next year I hid another £4, then £8 and so on.
    (i) How much did I have under the mattress in total on January 2nd 1994?
    (ii) How much did I have under the mattress in total on November 27th 1995?
    (iii) How much do I have under the mattress at present?
    (iv) When will I become a millionaire?

Investigation

The diagram shows a pile of cubical blocks, $5 \times 5 \times 5$. The pile is left outside and the faces that are exposed to the light become discoloured. How many cubes have (i) 3 (ii) 2 (iii) 1 (iv) 0 faces discoloured?

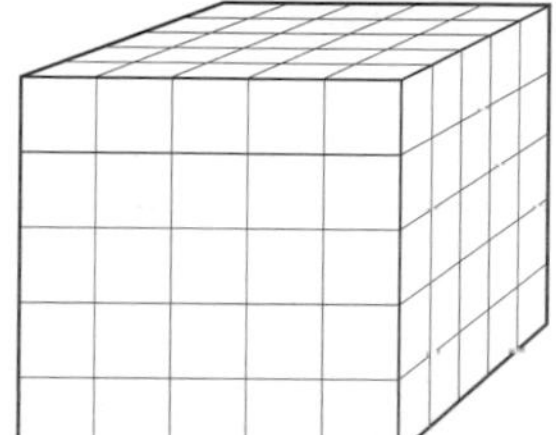

Squares and cubes

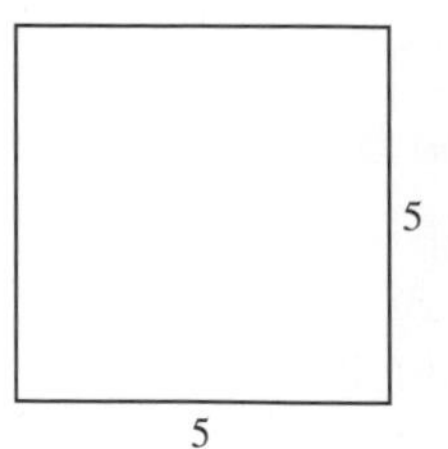

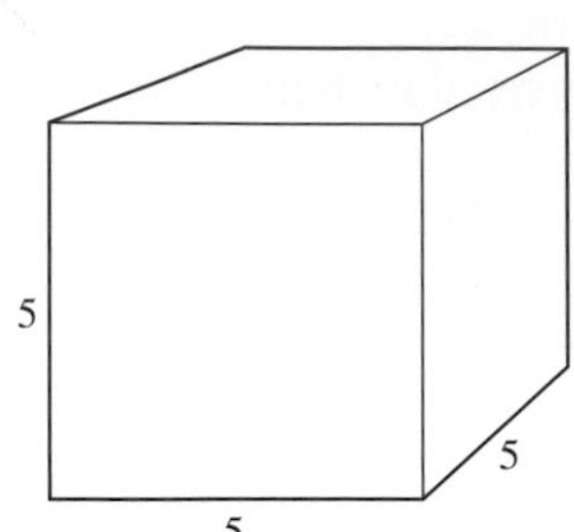

The powers of 2 and 3 are given special names.

The diagrams show a square and a cube, each with side 5 cm.

The area of the square is $5^2 \text{cm}^2 = 25 \text{cm}^2$.

The volume of the cube is $5^3 \text{cm}^3 = 125 \text{cm}^3$.

Because of this connection, a number to the power of 2 is called 'squared', and a number to the power of 3 is called 'cubed'.

There is no equivalent for 4 or higher powers. They are called 'to the power of 4' and so on.

Square roots and cube roots

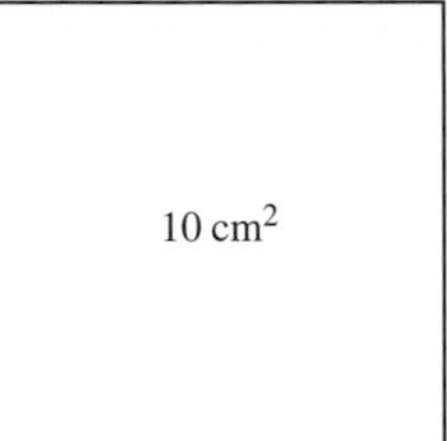

The diagram shows a square of area 10cm^2. How many centimetres long is each side?

The answer to this question is called the square root of 10, and is written $\sqrt{10}$. Since $3^2 = 9$ which is less than 10, and $4^2 = 16$ which is greater than 10, the answer must lie between 3 and 4.

You can find $\sqrt{10}$ on your calculator using the [√] key: 3.162... .

Since $(+3.162...) \times (+3.162...) = 10$ and $(-3.162...) \times (-3.162...) = 10$, the square root of 10 is usually written $\pm 3.162...$, the sign $\pm$ being read 'plus or minus'. You meet this when you use the formula for solving quadratic equations (page 68).

When finding the length of the side of a square, you are only interested in the positive value so you will discard the negative value.

To find the length of one side of a cube with volume 10cm^3 you need to find a number which, when raised to the power 3, gives the answer 10. This is called the cube root of 10 and is written $\sqrt[3]{10}$.

Some calculators have cube root keys, labelled [∛]. However, on most calculators you need to use the [$\sqrt[x]{y}$] key. Check that you can find cube roots on your calculator, e.g. that you get the cube root of 10 to be 2.154…

NOTE

It is not possible to find the square root of a negative number.

Any number, positive or negative, has just one cube root.
For example $\sqrt[3]{+8} = +2$ *and* $\sqrt[3]{-8} = -2$

Exercise

1. Work out the areas of the squares with the following sides: (Do not use a calculator.)
   (i) 4 cm (ii) 9 cm
   (iii) 10 cm (iv) 2.5 cm.
2. A landscape gardener draws a square patio on one of his sketches. He indicates that it will have 81 square paving slabs, each of side 60 cm. What are the dimensions of the patio?
3. Use your calculator to work out the area of a square which has a side of length 15.6 cm.
4. Work out the lengths of the sides of the squares with the following areas, using the $\sqrt{}$ key on your calculator:
   (i) 19 cm^2 (ii) 13 cm^2
   (iii) 50 cm^2 (iv) 83.7 cm^2.
5. Work out (without using your calculator) the volumes of the cubes with the following sides:
   (i) 6 cm (ii) 8 cm
   (iii) 10 cm (iv) 1.5 cm.
6. Use your calculator to find the lengths of the sides of the cubes with the following volumes:
   (i) 30 cm^3 (ii) 140 cm^3
   (iii) 65.5 cm^3 (iv) 211.16 cm^3.
7. The table shows the area of some squares with different side lengths. Copy and complete the table.

Side (cm)	1	2	3	4	5	6	7
Area (cm^2)				16			49

   Draw a set of axes, and plot 'length of side' along the horizontal axis (scale: 2 cm represents 1 cm) and 'area of square' up the vertical axis (scale 1 cm represents 2 cm^2).
   (i) From your graph estimate the length of the side of a square whose area is 43 cm^2;
   (ii) Use your calculator to find the accurate answer to part (i).

Activity

Choose an appropriate scale, and draw the following diagram accurately. Continue until you have at least three more triangles.

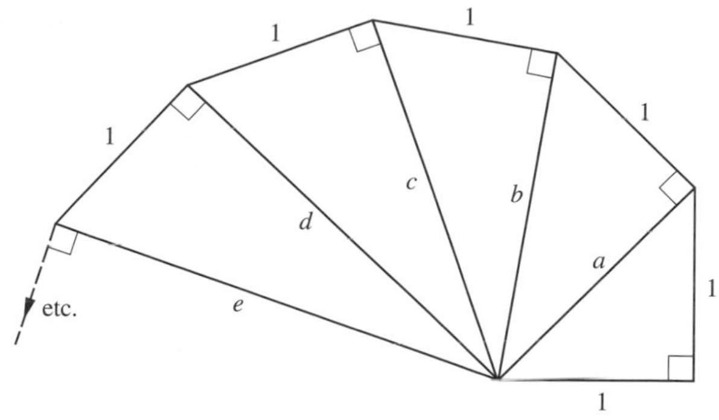

Measure the lengths a, b, c, d and so on, then copy and complete the following table.

Number	Square root
2	$\sqrt{2} = 1.414$
3	$\sqrt{3} - 1.732$
4	
5	
6	
7	
8	
9	

Compare the numbers in the right hand column with the lengths you measured in your diagram.
(i) What do you notice?
(ii) Explain why the diagram produces this pattern in the lengths.

Investigations

1. A red square has an area of 900 cm^2.
   A blue square has an area of 100 cm^2.
   Work out the side of:
   (i) the red square (ii) the blue square.

   Work out the following ratios:
   (iii) area of red square:area of blue square
   (iv) side of red square:side of blue square.
   (v) Compare your ratios.

2. A green cube has a volume of 512 cm^3.
   A white cube has a volume of 4096 cm^3.
   Work out the side of:
   (i) the green cube (ii) the white cube.

   Work out the following ratios:
   (iii) volume of green cube:volume of white cube
   (iv) side of green cube:side of white cube.
   (v) Compare your ratios.

SI units

When you measure a quantity, such as a length, a time interval or an electric current, you express the result as a number of units, such as 6 cm, 3 hours, or 0.5 ampères. There are many systems of units, but the one most widely understood today is the *Système Internationale* (SI). The SI units of mass, length and time are the kilogram, the metre and the second. In the UK the Imperial system (mass in pounds, length in feet, time in seconds) is still used in many situations, but it is gradually being phased out.

In the Imperial system, you need to know the definitions of many different units, for instance 1 inch is $\frac{1}{12}$ of 1 foot, 1 yard is 3 feet, 1 stone is 14 pounds, and so on. SI units are much easier, because they follow a simple pattern. You only need to know the names of a few basic units (grams, metres, litres and so on) and the meanings of a range of prefixes. To convert between the different units involves multiplying (or dividing) by powers of 10. The common prefixes and their meanings are given in the table.

Prefix	Meaning	Abbreviation	Example
mega-	$\times 1\,000\,000$	M	1 Mw = 1 000 000 W
kilo-	$\times 1000$	k	1 km = 1000 m
centi-	$\times \frac{1}{100}$	c	1 cm = $\frac{1}{100}$ m
milli-	$\times \frac{1}{1000}$	m	1 mg = $\frac{1}{1000}$ g

It is worth remembering the following relationships; you will often meet them.

- 1 km = 1000 m
- 1 m = 100 cm
- 1 cm = 10 mm
- 1 kg = 1000 g
- 1 tonne = 1000 kg
- 1 g = 1000 mg
- 1 l = 1000 ml
- 1 ml = 1 cm^3

A cubic centimetre of gold has a mass of 19.3 grams. You can say that the density of gold is 19.3 grams per cubic centimetre. The units here can be abbreviated as g/cm^3 or g cm^{-3}. Such units are called *compound units*. Compound units are used for many common quantities such as:

speed	metres per second	(ms^{-1} or m/s)
area	metres squared	m^2
engine speed	revolutions per minute	rpm
fuel consumption	miles per gallon	mpg

Converting between units

Sometimes you need to convert a quantity from one type of units to another. To do this you need to know the relationship between the two units. It is useful to memorise the following approximate relationships, since they come up very often.

- 1 inch ≈ 2.54 cm
- 1 metre ≈ 39 inches
- 1 km ≈ $\frac{5}{8}$ mile
- 1 litre ≈ 1.75 pints
- 1 gallon ≈ 4.5 litres.

For more accurate conversions, and for those not on this list, you need a conversion chart or table. Science data books and dictionaries often carry such tables.

Exercise

1. Convert the following quantities to the units indicated.
   (i) 357 cm to m;
   (ii) 3.45 m to cm;
   (iii) 1328 mg to g;
   (iv) 590 g to kg;
   (v) 3.4 l to ml;
   (vi) 1734 mm to m.

2. The diagram shows a non-standard window. Unfortunately the designer did not record all the measurements in mm. Copy out the dimensions but express them all in mm.

Dimensions	
Top panes	36.6 cm × 30 cm
Middle panes	37.6 cm × 30 cm
Lower panes	38.3 cm × 30 cm
Overall height	1 m 24.6 cm
Overall width	0.672 m
Glazing bar + frame	2.54 cm square section

3. In what compound units would you expect the following to be recorded?
   (i) the speed of a snail;
   (ii) the speed of an express train;
   (iii) the area of a sheet of A4 paper;
   (iv) the rate at which a bath empties;
   (v) the rate at which an oil droplet spreads on the surface of a puddle of water;
   (vi) the rate at which a drip feeds a solution into a patient's arm;
   (vii) the rate at which the temperature rises in a kettle.

4. Make sure you quote the correct units in your answers to each of the following.
   (i) What is the average speed of a car which travels 204 miles in $3\frac{1}{2}$ hours?
   (ii) A domestic oil tank holds 1200 l of oil and this is used up in 90 days. At what rate is the oil being used?
   (iii) A 3 m by 8 m sheet of copper is 5 mm thick. It weighs 1080 kg. What is its density?

5. Convert the following quantities to the units indicated, using the conversion formulae given (or those at the bottom of the opposite page).
   (i) 9.2 cm to inches;
   (ii) 14 km to miles;
   (iii) 12 gallons to litres;
   (iv) £50 to euros (£1 = 1.55 euros);
   (v) 9 litres to gallons;
   (vi) 24 km to miles (8 km = 5 miles).

Activities

inches	1	2	3	4	5	6	7	8	9
mm	25.4	50.8	76.2	101.6	127	152.4	177.8	203.2	228.6

1. The table above can be used to convert inches into millimetres. Use the table to construct a conversion graph, and then answer the following questions.
   (i) How many mm are there in 8.7 inches?
   (ii) How many inches are there in 200 mm?
   (iii) Which is larger, 160 mm or 6.35 inches?

2. Shelley needs to know how many kilometres her car will travel on a litre of petrol. She knows it travels 40 miles per gallon (mpg) on average.
   (i) Use the conversion factors given on the opposite page to convert 1 mpg into kilometres per litre.
   (ii) Draw a conversion graph for converting mpg into kilometres per litre.
   (iii) How many kilometres will Shelley's car travel on a litre of petrol (on average)?

Standard form

Large numbers

Look at the pattern of numbers below.

$$
\begin{aligned}
20 &= 2 \times 10 &&= 2 \times 10 \\
200 &= 2 \times 100 &&= 2 \times 10^2 \\
2000 &= 2 \times 1000 &&= 2 \times 10^3 \\
20000 &= 2 \times 10000 &&= 2 \times 10^4 \\
200000 &= 2 \times 100000 &&= 2 \times 10^5
\end{aligned}
$$

The numbers in this column are written in *standard form.*

You can see that standard form is a shorter way of writing very large numbers. It is very useful when you want to write huge numbers like 9 500 000 000 000 000 (which is the number of metres that light travels in one year). It is much quicker to write (and to read) the number in standard form: 9.5×10^{15}.

Example

(i) Write 2136 in standard form.

(ii) Write 4.56×10^5 in conventional form.

Notice that there is always just one figure before the decimal point in standard form.

Solution

(i) $2136 = 2.136 \times 1000 = 2.136 \times 10 \times 10 \times 10$
$= 2.136 \times 10^3$

(ii) $4.56 \times 10^5 = 4.56 \times 100\,000 = 456\,000$

Small numbers

Very small numbers can also be written in standard form. For example,

$$
\begin{aligned}
0.2 &= \tfrac{2}{10} &&= 2 \times 10^{-1} \\
0.02 &= \tfrac{2}{100} &&= 2 \times 10^{-2} \\
0.002 &= \tfrac{2}{1000} &&= 2 \times 10^{-3}
\end{aligned}
$$

These numbers are in standard form.

Again, this is very useful when you are writing extremely small numbers. The time taken for light to travel 1 metre is 0.000 000 003 3 seconds, or 3.3×10^{-9} seconds.

Example

(i) Write 0.000 417 6 in standard form.

(ii) Write 3.15×10^{-3} in conventional form.

Solution

(i) $0.000\,417\,6 = 4.176 \div 10\,000$
$= 4.176 \div 10^4$
$= 4.176 \times 10^{-4}$

(ii) $3.15 \times 10^{-3} = 3.15 \div 10^3$
$= 3.15 \div 1000 = 0.00315$

Make sure that you can use your scientific calculator to work with numbers in standard form. For example, check that $(4.6 \times 10^{28}) \times (3 \times 10^{-13}) = 1.38 \times 10^{16}$.

Exercise

1. Copy and complete:
   (i) $3000 = 3 \times 10 \times 10 \times ? = 3 \times 10^?$
   (ii) $343 = 3.43 \times ? = 3.43 \times 10^?$

2. Write the following numbers in standard form.
   (i) 400
   (ii) 60 000
   (iii) 3200
   (iv) 4 320 000
   (v) 237.6
   (vi) 991 000 000 000 000

3. The population of China is estimated to be 1 131 000 000, and that of the UK is roughly 57 600 000. Write both of these numbers in standard form.

4. Write the following numbers in conventional form.
   (i) 2×10^3
   (ii) 4×10^5
   (iii) 6.2×10^4
   (iv) 7.63×10^2
   (v) 7.97×10^2

5. The distance from London to New York is roughly 5.535×10^3 km and from London to Sydney is 1.7005×10^4 km.
   (i) By looking at the numbers in standard form, state which is the greater distance.
   (ii) Write both of the distances in conventional form.

6. Copy and complete:
   (i) $0.003 = \frac{3}{1000} = \frac{3}{10\times10\times10} = \frac{3}{?} = 3 \times 10^?$
   (ii) $0.000\,214 = \frac{2.14}{10000} = \frac{2.14}{10^?} = 2.14 \times 10^?$

7. Write the following numbers in standard form.
   (i) 0.003
   (ii) 0.000 17
   (iii) 0.001 59
   (iv) 0.000 017 26
   (v) 0.12

8. Write the following numbers in conventional form.
   (i) 4×10^{-3}
   (ii) 5.6×10^{-4}
   (iii) 4.89×10^{-2}
   (iv) 9.9989×10^{-1}

9. A blood cell is about 1.1×10^{-5}m long. What is this length in millimetres
   (i) in standard form;
   (ii) in conventional form?

10. Arrange the following numbers in order of increasing size.
    $30, \quad 3 \times 10^2, \quad 3, \quad 3 \times 10^{-2}, \quad 0.0003$

11. Use your calculator to work out the following.
    (i) $(3.5 \times 10^6) + (2 \times 10^5)$
    (ii) $(7 \times 10^{-5}) + (2.5 \times 10^{-4})$
    (iii) $(8 \times 10^{10}) - (7 \times 10^9)$
    (iv) $(6 \times 10^7) \times (4 \times 10^{15})$
    (v) $(3.69 \times 10^{18}) \div (3 \times 10^{27})$
    (vi) $(9 \times 10^{-8}) \times (7.2 \times 10^6)$
    (vii) $(2.4 \times 10^{-5}) \div (4.8 \times 10^{-17})$
    (viii) $(5 \times 10^{11})^2$
    (ix) $\dfrac{(6 \times 10^{-9}) \times (1.2 \times 10^{16})}{(1.8 \times 10^4)}$
    (x) $\dfrac{(1.6 \times 10^{25})}{(4 \times 10^7)^3}$

Activities

1. Write the following facts using standard form.
   (i) The mass of a planet is 10 000 000 000 000 000 tonnes.
   (ii) The diameter of the earth is 12 750 kilometres.
   (iii) Mount Everest is 8 850 metres high.
   (iv) The volume of the earth is 1 100 000 000 000 cubic kilometres.
   (v) The speed of light is 300 000 000 metres per second.
   (vi) The mass of an electron is 0.000 000 000 000 000 000 000 000 000 910 9 grams.
   (vii) Light takes 0.000 000 003 3 seconds to travel one metre.

   Find another 5 or 6 facts which involve either very large or very small numbers. You might use an encyclopaedia, a CD ROM, a data book or a textbook. Even some diaries contain lists of 'amazing facts'.

2. (i) Learn which keys to press on your scientific calculator in order to feed in numbers in standard form such as 1.6×10^6.
   (ii) Do a few simple calculations like $2.3 \times 10^5 \times 3$ on your calculator and check that your answers make sense.
   (iii) Work out the value of $10\,000\,609^2$ without using your calculator. Now do the same calculation on your calculator. Explain what the calculator has done.

Ratio

Day	Sausages	Bacon rashers
Mon	9	6
Tue	6	4
Wed	3	2
Thurs	12	8
Fri	18	12
Total	48	32

A café serves a weekday breakfast consisting of 3 sausages, 2 rashers of bacon, 1 egg and a portion of baked beans. The table shows the numbers of sausages and bacon rashers used during a particular week.

The number of breakfasts varied from day to day, but you can see that the *ratio* of sausages to bacon rashers was 3 to 2 each day. So for every 2 bacon rashers used, 3 sausages were used. Notice that the totals in the columns (i.e. 48 and 32) are also in the ratio 3 to 2.

A ratio, like a fraction, is in its simplest form when it is written using the smallest whole numbers possible. The ratio 50:10 is not in its simplest form. You can divide each number by 10 to obtain the ratio 5:1, which is its simplest form. Similarly the simplest form of the ratio 14:21 is 2:3.

When writing two quantities as a ratio, make sure they are both in the same units. The ratio of 50p to £1 is not 50 : 1, it is 50 : 100, which is 1 : 2.

Example

For apple and blackberry jam a manufacturer needs 3 kg of apples to every 2 kg of blackberries and 1 kg of sugar. The fruit buyer has ordered 15 tonnes of apples at a very good price. Find the amount of:

(i) sugar and blackberries needed (ii) jam produced.

Solution

(i)

Apples	: blackberries	: sugar
3 kg	: 2 kg	: 1 kg
15 tonnes	: 10 tonnes	: 5 tonnes

(ii) Total amount of jam (ignoring any reduction during cooking)

= 15 + 10 + 5 tonnes = 30 tonnes.

Converting between fractions and ratios

Example

In a Valentine's Day bouquet, the ratio of red to white roses is 3:1. What fraction of the bouquet are red roses?

Solution

The simplest bouquet would have 4 roses: 3 red and 1 white.
Fraction of roses in bunch that are red = $\frac{3}{4}$.

Example

Alf and Mo won £25 000 on the lottery. Alf paid 60p towards the £1 ticket and Mo paid 40p. They divided the winnings in the ratio of their contributions. How much did each receive?

Solution

Ratio of contributions = 60:40 = 3:2

The winnings were divided into 5 equal parts. Alf received 3 parts and Mo received 2.

£25 000 ÷ 5 = £5 000, so Alf's share = 3 × £5 000 = £15 000:
Mo's share = 2 × £5 000 = £10 000

Exercise

1. Write the following ratios in their simplest form.
   (i) 10 : 6 (ii) 24 : 16
   (iii) $7\frac{1}{2} : 2\frac{1}{2}$ (iv) $1 : \frac{1}{4}$
   (v) 50p : £1.50 (vi) 5 cm : 1 m
   (vii) 1 kg : 225 g (viii) 1 hour : 20 mins

2. In order to redress a gender imbalance in its workforce, a company needs to appoint men and women in the ratio 4:5. Over the next year 90 new jobs are expected to be advertised. How many of the appointees should be men, and how many women?

3. A football club has a ground capacity of 24 000 and the club's management makes the decision that the tickets sold to supporters should be in the ratio 5:1 in favour of the home supporters. How many tickets are allocated to the away supporters?

4. Megan and Greg set up a company and invest £5 000 and £10 000 respectively. They agree to share the profit between them in the ratio of their investments. After the first year the profit is £96 000. How much profit does each receive?

5. A map has a scale of 1:1000.
   (i) What distance on the ground (in metres) is represented on the map as
   (a) 2 cm (b) 7.6 cm?
   (ii) What distance on the map represents
   (a) 100 m (b) 2.6 km?

6. To make FoundaMix concrete I need cement, sand and aggregate mixed in the ratio 2:5:7. I need 7 tonnes of concrete for a driveway. How much of each material should I buy?

7. In a manufacturing process, the time allocated to machining, assembly, inspection, and packing is found to be in the ratio 6:3:1:2. Given that it takes 20 minutes to inspect the product, how long does the whole process take?

8. An alloy consists of lead, zinc and tin in the ratio 7:2:1 (by mass). How much of each metal is present in 70 kg of the alloy?

9. An amount of £32 000 is shared among three people, Leo, Mel and Nathan in the ratio 4:3:1. How much does each receive?

10. Here is a plan of the ground floor of Alexandra's house.

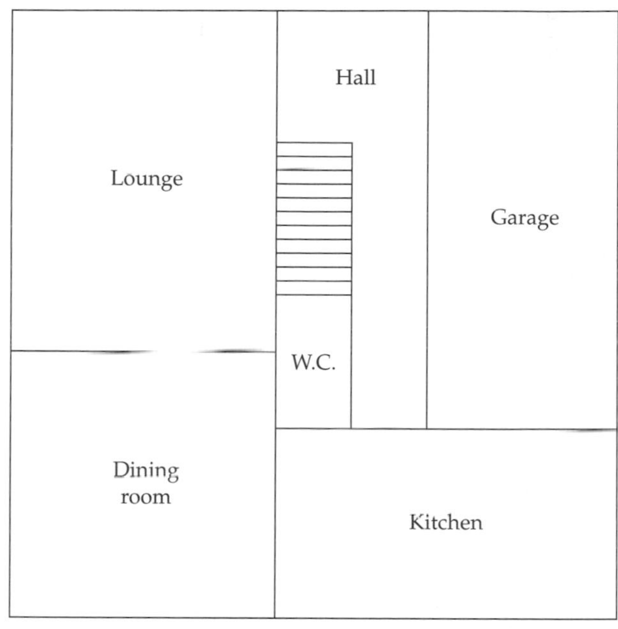

Scale 1:100

   (i) Work out the actual dimensions of
   (a) the garage, (b) the kitchen.
   (ii) Work out the floor area of
   (a) the lounge, (b) the dining room.

   A plan of the first floor is to be made using the same scale.

   (iii) Alexandra's bedroom is 4 m by 3 m. What lengths would be used to represent this?
   (iv) The area of the main bedroom is 15 m^2. What area on the plan will represent this?

Investigation

The rectangle in the diagram is special. The ratio of its length to its width is called the Golden Ratio and this is said to be the most pleasing to the eye.

Measure the sides and so express the ratio in the form x:1. To find an accurate value of x, calculate $\frac{1+\sqrt{5}}{2}$.

Now divide the rectangle into a square and rectangle. Show that the ratio of the length to the width of the new rectangle remains the same.

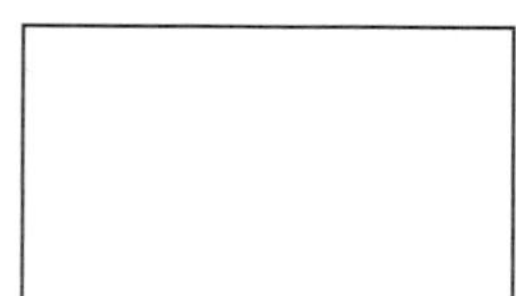

Proportion

Example

Given that 8 oranges cost 88p, how much would you expect to pay for 16 oranges?

Solution

This method is easy when the numbers are simple, e.g. when you are just doubling or trebling.

Method 1: scaling up

You are buying twice as many oranges, so you would expect to pay twice as much:

$$\text{price of 16 oranges} = 2 \times 88\text{p} = 176\text{p} = £1.76.$$

Method 2: unitary method

If 8 oranges cost 88p, price of 1 orange = 88p ÷ 8 = 11p.

If you are buying 16 oranges then you expect to pay 16 times as much as you pay for 1 orange:

$$\text{price of 16 oranges} = 16 \times 11\text{p} = 176\text{p} = £1.76.$$

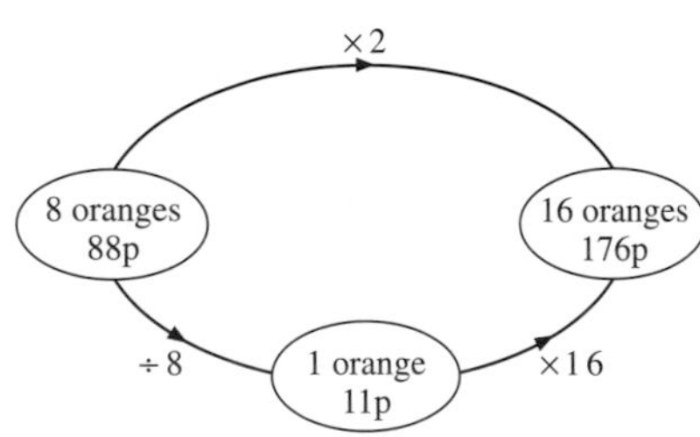

The diagram on the left shows how these two methods lead to the same answer.

Two quantities are said to *vary directly* or be in *direct proportion* if they increase or decrease at the same rate. In the example above, the price of the oranges is in direct proportion to the number you buy.

If two quantities are in direct proportion, the graph of one against the other is a straight line through the origin.

The next example differs fundamentally from the previous one. However, the unitary approach is once again the most reliable method of solution.

Example

Two people tidy a garden in 3 hours. How long will three people take to tidy a similar garden?

Solution

You can use the unitary method but you need to think carefully about it.

2 people take 3 hours
so 1 person takes 2 × 3 hours = 6 hours
so 3 people take 6 hours ÷ 3 = 2 hours

It will take one person longer.

In the last example the quantities *vary inversely* or are *inversely proportional*. The more people who work on the job, the less time it will take to complete. When two quantities vary inversely the graph of one against the other is *not* a straight line.

Exercise

1. (i) One gallon is the same as 4.55 litres. How many litres are there in 45 gallons?
   (ii) One inch is the same as 2.54 cm. How many centimetres are there in 12 inches?
   (iii) One kilogram is the same as 2.21 pounds. How many pounds are there in 15 kg?
   (iv) One mile is the same as 1.609 kilometres. How many kilometres are there in 30 miles?
   (v) £1 is the same as $1.57. How many dollars are there in £25?
2. (i) Which is the best buy in each of the following situations?
   (a) A net of 10 oranges at £1.20 or a net of 8 oranges at £1.05.
   (b) A 16 oz jar of strawberry jam priced at 76p or a 12 oz jar of the same jam at 58p.
   (c) A '4 Pack' of beer at £4.40 or a '10 Pack' at £10.10.

   (ii) Why do people buy smaller portions if they are more expensive?
3. Twelve bottles of a particular drink weigh 8 kg. How much do fifty bottles weigh?
4. Four people do a job in five days. How long will it take two people to do a similar job?
5. Seven apples cost 56p. How much would 12 cost?
6. A high-speed train travels 4 km in 2 minutes. How long will it take to go 5 km?
7. A journey takes 2 hours when travelling at 60 kilometres per hour. How long does the same journey take when traveling at 50 kilometres per hour?
8. A catering company provides two bottles of wine for every five people at a dinner. How many bottles are needed for 136 people?
9. A tomato grower reads on the side of a 125 g box of Tomogro tomato feed that it is sufficient for 15 plants. His glasshouse holds 800 plants, and Tomogro is available in commercial 10 kg packs. How many such packs should he buy?
10. A can of Synthaprufe waterproofer holds 5 litres and the recommended coverage is 1 litre per square metre for the first coat and 1 litre per 1.5 square metres for the second coat. How many cans will Mel require to give a 20 m^2 wall 2 coats if she uses the recommended coverage rate as a guide?

Activities

1. Go to a grocery shop or supermarket and find 6 products which are for sale in at least 2 sizes, such as cans of baked beans, packets of cornflakes or bottles of shampoo. Make a note of the pack price and of the contents for each one. Investigate whether the bigger pack always represents better value for money.
2. The table below shows the number of votes cast for each of the main political parties at the 1992 General Election. It also shows how many Members of Parliament for each party were elected as a result.

Party	No. of votes	No. of seats
Conservative	14 089 722	343
Labour	11 567 764	273
Liberal Democrat	6 027 552	18
Other	1 934 052	25

   (i) From the table, find
   (a) the number of Conservative votes cast per Conservative MP;
   (b) the number of Labour votes cast per Labour MP;
   (c) the number of Liberal Democrat votes cast per Liberal Democrat MP.

   (ii) If we had 'proportional representation', the 659 seats in Parliament would be allocated in proportion to the number of votes cast for each party. Under this system, how many seats would be held by each of these three parties?
3. A garage displays this graph to allow the customers to convert fuel consumption in miles per gallon into kilometres per litre.

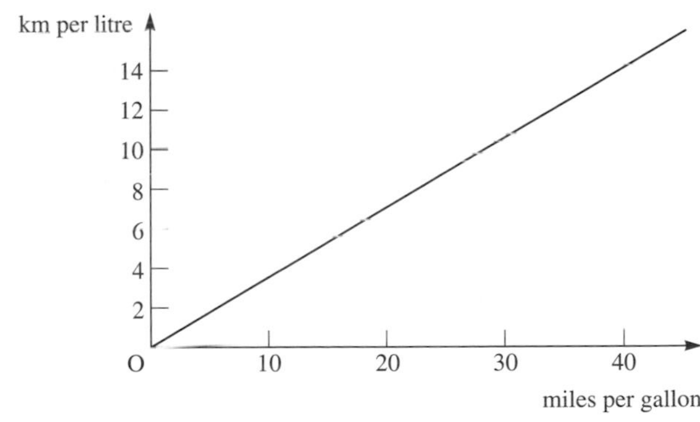

   (i) Use the graph to estimate how many kilometres per litre are equivalent to 36 miles per gallon.
   (ii) Given that 1 gallon = 4.55 litres and 1 mile = 1.609 km, calculate the answer you found from the graph in part (i).
   (iii) Explain why the graph is a straight line.

Accuracy

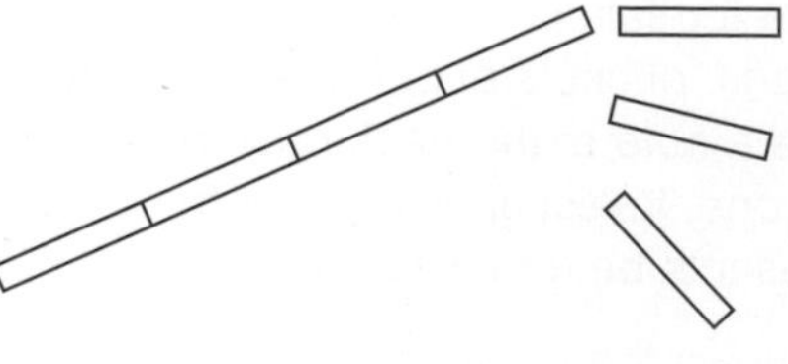

In a manufacturing process, a 1200 mm length of steel is cut into 7 equal pieces. The calculation 1200 ÷ 7 performed on a calculator gives the answer 171.42857. Although mathematically accurate, the answer is misleading: the degree of accuracy is far beyond what is attainable or needed in any engineering process.

A 14-year-old student would probably measure the bar with a plastic rule and cut it with a hacksaw. She might record her cut length as 171 mm. A precision engineer in a large manufacturing company might use the latest computer/laser technology and cut and record a length of 171.43 mm. In both cases the answer is given to an appropriate level of accuracy, taking account of who did the cutting and with what equipment.

When the student says her piece is 171 mm long, she is actually saying that it is closer to 171 mm than it is to 170 mm or 172 mm; that the length of the rod is between 170.5 mm and 171.5 mm. If its real length is 171.9 mm, then she is telling a lie! Look at the three cases below.

Actual length (mm)		Quoted length (mm)
171.29	→	171
171.7	→	172
171.5	→	172

This is nearer 171 than 172 because .29 is less than .5

This is nearer 172 because .7 is greater than .5

This is exactly half way between 171 and 172. In this case we usually round up, i.e. to 172.

This example involved rounding to the *nearest whole number.* Sometimes this is not accurate enough for your needs. Instead you may wish to specify a number of *decimal places.* In all rounding, look at the digit in the next place and compare it with 5.

Example

Write (i) 0.00216 (ii) 32.1567 to 3 decimal places.

Solution

You are asked for 3 decimal places, so look at the digit in the 4th place.

(i) $0.002\underline{1}6 \rightarrow 0.002$ (ii) $32.156\underline{8} \rightarrow 32.157$

1 is less than 5. 21 rounds down to 2.

8 > 5. 68 rounds up to 7.

Sometimes you need to take note of the size of a number when deciding how to round it. You do this by giving a number of *significant figures* as in the next example.

Example

Write (i) 23.418 (ii) 0.00213826 to 3 significant figures.

Solution

(i) 23.418
↓
23.4

Start at the left and find the first non-zero digit. That and the next 2 digits are the '3 significant figures'. You need to look at the 4th digit to decide whether to round digit 3 up or leave it unchanged.

(ii) 0.00213826
↓
0.00214

Exercise

1. (i) Round off the following numbers to the nearest whole number:
   (a) 7.1 (b) 6.243
   (c) 10.826 (d) 66.6.

   (ii) Round off the following numbers to the nearest 10:
   (a) 47 (b) 85
   (c) 201 (d) 174.5
   (e) 1837.

   (iii) Round off the following numbers to the nearest 100:
   (a) 139 (b) 873
   (c) 753 (d) 189.2
   (e) 165.

2. Which of the following statements are true and which are false? For each of the false statements, write down the correct version.
   (i) The number of patients treated in a hospital in one year was 20 963. This is 21 000 to the nearest hundred.
   (ii) The average length (measured with a micrometer) of the drawing pins in a box is 0.928 cm. This is 9 mm to the nearest millimetre.
   (iii) A pop-riveting machine produced 2379 rivets in half an hour. This is 4800 rivets per hour, to the nearest 50 rivets.
   (iv) The travel expenses claim form for a salesman said 'Norwich to Newcastle: 257 miles'. This is 260 miles, to the nearest 5 miles.

3. **New road bridge to cost £2,371,214.98**

   Comment.

4. (i) Write the following numbers correct to 1 significant figure:
   (a) 164 (b) 1739
   (c) 456.5 (d) 0.72
   (e) 14.11.

   (ii) Write the following numbers correct to 3 significant figures:
   (a) 1857 (b) 143.9
   (c) 76.32 (d) 5285
   (e) 16.439.

   (iii) Write the following numbers correct to 2 decimal places:
   (a) 0.579 (b) 0.445
   (c) 32.627 (d) 1.7837
   (e) 10.07.

5. Min has paid the phone bill, and is now calculating the amount of money owed to her by each of her flatmates for phone calls. She gets the following answers on her calculator. Write down what each person should actually pay.
   Simon owes £1.124
   Sharon owes £3.106
   Min owes £5.4555
   Chris owes £4.15921

6. A travelling salesman claims 29p per mile for his car journeys. What would he claim for a journey of 295 miles, to the nearest £1?

7. A restaurant bill came to £47 and although they had eaten different meals, three friends decided to share the bill equally between them. How much should each put into the kitty if a suitable tip is to be included? Show your reasoning.

Activity

The significant figures game
(for any number of players)

1. Feed an 8-figure number into your calculator.
2. Throw a die and divide the 8-figure number by the number shown on the die.
3. Throw the die again and round off the number on the display to the number of significant figures shown by the die.
4. Your score is the last non-zero digit of the number now displayed.

   Now it is the next person's turn. The player with the highest score wins.

Example:

Feed in 12 342 668

Throw a 2 → 12 342 668 ÷ 2 = 6 171 334

Throw a 3 → 6 170 000

Score: 7

Estimation

How many potatoes do you get in a 25 kg sack?

At first sight this question may seem rather silly. The answer obviously depends on how big the potatoes are and that involves knowing their type, and whether they are new or main crop potatoes. Even if you know that, there is no definite answer: two sacks of the same type are very likely to contain different numbers.

There are many situations like this where you want to know a rough answer; it does not matter that it may not be very accurate. This involves *estimation*. The original question would have been better worded as *Estimate the number of potatoes in a 25 kg sack.*

The work might then proceed as follows.

Description	Typical mass	Calculation	Estimate (sensibly rounded)
Main crop large	300 g	$25\,000 \div 300 = 83\frac{1}{3}$	80
Main crop medium	150 g	$25\,000 \div 150 = 166\frac{2}{3}$	170
New	50 g	$25\,000 \div 50 = 500$	500

You will often give your estimate not as a single number but as a range of values, say 150 to 200 for the main crop medium potatoes.

Estimation is very important in everyday life. It is your first check that things seem sensible, that the answer to a calculation has the right *order of magnitude*.

Sometimes you will do a rough estimate of a calculation to check that it is reasonable to continue, and that you have not made an obvious mistake.

Example

Give a rough estimate of $\dfrac{2.2 \times 39.95}{112}$

Solution

This can be written roughly as $\dfrac{2 \times 40}{100} = 0.8.$

NOTE

There is no right answer to a rough estimate. It depends how rough you are prepared for it to be.

Exercise

1. Estimate
   (i) how many cigarettes are smoked in Britain each year;
   (ii) how many times the average heart beats in a lifetime;
   (iii) the number of seconds you have been alive;
   (iv) the number of hot dinners you have eaten in your life;
   (v) the number of hours of sleep you have had this year.

   In each case, show your reasoning.

2. Estimate the answers to the following calculations, showing your reasoning.

(i) 7.4×1.95	(ii) 25.3×3.4
(iii) $396 \div 10.3$	(iv) $66.7 - 11.8$
(v) 1.96×800	(vi) $(19.1 \times 20.1) \div 49$
(vii) 611×1.07	(viii) $68.9893 + 0.99435$
(ix) $1520 \div 52$	(x) $99.56 \div 0.49$

3. Every day a business sends about 600 first-class letters and 1100 second-class letters.

   Estimate the cost of their postage for a year, at the current postage rates, showing your reasoning.

4. (i) Estimate the cost of Dave's electricity bill from the figures below. Show your working.

Standard quarterly charge:	£9.13
Units used:	1182
Price per unit:	9.33p
VAT (payable on total amount):	8%

   (ii) Compare answers with other students in your group. Who was the closest to the correct answer? Was his or her method better than yours?

5. Look at the calculation on the opposite page: $\frac{2.2 \times 39.95}{112}$. How might it have arisen?

6. A standard building brick measures 23 cm by 11 cm by 6.5 cm. How many bricks would you order to build the walls of the garage in the diagram, which are 2 m high and 11 cm thick? (A few of the bricks may be broken.)

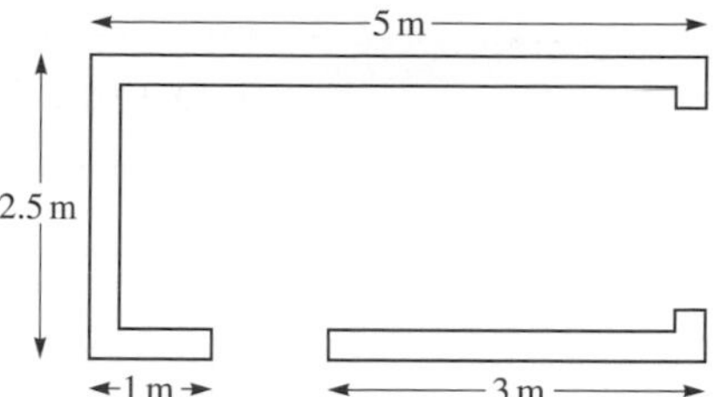

7. A naturalist is studying the number of eggs found in the birds' nests in a particular copse. His results are shown below.

1	5	4	16
20	1	5	17
2	5	20	1
2	15	4	6.

   How many different species do you think were involved in the survey? Why?

8. A weights and measures inspector is testing some drinks for alcohol content. The types of drink being tested are spirits, fortified wine, beer and shandy. Here are her results.

76.2	3.4	8.1	7.9
2.9	63.6	2.6	13
14.1	65.3	54	

   How many brands of each type of drink do you think she tested?

9. It is easy to press a wrong key on your calculator. Look at each of the following calculations, and the answers given. In which ones would you suspect immediately that a wrong key had been pressed?

	Calculation	Answer displayed
(i)	4×67	268
(ii)	400×67	2680
(iii)	$4 \times \pi \times 6^2$	75.3…
(iv)	2×16.2	32.4
(v)	$450 + 452 - 96$	998

Activities

1. (a) In a copy of your local paper find 5 cars whose prices are of the same order of magnitude. List their prices.
   (b) Write down the prices of 3 second-hand cars whose prices all have different orders of magnitude.

2. The approximate answer to a calculation was quoted as 100. Write down 5 possible calculations (preferably using the operators +, −, ×, ÷, √ or a mixture of these) that could give rise to this answer. Explain your reasoning.

Using rounded figures in calculations

Always be careful when using rounded figures in calculations. The errors involved may build up, as you will see in the two examples on this page.

Example

A rectangular piece of ground is described as 7 m long and 4 m wide, (to the nearest 1 m). Find

(i) the smallest and largest possible values of its area;

(ii) the greatest possible percentage error in taking the area to be $28\,\text{m}^2$.

Solution

The length (7 m) lies between 6.5 m and 7.5 m.
The width (4 m) lies between 3.5 m and 4.5 m.

(i) The smallest possible area is $6.5 \times 3.5 = 22.75\,\text{m}^2$
The largest possible area is $7.5 \times 4.5 = 33.75\,\text{m}^2$

(ii) Smallest area (in m^2): Possible actual error $= 28 - 22.75 = 5.25$

Possible % error $= \dfrac{5.25}{28} \times 100 = 18.75\%$

Largest area (in m^2): Possible actual error $= 33.75 - 28 = 5.75$

Possible % error $= \dfrac{5.75}{28} \times 100 = 20.54\%$

The greatest possible error is 20.54%, a large figure.

Example

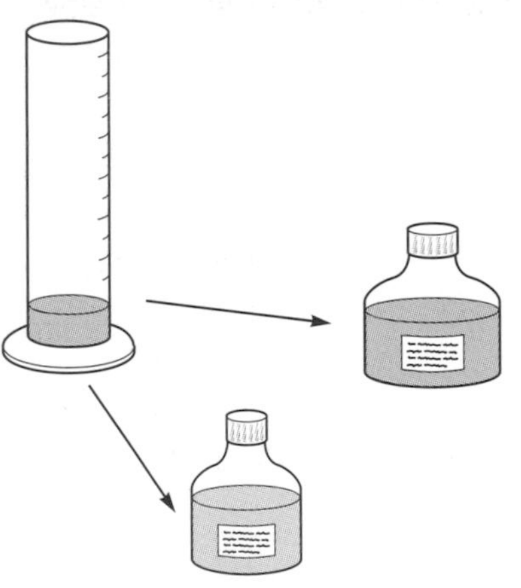

A pharmacist is supplied with a chemical in a bottle that contains 10.0 ml (correct to the nearest 0.1 ml). He uses it to make up two different drugs: the first requires 9.5 ml and the second 0.5 ml. He starts by taking 9.5 ml (correct to 0.1 ml) for the first drug and then uses what remains for the second drug. Find the largest possible percentage error in the amount used in the second drug.

Solution

The volume supplied (10.0 ml) lies between 9.95 ml and 10.05 ml.

The volume taken for the first drug lies between 9.45 ml and 9.55 ml.

Smallest possible amount remaining (ml) $= 9.95 - 9.55 = 0.40$

Possible actual error (ml), for the second drug $= 0.50 - 0.40 = 0.10$

Possible % error $= \dfrac{0.10}{0.50} \times 100 = 20\%$

The greatest possible amount remaining (ml) $= 10.05 - 9.45 = 0.60$

Possible actual error (ml) $= 0.60 - 0.50 = 0.10$

Possible % error $= \dfrac{0.10}{0.50} \times 100 = 20\%$

So, despite the high level of accuracy of the measurements involved, the volume used in the second drug could be as much as 20% in error, either too much or too little (sometimes written as ± 20%). How could the pharmacist improve his procedures?

Exercise

1. State the maximum and minimum possible values for these measurements:
   (i) the length of a car journey, recorded as 15 miles to the nearest mile;
   (ii) the weight of a can of tomato soup, recorded as 405 g to the nearest gram;
   (iii) the time to boil an egg, recorded at 4 minutes 20 seconds to the nearest 10 seconds;
   (iv) the volume of medicine in a dose, recorded as 7.5 ml to the nearest 0.1 ml;
   (v) the temperature at a holiday resort, recorded as 17.3°C to the nearest 0.1°C;
   (vi) the time for a 100 m race, recorded as 10.16 seconds to the nearest $\frac{1}{100}$ of a second.

2. An architect is measuring the width of a bungalow when her tape-measure snaps. She decides to use what remains of the tape-measure, as shown in the diagram. The readings are to the nearest 1 cm. She subtracts 12 cm from 729 cm to find the width.
   (i) What is the maximum possible width of the bungalow?

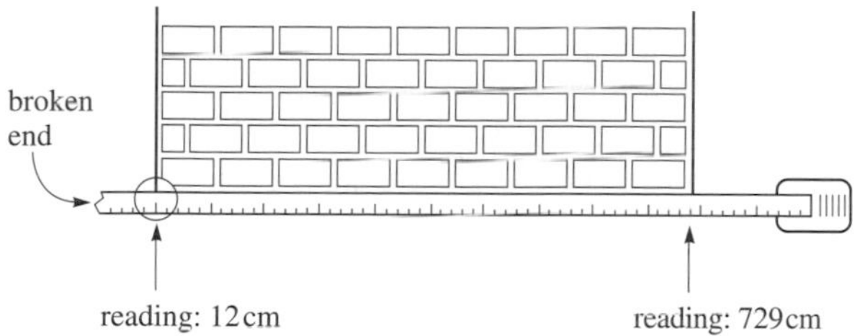

   (ii) What is the maximum possible percentage error in taking the width as 717 cm?

3. In the diagram below, all measurements are given to the nearest 10 cm.

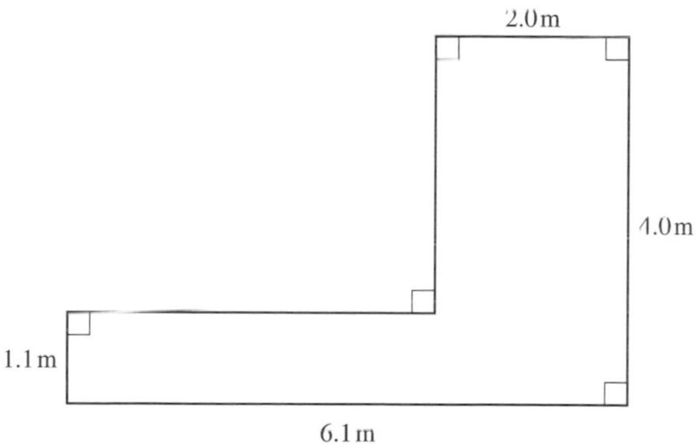

Find the smallest and largest possible values of the area, and the greatest possible percentage error in taking the area as 12.51 m^2.

4. Gold is a very expensive metal to buy. Assume that it costs £12.95 a gram. Find
   (i) the maximum value (to the nearest 1p) of the gold in a ring which weighs 18 grams to the nearest gram
   (ii) the maximum value (to the nearest 1p) of the gold in a ring which weighs 18 grams to the nearest 0.1 gram
   (iii) the maximum value (to the nearest 1p) of the gold in a ring which weighs 18 grams to the nearest 0.01 gram

   Using your answers to this question, explain why a jeweller would want to weigh gold as accurately as possible.

5. (i) You are to time a relay race. Is it more accurate to time it from start to finish, or to time each leg separately and add the four answers together? (You should make up some realistic figures to help you with your argument.)
   (ii) A nurse is to give 20 ml of medicine to a patient. Is it more accurate to give one 20 ml dose (measured to the nearest ml) or four 5 ml doses (measured to the nearest ml)?

6. A Premier League football club decided to reseed its pitch. The groundsman knew that the dimensions of the pitch were 110 m by 65 m. His measurements were taken to the nearest metre. The grass was supplied in 10 kg bags with information that 1 bag would cover 250 square metres.

   The groundsman's calculation was as shown below:

   $$\text{Area of pitch (m}^2) = 110 \times 65 = 7150.$$

   $$7150 \div 250 = 28.6.$$

   He ordered 29 bags of seed.

   If he applied the seed at the recommended rate of coverage, could he guarantee that he had enough grass seed?

7. In an experiment the speed of a mouse is estimated by timing how long it takes to run along a measured length. The speed is given by $v = \dfrac{d}{t_2 - t_1}$.

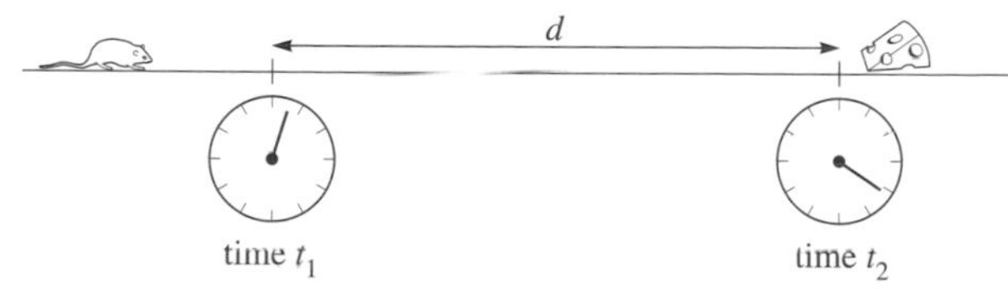

In one such experiment, $d = 1.000$ m (to nearest mm)
$t_2 = 1.4$ s (to nearest 0.1s)
$t_1 = 0.2$ s (to nearest 0.1s)

Between what values does the speed of the mouse lie?

Perimeter and area

The **perimeter** of a shape is the distance around it. The perimeter is measured in units of length.

The **area** of a shape is the amount of surface contained within it. Units of area include square centimetres (cm^2) and square metres (m^2).

Areas of common shapes

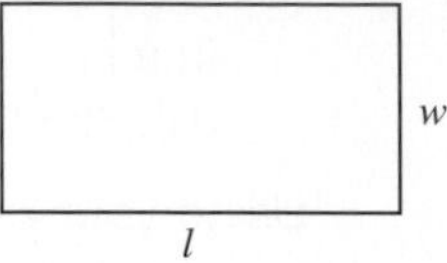

Rectangle: area = length × width

$A = lw$

Scalene triangle: area = $\frac{1}{2}$ base × height

$A = \frac{1}{2}bh$

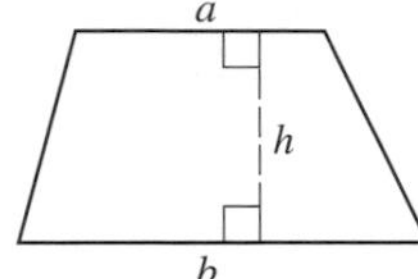

Trapezium:
area = $\frac{1}{2}$ sum of parallel sides × distance between them

$A = \frac{1}{2}(a + b)h$

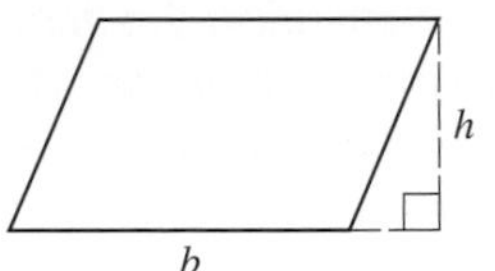

Parallelogram:
area = one side × distance to parallel side

$A = bh$

You may need to break down a shape into rectangles, triangles and trapezia. Sometimes it may be easier to think of the shape as a large rectangle with pieces removed, as in the next example.

Example

What is the area of card required to make the net of the cuboid shown below, which is to be used for packaging a clock?

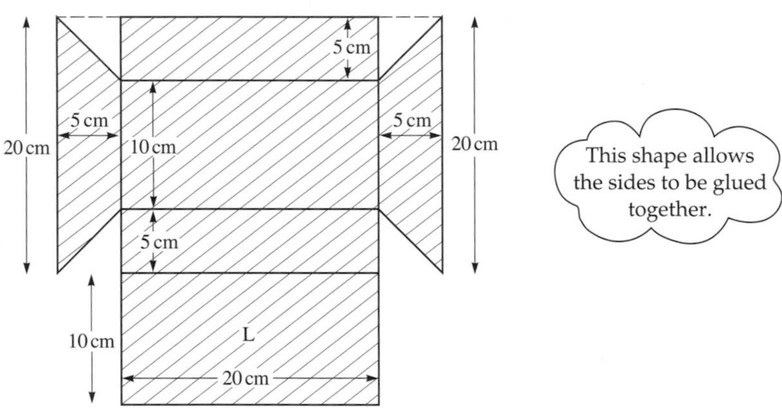

Solution

The shape can be treated as two parts.

1. The lid, L, is a rectangle 20 cm by 10 cm. Its area is 200 cm^2.
2. The rest is a rectangle, 20 cm by 30 cm, an area of 600 cm^2, with 4 triangles removed. The area of each triangle in cm^2 is $\frac{1}{2} \times 5 \times 5$, so the total area removed is $4 \times \frac{1}{2} \times 5 \times 5 = 2 \times 25 = 50$. The area of the rectangle with these triangles removed is therefore 550 cm^2.

   The total area of the net is $(200 + 550)\,cm^2 = 750\,cm^2$.

Exercise

1. A square has a side of 6 cm. Calculate
   (i) the perimeter; (ii) the area.

2. A rectangle is 8 m long and 5 m wide. Calculate
   (i) the perimeter; (ii) the area.

3. Find the areas of these shapes.

(i)

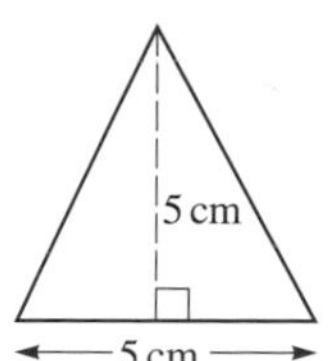

(ii)

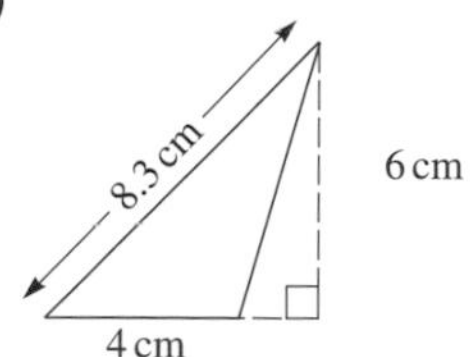

(iii)

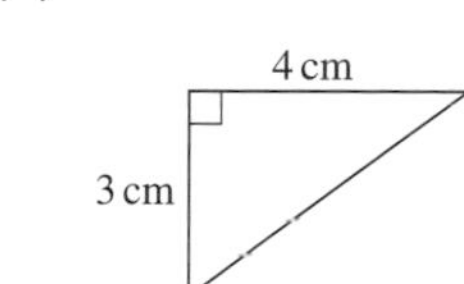

(iv)

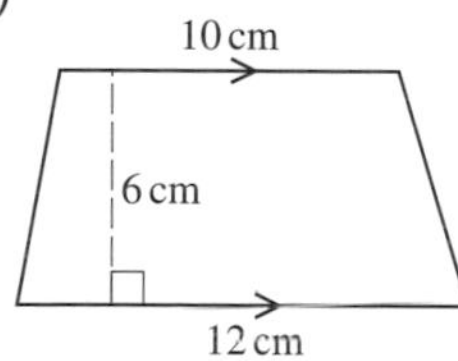

(v)

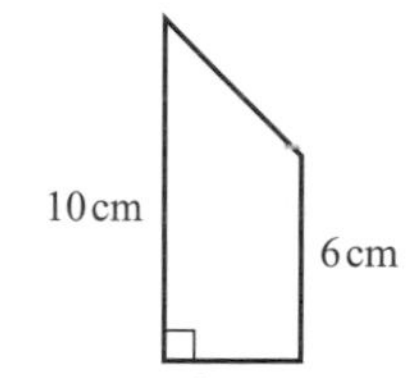

(vi)

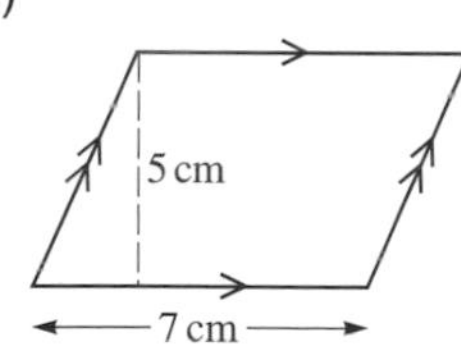

4. Work out the areas of these shapes.

(i)

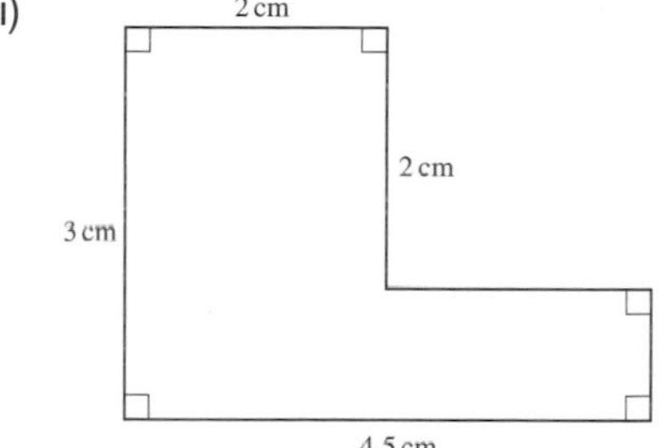

(ii)

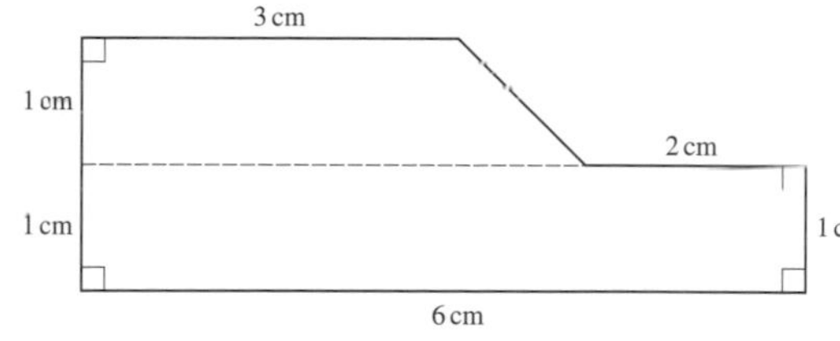

5. The diagram shows an aluminium panel used by an air-conditioning company. Its front surface has to be spray-painted before it is installed. Work out the surface area (shaded) that is to be painted.

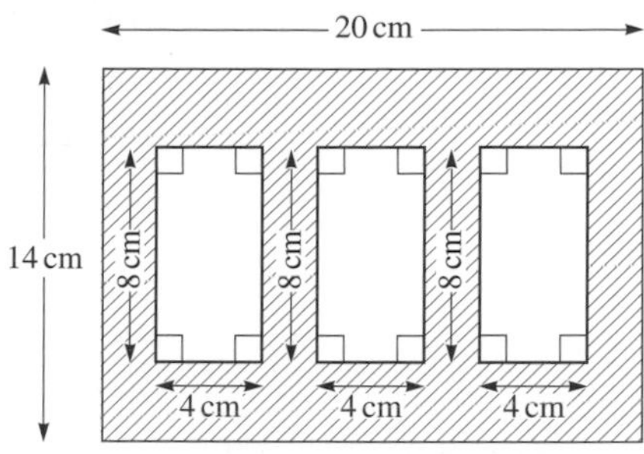

6. What area of gold leaf is needed to coat this brooch?

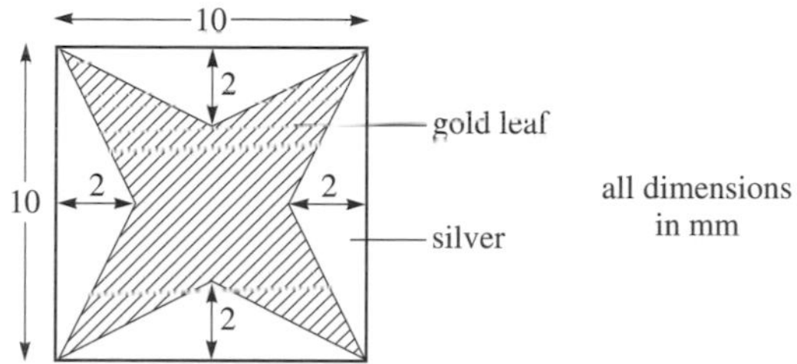

7. The diagram shows the plan of a small bungalow.
   (i) How much carpet is needed to carpet the lounge, dining room and hall?
   (ii) How many 15 cm by 15 cm vinyl tiles will be needed to cover the kitchen floor? (You can assume the thickness of the walls to be negligible).

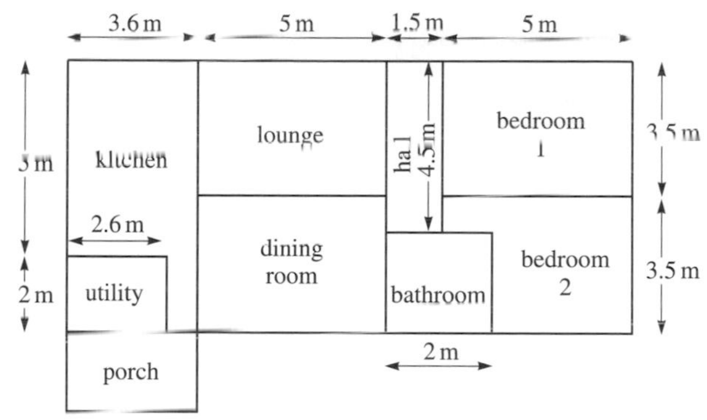

Activity

Choose an irregular-shaped object, such as a fishing reel, a teapot or a large brandy glass, and design a box to package it in. Any tabs you use to hold it together should be triangles.

Draw out the net of your box on a piece of stiff card, then construct the box. How much of the card did you waste? Could you improve the design?

The circle

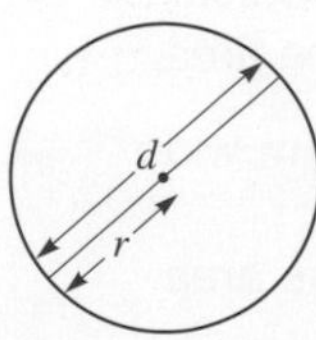

The diameter of a circle is twice, the radius $d = 2r$.

The perimeter of a circle is called its *circumference* and is given by:

$$C = \pi d \quad \text{or} \quad C = 2\pi r.$$

The value of π is approximately 3.14. Your calculator gives the value more accurately as 3.141592…

Example

How far does a bicycle travel for each full rotation of the front wheel, given that the wheel has a radius of 31 cm?

Solution

The distance travelled (in cm) is given by the circumference of the wheel.

$$\begin{aligned} C &= 2\pi r \\ &= 2 \times \pi \times 31 \\ &= 194.7787\ldots \end{aligned}$$

So the bicycle goes forward by 195 cm or 1.95 m (to 3 significant figures) every time the wheel goes round.

To work out the area of a circle, you again need to know either its radius or its diameter. The area of a circle is given by

$$A = \pi r^2$$

Example

What is the area of a potter's wheel whose radius is 15 cm?

Solution

The area (in cm^2) is given by $A = \pi r^2$: $\quad A = \pi \times 15^2 = 706.858\ldots$

The area is 707 cm^2.

Example

What area of gold leaf is needed to cover the shape on the left?

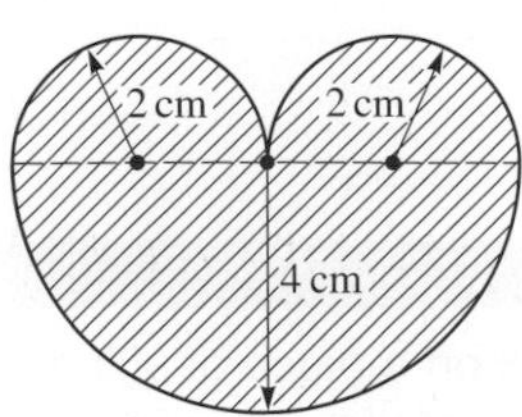

Solution

The large semi-circle has half the area of a circle with radius 4 cm:

$$\text{area (in cm}^2) = \tfrac{1}{2}\pi r^2 = \tfrac{1}{2} \times \pi \times 4^2 = 25.13.$$

Each small semi-circle has radius 2 cm. The two together are equivalent to a whole circle of radius 2 cm:

$$\text{area (in cm}^2) = \pi r^2 = \pi \times 2^2 = 12.57.$$

Total area of gold leaf = (25.13 + 12.57) cm^2 = 37.7 cm^2 (to 3 sig. figs).

Exercise

In all of these questions, use the π key on your calculator and round your answer to 2 decimal places.

1. Work out the circumference and area of each of these circles.

(i)

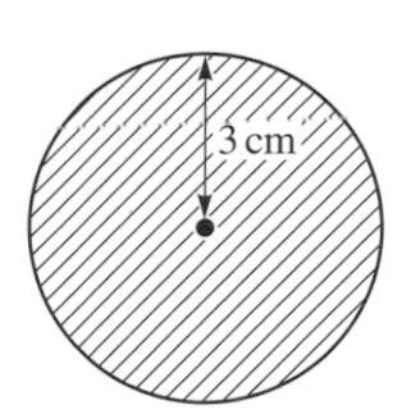

(ii)

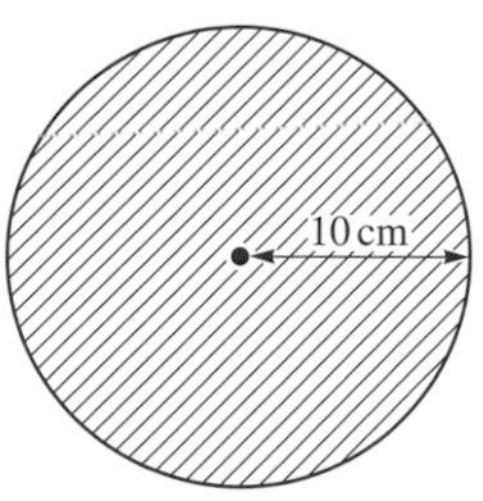

2. Work out the area and perimeter of these shapes.

(i)

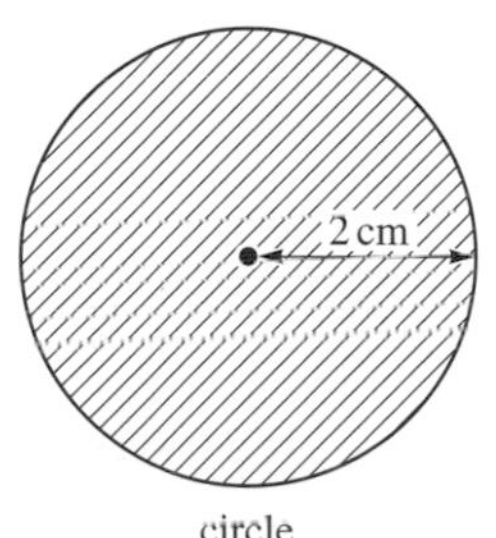

circle

(ii)

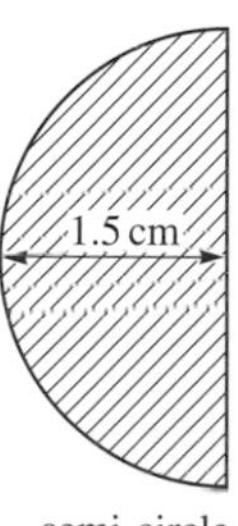

semi-circle

(iii)

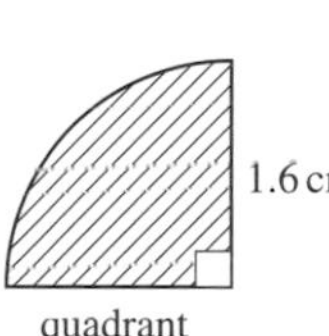

quadrant

(iv)

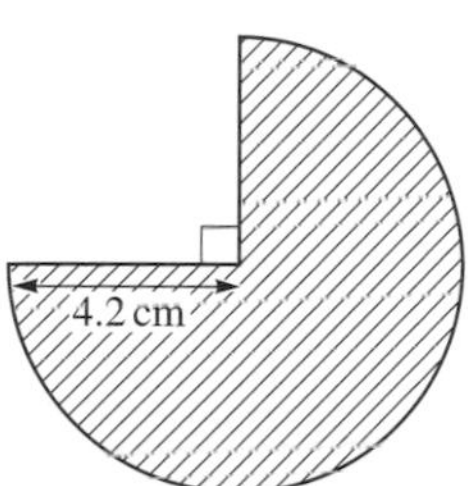

3. Work out the area and perimeter of these shapes.

(i)

85 m

82.6 m

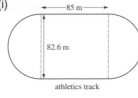

athletics track

(ii)

3 cm

3 cm

3 cm

'tulip' design

4. The diagram shows the cylinder-head gasket of a car. What is its surface area?

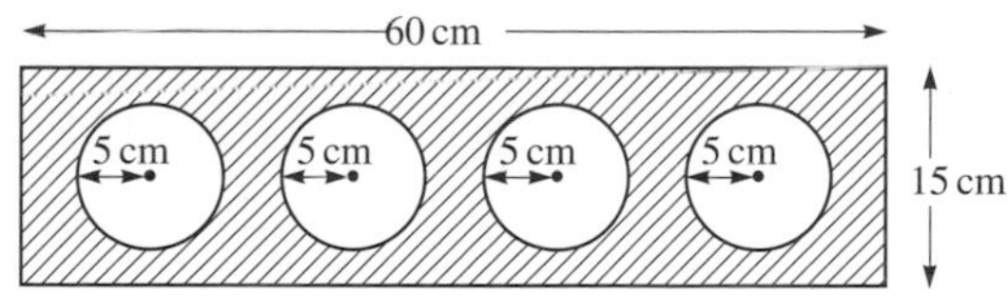

5. A piece of copper pipe has an outside diameter of 40 mm and a bore of 30 mm. Work out the cross-section area of the copper.

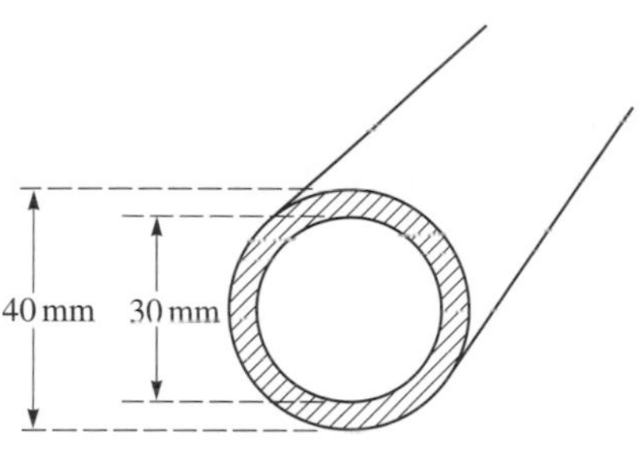

Investigation

Take 5 different cylindrical cans (e.g. baked beans, tuna fish, tomato purée etc.)

For each one,

(i) measure the circumference of its lid with a tape measure or a piece of thread, to the nearest mm;
(ii) measure the diameter of its lid with a ruler, to the nearest mm.

Record your measurements in a copy of the table, then complete the final column

	Circumference (C)	**Diameter (d)**	$C \div d$
Can 1			
Can 2			
Can 3			
Can 4			
Can 5			

What do you notice about the numbers in the last column?

Activity

The value of π is a never-ending decimal. The first twelve digits are

3.14159265359

Make a sentence whose first word has 3 letters, second word has 1 letter, third has 4 and so on, to help you remember these digits. You could start with "How I wish…".

Volume

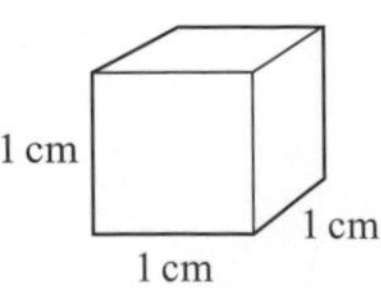

The **volume** of an object is the amount of space it occupies. Units of volume include cubic centimetres (cm^3) and cubic metres (m^3).

The volume of a cube of side 1 cm is 1 cm^3. Note that 1000 cm^3 = 1 litre.

Similarly, the volume of a cube of side 1 m is 1 m^3. This is a surprisingly large volume. A cubic metre of water is 1 000 000 cm^3, or 1000 litres. Its mass is 1 tonne.

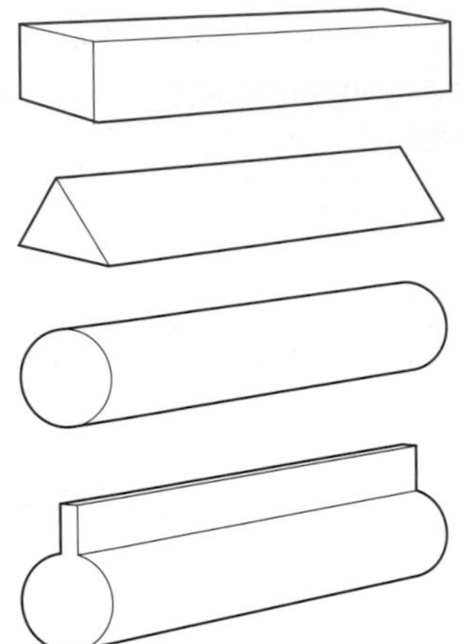

You will sometimes need to calculate the volume of a solid shape. You have already seen how to do this for a cube, but it is also helpful to be able to do it for other shapes. It is straightforward to work out the volume of a body which has uniform cross section. Such shapes are called prisms, and a selection of them are shown on the left.

For any prism, the volume is the cross-section area × the length.

The simplest type of prism is a cuboid. Its cross-section area is given by

$$\text{width} \times \text{height},$$

so its volume is simply

$$\text{width} \times \text{height} \times \text{length}.$$

Another familiar shape is the triangular prism. You start by working out the area of the triangle, as in the next example.

Example

Find the volume of the loft space shown below.

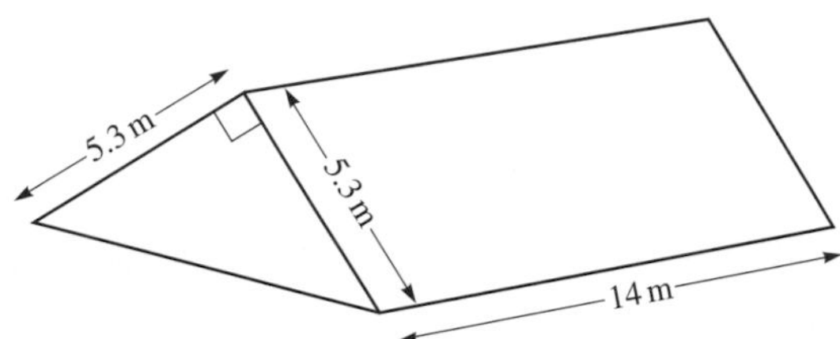

Solution

The cross section is a right-angled triangle, with 'base' 5.3 m and 'height' 5.3 m.

Area of cross section (in m^2) = $\frac{1}{2} \times 5.3 \times 5.3 = 14.045$.

Volume of loft space (in m^3) = cross-section area × length

$= 14.045 \times 14 = 196.63$

Volume of loft = 197 m^3 to 3 significant figures.

Another very common type of prism is the cylinder, which has a circular cross section. You will learn more about cylinders on the next spread.

Exercise

1. Calculate the volumes of these cuboids.

(i)

4 cm, 5 cm, 20 cm

(ii)

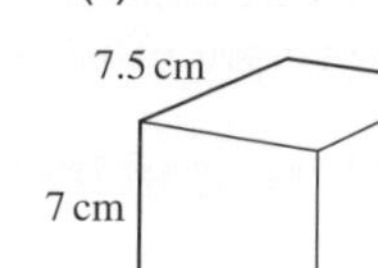

2. The solids below are made up of cuboids. Calculate the volume of each one.

(i)

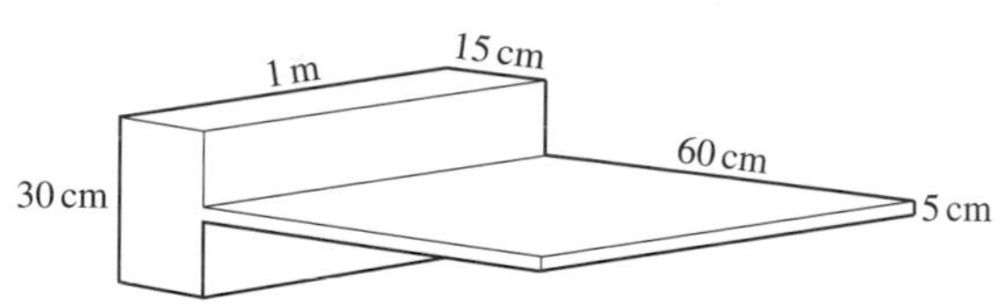

(ii)

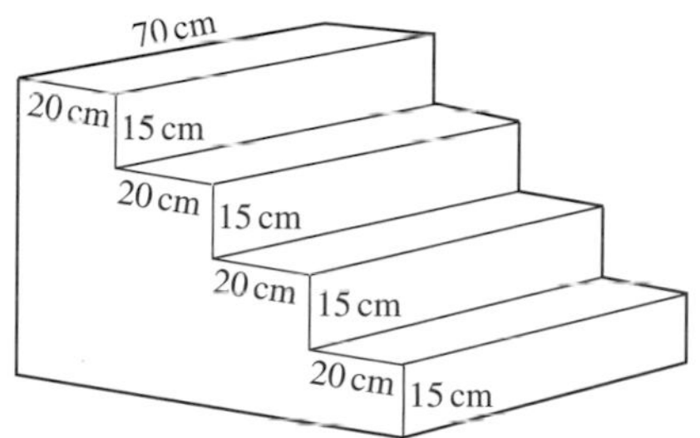

3. The cross-section areas of the solids below are as shown. Calculate their volumes.

(i)

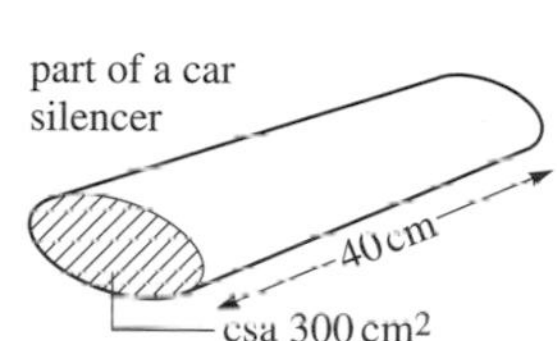

(ii)

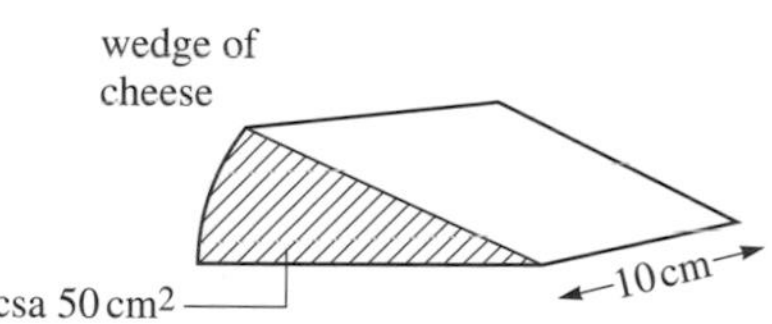

4. The diagram shows the plan view of a triangular patio area. It is to be covered with concrete $15\,\text{cm}$ deep. What volume of concrete should be ordered?

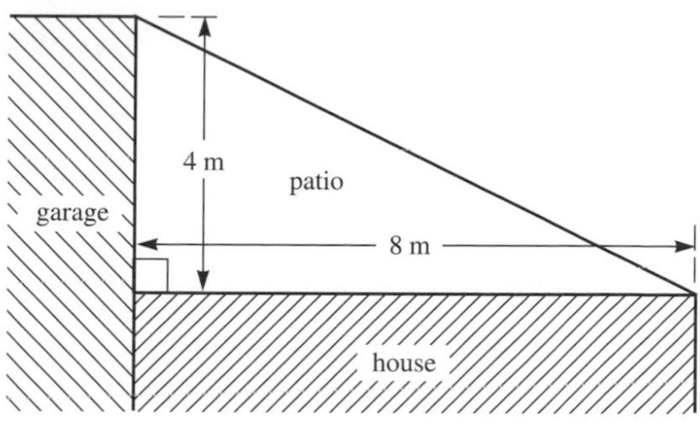

5. The swimming pool shown is to be filled with water right to the rim. What volume of water is required?

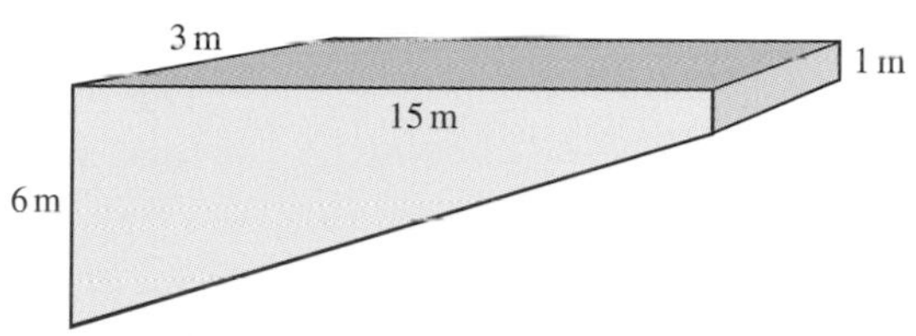

6. A box has dimensions $3\,\text{m} \times 2\,\text{m} \times 1\,\text{m}$.
   (i) State these dimensions in (a) cm and (b) mm.
   (ii) Find the volume of the box in
   (a) m^3 (b) cm^3 (c) mm^3
   (iii) How many cubic centimetres equal one cubic metre?

Activities

1. Find a number of 1 litre containers that are prisms, such as a fruit juice carton, a milk carton and a washing-up liquid bottle (the latter may only approximate to a prism). Measure their dimensions.
   (i) How much 'air-space' is allowed within each container?
   (ii) Which container has the smallest surface area?

2. *If I double the dimensions of a cuboid, I double its volume.* True or false?
   Without using a calculator, write down the factor by which you must increase the dimensions of a cuboid in order to
   (i) double its volume;
   (ii) treble its volume;
   (iii) create a volume 27 times as large.

Cylinders

A cylinder is a prism with a circular cross section. The cylinder and the cuboid are the prisms that we see most often in everyday life.

Surface area of a cylinder

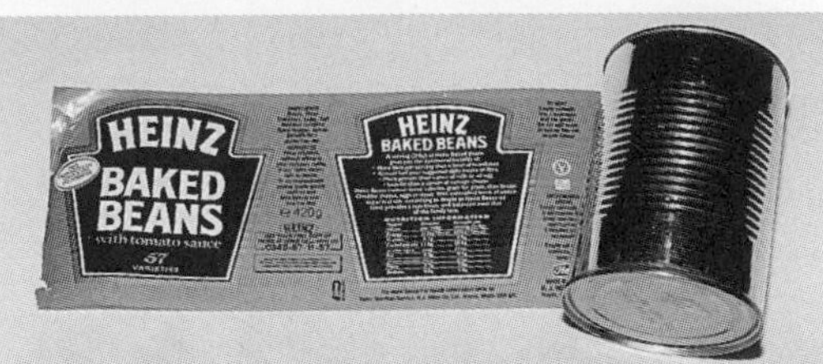

Look at the can of beans and its unrolled label. The height of the label is the same as the height of the can. The width of the label is the same as the circumference of the lid of the can.

Area of label = width × height = $2\pi r \times h$.

Area of curved surface of cylindrical can = $2\pi rh$.

The lid and base of the can are both circles, each with area πr^2, so the total surface area of a cylinder is

$$A = 2\pi rh + 2\pi r^2.$$

Volume of a cylinder

As for other prisms, the volume of a cylinder is its cross-section area multiplied by its height (or length).

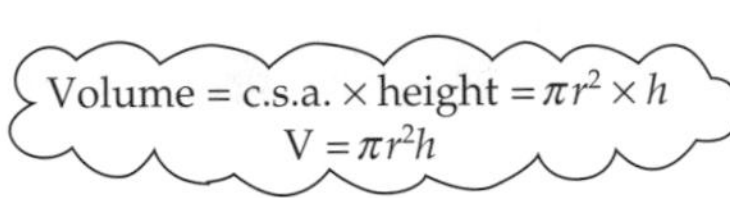

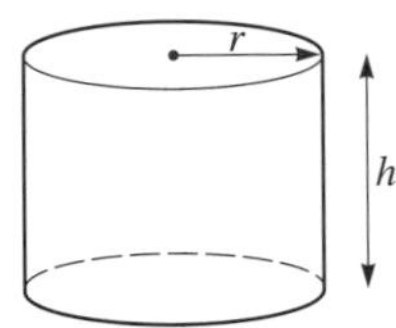

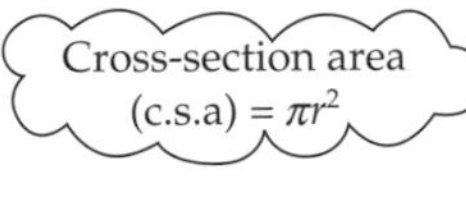

Example

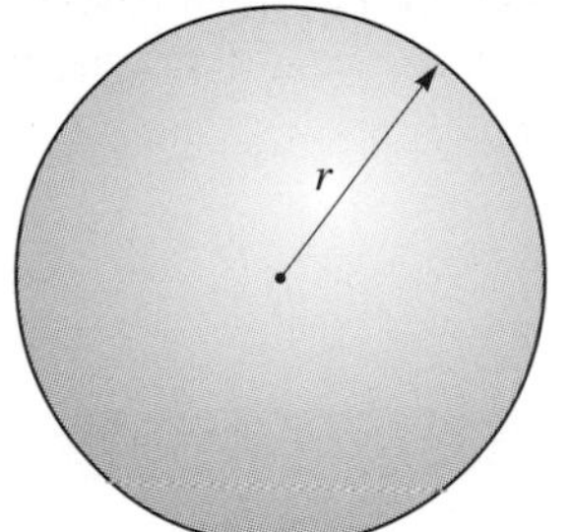

What is the volume of a can of baked beans of radius 3.7 cm and height 10.7 cm?

Solution

$$V = \pi r^2 h$$
$$= \pi \times 3.7^2 \times 10.7$$
$$= 460.189\ldots$$

When using these formulae, make sure all the quantities are in the same units. Here, r and h are in cm, so V will be in cm^3.

The volume is 460 cm^3 to 3 significant figures.

Spheres

The surface area of a sphere is given by the formula $A = 4\pi r^2$

The volume of a sphere is given by the formula $V = \frac{4}{3}\pi r^3$

The mathematics needed to prove these formulae is quite complicated and at this stage you are only expected to use them rather than prove them.

Exercise

1. Work out the volume of these cylinders, giving your answer in the appropriate units.
   (i)

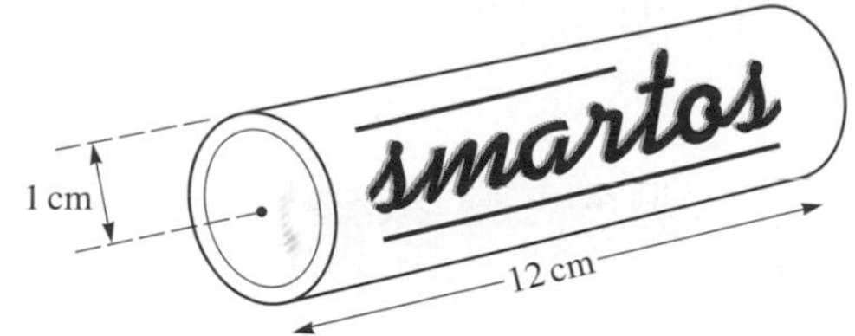

   (ii)

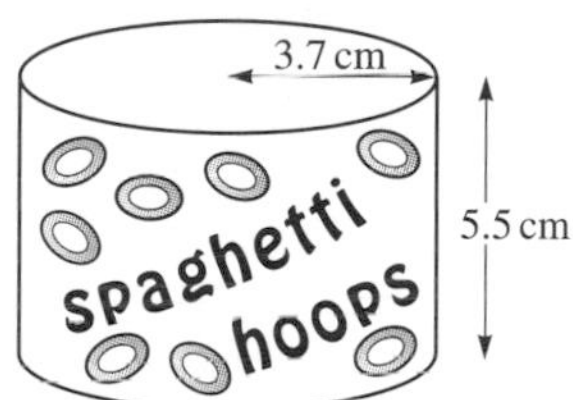

   (iii)

2. The pistons in a car engine move back and forth inside cylinders. The volume swept out by each piston is called the engine 'capacity' and is an indicator of its power.
   In a particular engine, the cylinder diameter is 93 mm and the stroke (the distance travelled by the piston) is 23 cm. What is the capacity of this engine in
   (i) cm^3 (or cc) (ii) litres?

3. (i) Blood is taken from a patient using a syringe which has a radius of 0.5 cm and can be drawn back a maximum distance of 10 cm from the empty position. What is the maximum amount of blood that can be taken from a patient using this syringe?
   (ii) Another syringe has double the radius but the same length. What is the maximum amount for this syringe?

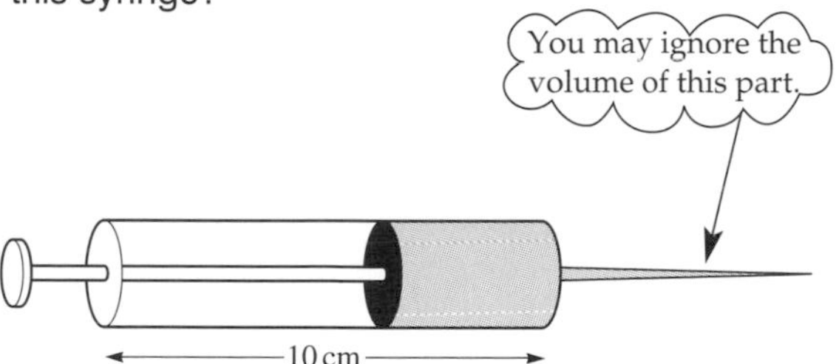

4. What is the volume and surface area of each of the following spheres?
   (i)

ball bearing, radius 1 mm

   (ii)

football, radius 12.5 cm

5. A lump of lead in the shape of a cube whose sides are 3 cm long is melted down to make spherical shotgun pellets which have a radius of 1 mm. How many pellets can be made?

Activity

Work out the volume and surface area of the Earth assuming its radius to be 6370 km.

Convert the volume of the Earth from cubic kilometres to cubic metres.

The mass of the Earth is approximately 5.98×10^{24} kg. Divide the mass in kilograms by the volume in cubic metres to find the average mass of 1 cubic metre of the Earth.

The mass of a cubic metre of soil is approximately 1000 kg. Compare this with the mass that you calculated. Why do you think the two values are different?

Pyramids

A solid with a polygon-shaped base and triangular faces which meet at a vertex above the base is called a pyramid. The diagram shows a square-based pyramid.

The volume of a pyramid is

$\frac{1}{3}$(area of base) × height.

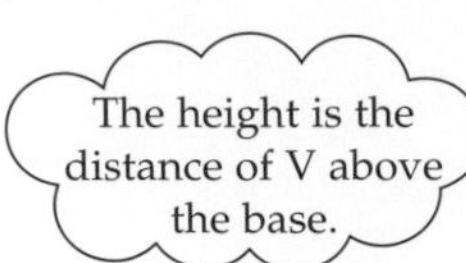

The height is the distance of V above the base.

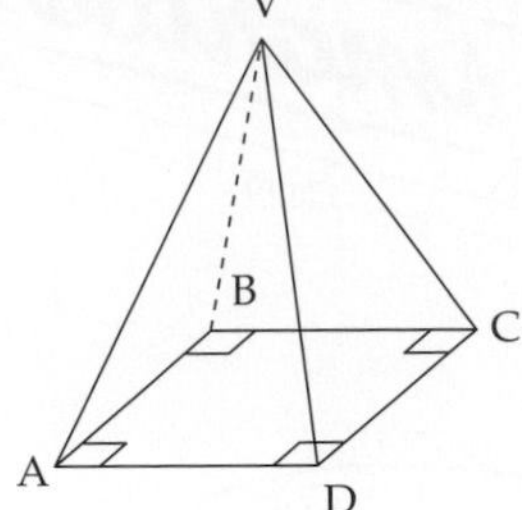

The surface area of this pyramid can be found by adding the area of the base and the areas of the four triangular faces.

Example

Calculate the volume of a pyramid which has a square base of side 12 cm and a height of 20 cm.

Solution

$V = \frac{1}{3}$(area of base) × height

$= \frac{1}{3}(12 \times 12) \times 20 = 960$

The volume is 960 cm^3.

Cones

A solid with a circular base and a vertex above the centre of the base is a called a cone.

The volume of a cone is

$\frac{1}{3}$(area of base) × height

$= \frac{1}{3}(\pi r^2) \times h$

The base is a circle of radius r.

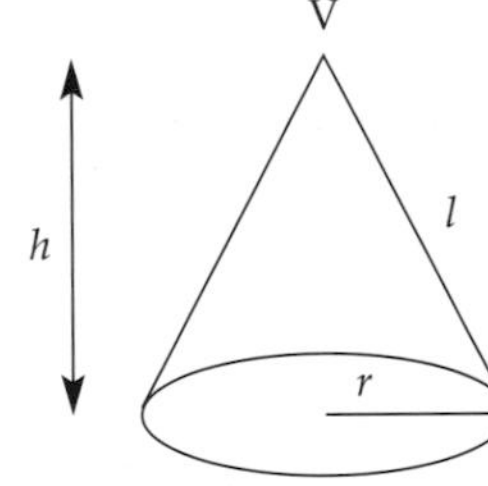

The curved surface area of a cone is

πrl

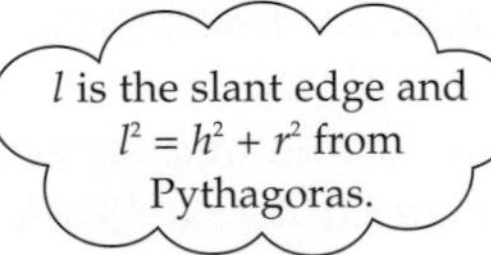

l is the slant edge and
l² = h² + r² from
Pythagoras.

The total surface area is

$\pi r^2 + \pi rl$

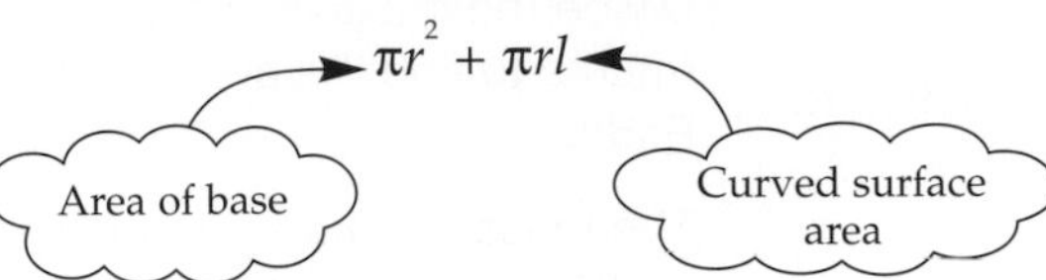

Exercise

1. Calculate the volume of these pyramids.

(i)

(ii)

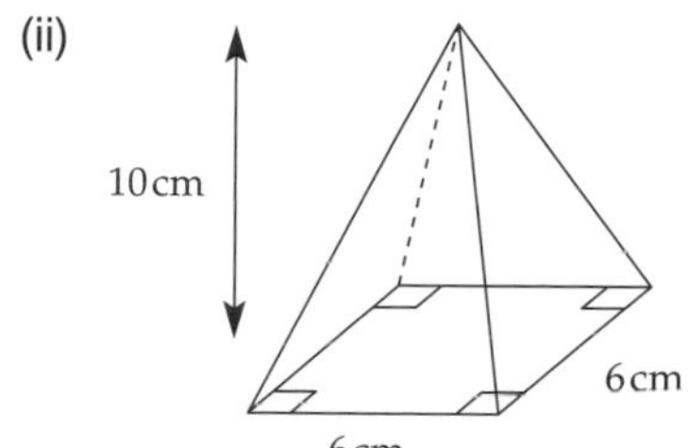

2. Calculate the surface area of the pyramid formed from this net.

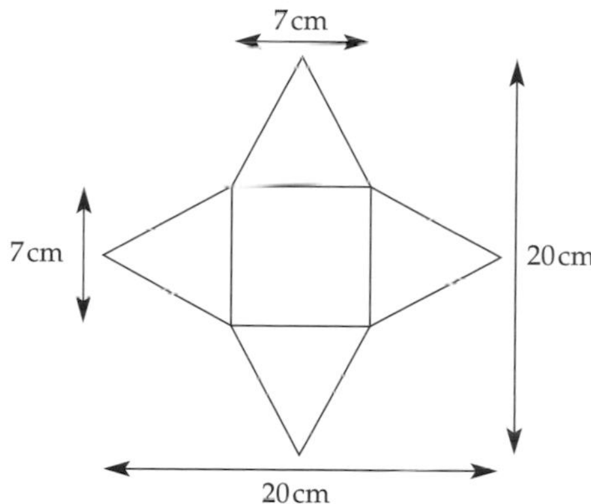

3. A square-based pyramid has a volume of 12 000 mm³. Its height is 90 mm. Calculate the size of the base.

4. Calculate the volume of these cones.

(i)

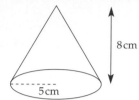

(ii)

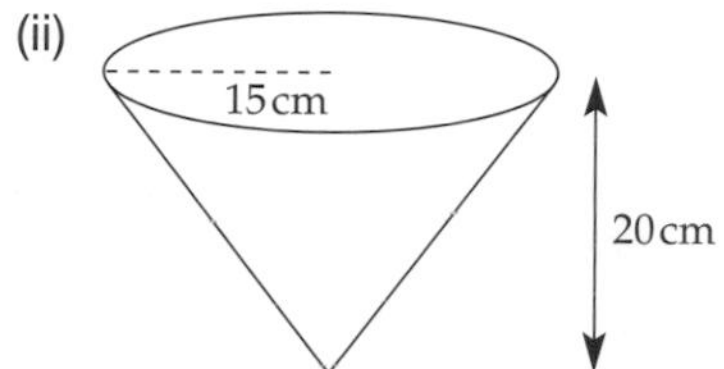

5. A solid cone has a radius of 8 cm and a slant edge of 17 cm. Calculate
   (i) the curved surface area;
   (ii) the total surface area.

6. The volume of a cone is 254 cm^3. Its radius is 2.7 cm. Calculate the height.

7. A cone has a volume of 453 cm^3 and a height of 12 cm. Calculate the radius.

8. A solid metal cone has a radius of 2.5 cm and a height of 6.3 cm. The metal has a density of 7.8 g/cm^3. Work out the mass of the cone.

9. The diameter and height of a cone are equal. The volume is 579 cm^3. Calculate the radius.

Investigations

Cone A has a radius of 4 cm and a height of 7 cm.
Cone B has a radius of 8 cm and a height of 14 cm.

1. Work out the following ratios.
   (i) radius of cone A: radius of cone B
   (ii) height of cone A: height of cone B.

2. Work out the following ratios involving the total surface area (t.s.a.). (Do not substitute a value of π.)
   (i) t.s.a. of cone A: t.s.a. of cone B
   (ii) volume of cone A: volume of cone B.

3. Explain what happens to
   (i) the t.s.a. and
   (ii) the volume

   when a cone is enlarged by a scale factor of 2.

Chapter 

Using Algebra

A car hire company's rates are as follows:

Fixed charge £10
\+ Daily charge £25

Knowing this, you can work out the cost for different numbers of days.

Number of days	Fixed charge (£)	+	Daily charge (£)	=	Cost (£)
1	10	+	25×1	=	35
2	10	+	25×2	=	60
3	10	+	25×3	=	85
4	10	+	25×4	=	110
...	...		...		...

In each case the calculation has the same pattern:

$$\text{cost} = 10 + 25 \times \text{number of days}.$$

This can be written more neatly by abbreviating cost to C and number of days to d so that it becomes

$$C = 10 + 25 \times d.$$

The rule for working out cost is now written as a *formula* using algebra. The right-hand side is called an *expression*. Because the letters C and d can take different values they are described as *variables*; the numbers 10 and 25 are *constants* because they are the same in all the calculations.

Once you have seen how to do a calculation with numbers it is a straightforward process to write it using algebra, with letters to represent any variables involved.

All the common symbols have exactly the same meaning except that in algebra we usually miss out the × sign; $5 \times x$ is written as $5x$. (So the formula for the cost of hiring cars is given as $C = 10 + 25d$.)

Look at the example below and make sure you follow the steps involved.

Example

(i) I buy 3 lb apples at 48p per pound and 6 oranges at 12p each. What change do I get from a £5 note?

(ii) I buy m lb apples at 48p per pound and n oranges at 12p each. What change do I get from a £P note?

Solution

	(i)	(ii)
Cost of apples (pence)	$3 \times 48 = 144$	$m \times 48 = 48m$
Cost of oranges (pence)	$6 \times 12 = 72$	$n \times 12 = 12n$
Total cost (pence)	$144 + 72 = 216$	$48m + 12n$
Change (pence)	$500 - 216 = 284$	$100P - (48m + 12n)$

Exercise

1. Write an algebra expression which means
   (i) add 2 to N;
   (ii) half of N;
   (iii) four times N;
   (iv) two-thirds of N;
   (v) three less than N;
   (vi) 6 more than N;
   (vii) multiply N by 3;
   (viii) subtract 4 from N;
   (ix) double N and add 2;
   (x) 1 less than N;
   (xi) divide 12 by N;
   (xii) halve N and then add 2;
   (xiii) multiply N by 4 and subtract 2 from the result.

Each of the questions 2 – 7 consists of an arithmetic problem and its algebra equivalent.

2. (i) A secretary buys 12 stamps at 25p each, 2 rolls of sellotape at 72p each and a packet of envelopes at £1.32. How much change is there from £10?
   (ii) A secretary buys x stamps at 25p each, y rolls of sellotape at 72p each and a packet of envelopes at £1.32. How much change is there from £d?

3. (i) A hotel charges £30 for the first night and £25 for each additional night. What is the cost of staying 10 nights?
   (ii) A hotel charges £m for the first night and £p for each additional night. What is the cost of staying 7 nights?

4. (i) Fiona is paid £6 per hour for a normal week of 40 hours and £8 per hour for overtime. Find her wage for a week of 48 hours.
   (ii) Jenny is paid £x per hour for a normal week of 40 hours and £z per hour for overtime. Find her wage for a week of 48 hours.

5. (i) The cost of running a coach to the coast is £60. If there are 28 passengers each paying £3.50, find the profit made by the bus company.
   (ii) The cost of running a coach to the coast is £c. If there are 36 passengers each paying £x find the profit made by the bus company.

6. (i) A boat travels 12 miles upstream at 4 miles per hour and returns at 6 miles per hour. Find the total time taken.
   (ii) A boat travels x miles upstream at 4 miles per hour and returns at 6 miles per hour. Find the total time taken.

7. (i) A room measuring 13 feet by 9 feet is 10 feet high. Find the total surface area of the four walls.
   (ii) A room measuring a feet by b feet is h feet high. Find the total surface area of the four walls.

Here are some algebra problems without their arithmetic equivalents. If you have difficulty with any of these, try substituting some numbers to create your own arithmetic equivalent.

8. Hilary has £x and Nick has £y. Express the following as equations:
   (i) Nick has twice as much as Hilary;
   (ii) Nick has half as much as Hilary;
   (iii) together Nick and Hilary have £100;
   (iv) Hilary has £20 less than Nick;
   (v) Nick has £15 more than Hilary;
   (vi) if Hilary gives Nick £10 they will then have the same;
   (vii) if Nick gives Hilary £20 she will then have twice as much as Nick;
   (viii) Hilary has spent all her money and has none left;
   (ix) Nick is £10 in debt.

9. Rae is in charge of p patients and Martin looks after q patients. Express the following as equations:
   (i) Rae and Martin have 40 patients altogether;
   (ii) Rae has 3 times as many patients as Martin;
   (iii) if Martin transfers 10 patients to Rae's ward, he will then have twice as many as Rae;
   (iv) if Martin discharges 5 of his patients Rae will then have half as many as he does;
   (v) if Rae discharges half of her patients they will then have the same number each;
   (vi) Martin's ward is empty.

10. A school concert is performed on Thursday, Friday and Saturday. Tickets cost £x each.
    The school sells t tickets for Thursday, f tickets for Friday and s tickets for Saturday. Explain, in words, what the following expressions represent.
    (i) tx (ii) fx
    (iii) sx (iv) $fx + sx$
    (v) $t + f + s$ (vi) $\frac{t + f + s}{3}$
    (vii) $s - f$ (viii) $f - t$

11. An apple costs a pence and a banana costs b pence. Jangir buys m apples and n bananas. Explain, in words, what the following expressions represent.
    (i) $m + n$ (ii) ma
    (iii) nb (iv) $ma + nb$
    (v) $\frac{ma}{100}$ (vi) $\frac{nb}{100}$
    (vii) $0.01(ma + nb)$

Substitution

As you have seen, algebra allows you to set up general expressions or formulae, and to work with them. This will often involve you in substituting particular values for the variables.

Example

The formula for converting a temperature, f, in degrees Fahrenheit to degrees Celsius, c, is given by

$$c = \frac{5}{9}(f - 32).$$

Find the Celsius equivalents of

(i) 32°F (freezing point for water); (ii) 86°F (a hot day);
(iii) 98.4°F (body temperature).

Solution

(i) $c = \frac{5}{9} \times (32 - 32) = 0$ So 32°F = 0°C

(ii) $c = \frac{5}{9} \times (86 - 32) = 30$ So 86°F = 30°C

(iii) $c = \frac{5}{9} \times (98.4 - 32) = 36.9$ (to 1 decimal place)

So 98.4°F = 36.9°C (to 1 decimal place)

Example

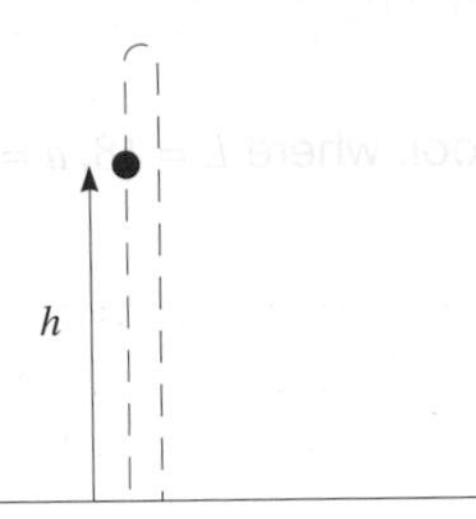

A ball is thrown into the air from ground level with speed u ms^{-1}. Its height, h m, after t s is approximately given by

$$h = ut - 5t^2.$$

(i) Find the height of the ball when
 (a) $u = 20$ and $t = 2$
 (b) $u = 20$ and $t = 4$

(ii) What happens in the case when $u = 20$ and $t = 5$?

Solution

(i) (a) $h = 20 \times 2 - 5 \times 2^2 = 40 - 20$. The ball is 20 m high.
 (b) $h = 20 \times 4 - 5 \times 4^2 = 80 - 80 = 0$. The ball is at ground level.

(ii) In the case when $u = 20$ and $t = 5$, $h = 20 \times 5 - 5 \times 5^2 = 100 - 125 = -25$.

The minus sign means that the ball has already fallen back to the ground. Unless it happened to go down a well or a mine-shaft, its height is probably zero when $t = 5$.

Notice the order in which the operations are done: squared first, then ×, then −.

Part (ii) of the last example illustrates an important point. Many formulae are only valid for certain ranges of the variables involved; in this case t can only take values from 0 to 4. We express this as $0 \le t \le 4$.

Further, in some cases the variables may only take some values within the range; for example, if x represents a number of people it must be a whole number, but that is not the case if it is the length of a line in cm.

Exercise

1. To convert the temperature f in degrees Fahrenheit to c in degrees Celsius, the following formula is used:

$$c = \frac{5}{9}(f - 32).$$

Convert the following temperatures to degrees Celsius:
(i) 113°F,
(ii) 122°F,
(iii) 41°F,
(iv) –13°F

2. The distance s m travelled by an ambulance with speed u ms^{-1} which accelerates to a speed v ms^{-1} is given by

$$s = \frac{v^2 - u^2}{2.5}.$$

Find s when
(i) $u = 12$ and $v = 18$;
(ii) $u = 10.5$ and $v = 15.5$.

3. When a sum of money, £P, is invested at a compound interest rate of r% per annum, the amount in the account after 2 years is given by

$$A = P\left(1 + \frac{r}{100}\right)^2.$$

Find the amount in an account after 2 years if the initial sum is £2600 and the interest rate is 13.5% per annum.

4. The cost, £C, of printing n posters for a concert is given by

$$C = 7.2 + 0.14n.$$

(i) Find the cost of printing
(a) 100 posters;
(b) 1000 posters.

(ii) Find the cost per poster when 400 are printed.

5. The height of a trapezium can be found by using the formula

$$h = \frac{2A}{a+b}$$

where A is the area of the trapezium and a and b are the lengths of the parallel sides.

Find h if
(i) $A = 90$, $a = 5$ and $b = 13$;
(ii) $A = 44$, $a = 6$ and $b = 5$.

6. The volume of medicine, V cm^3, in a bottle of height h cm and glass thickness t cm is given by:

$$V = 3h(h - t)^2.$$

Find the volume in a bottle for which
(i) $h = 8$ and $t = 0.2$;
(ii) $h = 9$ and the glass is 5 mm thick.

7. The cost, £C, of transporting a racehorse to a racecourse is estimated to be

$$C = 100 + 0.75m + 0.25s,$$

where m miles is the distance to be travelled by road and s miles is the distance to be travelled by sea. A racehorse owner in Surrey wants a horse delivered to Longchamps. Delivery will involve a road distance of 160 miles and a sea distance of 120 miles. Find the cost.

8. The local leisure centre is refurbishing its pool area, and this involves painting the cubicles and tiling the pools.
(i) The surface area S to be painted in each cubicle is given by

$$S = d(3h + 2)$$

where h is the height in metres and d is the width in metres. Find the surface area of a cubicle 2 metres high and 1.5 metres wide.
(ii) The 2 trapezium shaped sides of each pool need to be tiled. The area A of each side is given by

$$A = L\left(\frac{a+b}{2}\right),$$

where L is the length of the pool, a is the minimum depth and b is the maximum depth (all in metres). Find the area to be tiled
(a) in the large pool, where $L = 25$, $a = 1$ and $b = 3$;
(b) in the small pool, where $L = 18$, $a = 0.5$ and $b = 1$.

(iii) After the tiling is completed, the pools are to be refilled with water. The volume of water, V (in cubic metres) required for each pool is given by

$$V = \frac{L}{2}w(a + b),$$

where a and b are as above, and w is the width of the pool in metres. Find the volume of water in
(a) the large pool, where $L = 25$, $a = 1$, $b = 3$ and $w = 20$;
(b) the small pool, where $L = 18$, $a = 0.5$, $b = 1$ and $w = 10$.

9. Find the value of $x^2 - 6x - 7$ when
(i) $x = 7$,
(ii) $x = 2$,
(iii) $x = -4$.

10. A professional photographer knows that if the bridal couple is a distance d cm from her camera, she must use a focal length, f cm, given by

$$f = \frac{10d}{d+10}.$$

What focal length should she set if
(i) $d = 200$ cm, (ii) $d = 10$ m?

Simplifying expressions

Addition and subtraction

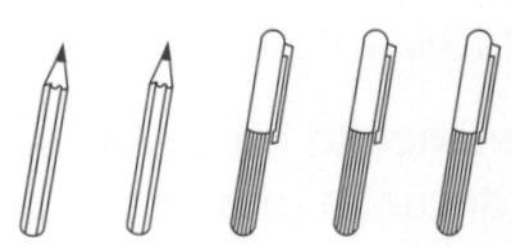

If you have these pencils and pens, the neatest way of saying it is not 'pencil + pencil + pen + pen + pen', but 2 pencils and 3 pens. Similarly in algebra,

$$a + a + b + b + b = 2a + 3b.$$

Multiplication and division

In algebra, $a \times b$ is written as ab. So $21 \times f \times g$ is written as $21fg$.

When you want to divide one quantity by another in algebra, you write one over the other as in arithmetic. You then cancel numbers in the usual way. You can also cancel letters in the same way:

$$\frac{21fg}{7f} = \frac{{}^{3}\cancel{21}\cancel{f}g}{{}_{1}\cancel{7}\cancel{f}} = 3g$$

Powers

In arithmetic, you write $4 \times 4 \times 4$ as 4^3.

Similarly in algebra, $a \times a \times a = a^3$

When you meet an expression like $p^4 \times p^3$, you add the powers:

$$p^4 \times p^3 = (p \times p \times p \times p) \times (p \times p \times p) = p^7.$$

4+3=7

Similarly when you meet an expression like $\frac{p^5}{p^3}$ you subtract the powers:

$$\frac{p^5}{p^3} = \frac{p \times p \times p \times p \times p}{p \times p \times p} = p^2 .$$

5−3=2

When you meet an expression like $(p^3)^2$ you multiply the powers:

$$(p^3)^2 = p^3 \times p^3 = p^6$$

$3 \times 2 = 6$

Sometimes you will meet combinations of letters.

For example $\frac{a^3b^4}{ab^3} = a^2b$, and $\frac{x^2}{x^3y} = \frac{1}{xy}$.

a^3b^4 means $a \times a \times a \times b \times b \times b \times b$

Brackets

In arithmetic, brackets are used to say 'do this first'. For example,

$$2 \times 3 + 5 = 6 + 5 = 11,$$

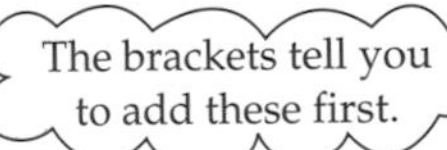

The brackets tell you
to add these first.

but $2 \times (3 + 5) = 2 \times 8 = 16$.

You will notice that in the second case the calculation could be carried out as $2 \times 3 + 2 \times 5$: each term inside the bracket has been multiplied by the 2.

The same idea is used in algebra, so

$$2(x + y) = 2x + 2y.$$

If the term outside the bracket is negative, all the signs inside the bracket will be changed when the bracket is removed:

$$-3(c - d + 2e) = -3c + 3d - 6e$$

Exercise

1. Simplify these expressions.
   (i) $7a - 3a + 2a$
   (ii) $5a + a - 8a$
   (iii) $4c + d + 3c$
   (iv) $2e - 5e + 4f$
   (v) $4a + 3x + 5a + 2x$
   (vi) $2k + 3n + k - 5n$
   (vii) $5r - 7t - 2r + 3t$
   (viii) $x + 3y - 4x + 2y$
   (ix) $3q - r - 5q - 2r$
   (x) $2a + 3c - 4f + 2a + f$
   (xi) $a - 4t - 3a + t + 4a$
   (xii) $2c + 3d + c - 7d - 3c - 5d$

2. Simplify these expressions.
   (i) $2x \times 3y$
   (ii) $(-7x) \times 4y$
   (iii) $6x \div 3$
   (iv) $10x \div (-2x)$
   (v) $(-5k) \times (-6k)$
   (vi) $4t \times 3u \times (-2v)$
   (vii) $5c \times (-4d) \div 10c$
   (viii) $8e \div (-24ef)$

3. Simplify these expressions.
   (i) $x^5 \times x^2$
   (ii) $y^6 \div y^3$
   (iii) $(z^2)^4$
   (iv) $a^4 \times a^5 \div a^6$
   (v) $b^2 \times b^4 \times b^2 \div b^3$
   (vi) $c^3 \times c^3 \times c^2 \div c^4$
   (vii) $7a^3 \times 6a^4 \div 3a^5$
   (viii) $9c^4 \times 4c^3 \div 6c^5$
   (ix) $a^2 \times a^3 \times b^2$
   (x) $a^4 \times b^2 \times a^3 \div a^2$
   (xi) $(a^3)^2 \div a^2$
   (xii) $(3c)^2 \div 3c^2$

4. Remove the brackets from the following expressions.
   (i) $2(3a - 4)$
   (ii) $-1(4n - 2)$
   (iii) $2p(p + q)$
   (iv) $-2(6x - 9y)$
   (v) $4(6b + c)$
   (vi) $7c(c - d)$
   (vii) $3(2r - s - t)$
   (viii) $-5(3x - 7y)$

Activities

1. Take twelve ordinary playing cards, rejecting court cards (jacks, queens and kings). Lay the cards face up, one by one. Record the suit of each card as h (for a heart), c (for a club), d (for a diamond) or s (for a spade). So if the first two cards are clubs and the third is a heart you will write $c + c + h + \ldots$

   When you have recorded the suit of all twelve cards, simplify your expression.

   Check your answer by sorting your cards into suits and seeing how many you have of each.

2. Shuffle your cards thoroughly and lay them out again but this time record the number as well as the suit. For example record the eight of hearts as $8h$. (An ace counts as 1.) Simplify this expression and check your answer.

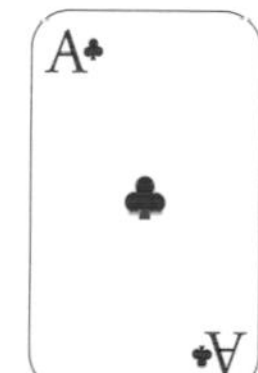

Investigations

Cubes

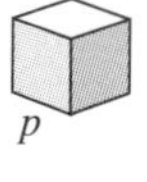

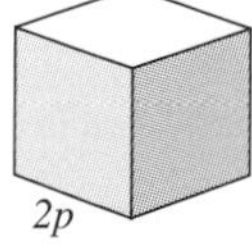

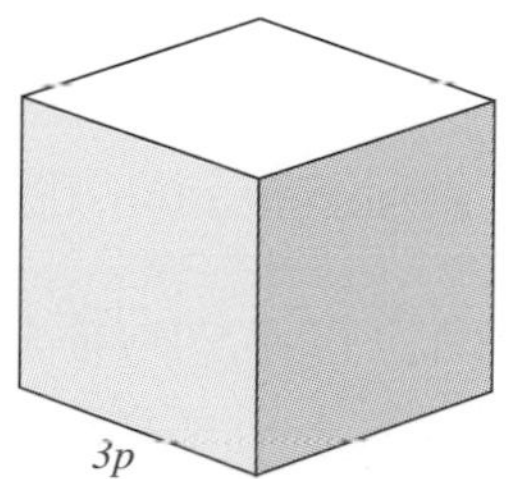

1. Calculate the surface area and the volume of each of these cubes, in terms of p.

2. (i) How many of the smallest cubes would fit inside the medium cube?
   (ii) How many of the smallest cubes would fit inside the large cube?

3. How many of the medium cubes would fit inside the large cube? How many small cubes would you need to fill the gaps?

4. If the cubes are hollow and can be filled with water,
   (i) how many full medium cubes can be emptied into the large cube without it overflowing?
   (ii) how many full small cubes can then be emptied into the large cube without it overflowing?

Equations

An equation is a mathematical statement containing an = sign. The statement will remain true provided you do the same thing to both sides of the equation.

Often you will want to find the value of the variables in an equation. This is called *solving* the equation. The examples on this page illustrate how you do this for an equation which contains only one variable. (If you need to find the values of two variables you need two *simultaneous* equations, for three variables you need three simultaneous equations, and so on. You will see how to solve simultaneous equations later in this chapter.)

An equation is like balancing scales. If you add the same weight to each side they still balance.

Example

(i) Solve the equation $2x - 5 = 7$

(ii) Check your answer.

Solution

(i) Write down the equation given: $2x - 5 = 7$

Now add 5 to both sides: $2x - 5 + 5 = 7 + 5$

Tidy up: $2x = 12$

Divide both sides by 2: $x = 6$

(ii) Substituting $x = 6$ into the left-hand side (LHS) of the original equation gives $2 \times 6 - 5 = 7$, so the solution is correct.

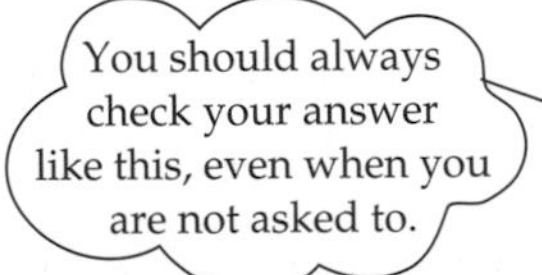

Example

Solve the equation $5(x + 1) = 7$

Solution

Write down the equation given: $5(x + 1) = 7$

Multiply out the bracket: $5x + 5 = 7$

Subtract 5 from both sides: $5x = 2$

Divide both sides by 5: $x = \frac{2}{5}$ (= 0.4)

Check: substituting $x = \frac{2}{5}$ into the LHS of the original equation gives $5(\frac{2}{5} + 1) = 5 \times \frac{7}{5}$, so the solution is correct.

Example

Solve the equation $3(3x - 2) = x + 2$

Solution

Write down the equation given: $3(3x - 2) = x + 2$

Multiply out the brackets: $9x - 6 = x + 2$

Subtract x from both sides: $8x - 6 = 2$

Now add 6 to both sides: $8x = 8$

Divide both sides by 8: $x = 1$

You should now check, by substitution, that $x = 1$ is the correct answer.

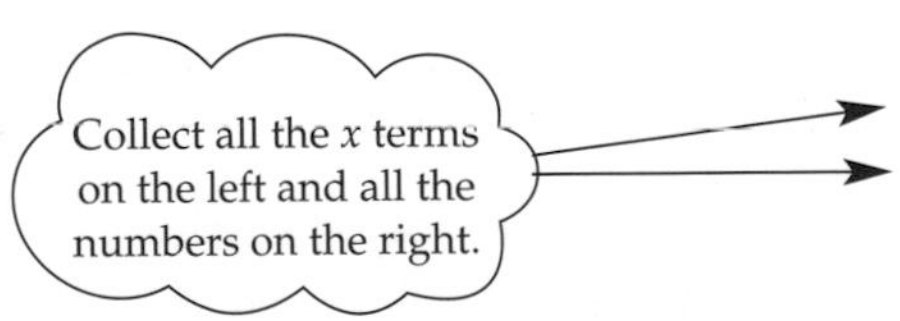

Collect all the x terms on the left and all the numbers on the right.

What happens when you try to solve $3(x + 6) = 3x + 18$?

Multiply out the brackets: $3x + 18 = 3x + 18$

This is true for all values of x! So $3(x + 6) = 3x + 18$ is an example of an *identity*.

Exercise

1. Solve the following equations, remembering to check each answer.
   (i) $x + 4 = 13$
   (ii) $y - 5 = 7$
   (iii) $4d = 16$
   (iv) $\frac{n}{2} = 16$
   (v) $3c - 10 = 5$
   (vi) $2x + 5 = 17$
   (vii) $15 = 7 + 4x$
   (viii) $7 = 17 - 2n$
   (ix) $4d - 3 = 7$
   (x) $10 = 3x + 22$
   (xi) $5(y - 1) = 30$
   (xii) $3(x + 2) = 18$
   (xiii) $2(n - 5) = 11$
   (xiv) $\frac{d + 5}{4} = 12$
   (xv) $5x - 9 = 2x + 3$
   (xvi) $3x + 4 = 5x - 2$
   (xvii) $1 - 7x = 4 - 2x$
   (xviii) $5x - 7 = 4(x + 2)$
   (xix) $7(y - 4) = 3(y + 8)$
   (xx) $3(n + 2) = 2(1 - n)$

2. Ian Ventor designs a new board game, and offers the design to Paddington Games p.l.c. for £25 000. The company offers an alternative deal: they will pay Ian £2000, plus a royalty of 50p for each game sold. How many games would need to be sold to make this offer attractive to Ian?

3. Warmawear produce ski-pants at a cost of £12 per pair. The ski-pants sell for £27 per pair.
   (i) Write down the profit on each pair.
   Warmawear hope to sell x pairs. This will cover the overheads of £10 000 and produce an overall profit to the company of £20 000.
   (ii) Form an equation in x and solve it.

4. At Christmas, a firm gives out calendars and diaries to its regular customers. The cost of a calendar is 6 times the cost of a diary which costs x pence. Write down, in terms of x,
   (i) the cost of a calendar;
   (ii) the cost of 40 calendars;
   (iii) the cost of 120 diaries.
   (iv) The firm sends out 40 calendars and 120 diaries for a total of £252. Form an equation in x and use it to find the cost of a diary.

5. Babies are weighed at birth to monitor their progress. Martin weighs 0.6 kg less than Daniel who weighs 0.4 kg less than Kevin. Martin weighs x kilograms and the total weight of the three boys is 10.3 kg. Form an equation in x, solve it and state the weight of each baby.

6. Photocopying a sale leaflet costs x pence and photocopying a full brochure costs 25 pence more. Write down, in terms of x:
   (i) the cost of photocopying 80 sale leaflets;
   (ii) the cost of photocopying a brochure;
   (iii) the cost of photocopying 50 brochures.

   Estate agents, Howes & Holmes, photocopy 80 leaflets and 50 brochures for £25.50.
   (iv) Write down an equation in x, solve the equation to find x, and find the cost of photocopying one leaflet.

7. Ainsley regularly attends the local theatre.
   (i) For x plays he buys a seat in the balcony. Each seat costs £9. Write down an expression for the cost of seeing these x plays.
   (ii) For each of the remaining plays he buys a ticket costing £5, for the stalls. Given that he attends 20 plays in a season, write down an expression in x for the total cost of his stalls tickets.
   (iii) Using your answers to (i) and (ii), write down and simplify an expression for the total cost of his tickets for the season.
   (iv) The cost for the season is £148. Write down an equation in terms of x.
   (v) Solve the equation to find the number of plays he watches from the balcony.

8. The manager of the Freeway Department Store keeps a close eye on sales of three different colours of suitcases: red, green and black. He notices that the green sells least well. Twice as many red suitcases are sold as black and, in a three month period, the store sells 25 more black suitcases than green. Let x be the number of green suitcases sold in the store.
   (i) Write down in terms of x:
      (a) the number of black suitcases sold;
      (b) the number of red suitcases sold.

   The store sells 135 suitcases in the three-month period.
   (ii) Form an equation in x and solve it. State the number of red suitcases sold.

9. A toothbrush costs 25p less than a flannel. Twelve toothbrushes and ten flannels are bought. The cost of a toothbrush is x pence.
   (i) Write down, in terms of x
      (a) the cost of a flannel;
      (b) the cost of 12 toothbrushes;
      (c) the cost of 10 flannels.

   The total cost of the 12 toothbrushes and 10 flannels is £20.10.
   (ii) Using this information, write down an equation in terms of x, solve it to find x, and find the cost of each item.

Inequalities

In an equation the two sides are equal. You will also meet situations in which two things are *not* equal. For example, someone might say 'It is less than 100 miles to Dundee' or 'At least half our employees are female'. You can express ideas like this using algebra, with the following symbols.

- $h < 6$ means h is less than 6;
- $x > y$ means x is greater than y;
- $p \leq 15$ means that p is less than or equal to 15;
- $ab \geq 8$ means that a times b is greater than or equal to 8.

There are two other symbols that you will find useful:

- $x \approx y$ means x is approximately equal to y.
- $x \neq y$ means x is not equal to y.

Example

(i) Use symbols to write 'the number N is smaller than the number P'.

(ii) Write 'the number n is at least 4 but less than 10' in symbols.

Solution

(i) $N < P$

(ii) $4 \leqslant n < 10$ (or $10 > n \geqslant 4$).

When you *solve* an inequality you end up with a range of possible values rather than a single value for the variable. The next example can be solved with the rules you used to solve the equations.

Example

An inequality is like unbalanced scales. If you add the same to each side the imbalance remains.

Solve the inequality $3x - 35 > 2x - 5$

Solution

Write the inequality as it was given: $3x - 35 > 2x - 5$

Subtract $2x$ from both sides: $x - 35 > -5$

Add 35 to both sides: $x > 30$

This is the solution: any number larger than 30 will satisfy the inequality.

Multiplying and dividing by negative numbers

Although inequalities are treated almost exactly the same as equations, there is one exception. Look at the inequality $2 < 3$, which is clearly true. If you multiply both sides of this by -2 you make it into $-4 < -6$, which is false.

To overcome this problem, you apply the rule that

- when multiplying or dividing by a negative number, you reverse the inequality sign.

Example

Solve $2 + 3x > 12 + 8x$

Solution

$$3x - 8x > 12 - 2$$
$$-5x > 10$$

Divide each side by -5:

$$x < -2$$

Notice that we have reversed the sign.

Exercise

1. State the following in words.
   - (i) $7 > 5$
   - (ii) $6 < 8$
   - (iii) $3 < 4 < 6$
   - (iv) $10 > 8 > 7$
   - (v) $2^{10} > 1000$
   - (vi) $N > 3$
   - (vii) $A < 8$
   - (viii) $x = y = 3$
   - (ix) $a \neq 2$
   - (x) $x \neq 0$
   - (xi) $\pi^2 \approx 10$
   - (xii) $a \neq b$

2. Write the following in symbols.
   - (i) 32 is greater than 3
   - (ii) 0.5^2 is less than 0.5
   - (iii) N is equal to 8
   - (iv) A is greater than 6
   - (v) z is not equal to zero
   - (vi) p is less than 3.5
   - (vii) A and B are equal
   - (viii) x is greater than y
   - (ix) the numbers a and b each equal 7
   - (x) y lies between 10 and 20

3. List all the whole numbers that n could be if
   - (i) $2 < n < 7$
   - (ii) $4 > n > -3$
   - (iii) $-2 < n < 2$
   - (iv) $50 > n > 46$
   - (v) $4 \leq n \leq 7$
   - (vi) $-2 < n \leq 0$
   - (vii) $5 < 2x < 11$
   - (viii) $12 \geq 3x \geq 9$
   - (ix) $7 < x + 3 < 12$
   - (x) $2 \leq x - 1 \leq 4$

4. Solve these inequalities.
   - (i) $R + 4 < 32$
   - (ii) $4t < 36$
   - (iii) $x + 5 < 20$
   - (iv) $5x > 25$
   - (v) $2a + 9 < 15$
   - (vi) $5b - 7 < 8$
   - (vii) $2p - 14 > 6$
   - (viii) $7 - h < 4$
   - (ix) $15 - 2x > 7$
   - (x) $21 - 3x < 12$
   - (xi) $6 - 2b > 8$
   - (xii) $3(3 - c) > 15$
   - (xiii) $2(4 + x) > 18$
   - (xiv) $5(x - 1) \leq 10$
   - (xv) $3x - 4 > 2x + 1$
   - (xvi) $5x + 1 < x + 13$
   - (xvii) $2x + 7 \leq 5x - 2$
   - (xviii) $9 - 3x \leq x + 7$

5. A team of six inspectors checks products for faults. The team is expected to check at least 720 items per hour.
   - (i) Write down inequalities for the number of items, x, expected to be checked per hour by:
     - (a) the team of six inspectors;
     - (b) each inspector.
   - (ii) Find the maximum time per item.

6. It is recommended that a company employs one telephone operator for every 200 customers per day. The number of customers per day at Innotech is from 1200 to 1600.
   - (i) Write this statement as an inequality in x, where x is the number of telephone operators needed at Innotech.
   - (ii) Solve the inequality to find possible values of x.

7. On a 14-day holiday a courier receives at least one complaint every day but no more than 4.
   - (i) Write down an inequality to express this statement.
   - (ii) Write inequalities to express the number of complaints the courier receives
     - (a) per week;
     - (b) per holiday.

8. Dennis travels on holiday to Germany. He wishes to buy several bottles of wine to take home to friends. The bottles weigh 1250 grams each. His suitcase, when packed, already weighs 24 kg. Let x be the number of bottles of wine he buys.
   - (i) Write down an expression, in terms of x, for the total weight of case and wine.

   Dennis does not wish to exceed the weight allowance for luggage, which is 35 kg.
   - (ii) Write down an inequality and solve it to find the maximum number of bottles of wine that Dennis can take home.

9. A company with 15 employees wishes to give each one a Christmas bonus of £b. The management decides that they must not spend more than £3000 in total.
   - (i) Write this statement as an inequality.
   - (ii) What is the largest bonus they can afford per employee?

10. Graeme has taken £10 from petty cash to buy a ream of paper and some envelopes. The paper costs £6.52 and envelopes cost 62p per packet. Let x be the number of packets of envelopes he buys.
    - (i) Write down an expression for the total cost, in pence, of paper and envelopes.
    - (ii) Write down an inequality in x. Solve the inequality to find possible values of x.

Activity

You will need graph paper and red, green and blue pens. Draw x and y axes numbered from 0 to 5.
Put a coloured cross on each point whose x and y co-ordinates are whole numbers, by following the steps below. Start at the top left, at point $(0, 5)$, and work systematically. Put a blue cross on all points at which $x + y < 5$, a red cross on points at which $x + y = 5$ and a green cross on points at which $x + y > 5$.

When you have finished,
- (i) join the red points with a line. What does this line represent?
- (ii) shade blue the region which has blue crosses, and shade green the region with green crosses. What do these regions represent?

Sequences

Clone Copying

Sheets	**Cost** (p)
1	15
2	20
3	25
4	30

and so on.

Quick-Copy

Copies	**Cost** (£)
100	4
200	7
300	10
400	13

and so on.

You need to have some photocopying done and are given rates from these two companies. How would you decide which company to use?

Patterns of numbers like these are called *sequences*. To find the formula for a sequence start by looking at the differences.

Clone Copying	**No. of copies n:**	1	2	3	4
	Cost C (pence):	15	20	25	30
			5	5	5

This shows that the cost goes up 5 pence for every copy, so each copy costs 5 pence and n copies cost $5n$ pence. However the first copy costs 15 pence and not 5 pence: there is a fixed charge of 10 pence as well.

So $C = 10 + 5n$.

Quick-Copy	**No. of copies n:**	100	200	300	400
	Cost C (pence):	400	700	1000	1300
	Difference:		300	300	300

In this case the cost goes up 300 pence for every 100 copies. This is 3 pence per copy, or $3n$ pence for n copies. Since the first 100 copies cost 400 pence rather than 300 pence, the fixed charge is 100 pence (i.e. £1).

In this case $C = 100 + 3n$.

For small numbers of copies, Clone Copying is cheaper.
For example when $n = 5$

Clone Copying: $C = 35$; Quick-Copy: $C = 115$.

However for large numbers, Quick-Copy is cheaper.
For example when $n = 500$,

Clone Copying: $C = 2510$; Quick-Copy: $C = 1600$.

To find the break-even, solve $10 + 5n = 100 + 3n$.

Collecting terms gives

$$
\begin{aligned}
5n - 3n &= 100 - 10 \\
2n &= 90 \\
n &= 45.
\end{aligned}
$$

For fewer than 45 copies it is cheaper to go to Clone Copying.
For more than 45 copies it is cheaper to go to Quick-Copy.
For exactly 45 copies, both companies charge the same amount, £2.35.
You can see this on this graph of C against n for the two companies.

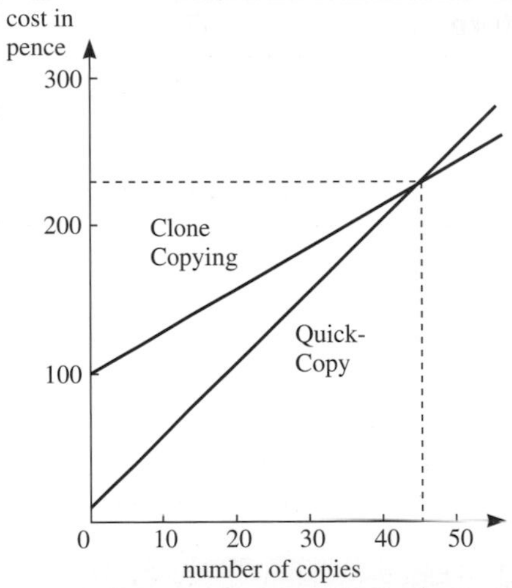

Exercise

1. For each of these sequences write down the next term and an expression for the nth term.
   (i) 3, 6, 9… (ii) 4, 7, 10…
   (iii) 6, 9, 12… (iv) 5, 11, 17…
   (v) 1.2, 1.4, 1.6… (vi) 18, 14, 10…

2. The time taken for a computer to scan and then print a copy of a document is
   120 seconds for 1 copy,
   195 seconds for two copies,
   270 seconds for three copies, and so on.

   Find the pattern in the numbers, and express the time taken in terms of n, where n is the number of copies required.

3. After a party the amount of alcohol in Dave's bloodstream is measured every hour. At 2 am when he stops drinking, the amount is 150 mg/cl. At 2.30 am it is 145 mg/cl, and at 3 am it is 140 mg/cl. Assuming a steady decrease, express the amount in terms of t, the number of hours after the end of the party. Hence find the time at which the amount reaches 50 mg/cl, which is the maximum amount at which it is safe for Dave to drive.

4. The entry charge for a safari in a wildlife park varies according to the number of people in the party, as shown below.

Wildlife Safaris

Number in party	Charge per person
30	£270
20	£320
10	£370
1	£415

   Express the charge in terms of n, the number of people in the party.

5. The cost of hiring a self-drive van can be calculated using the following scale.
   £35 for 10 miles,
   £45 for 20 miles,
   £55 for 30 miles.

   State the cost of hiring the van to go
   (i) 5 miles;
   (ii) 50 miles;
   (iii) n miles.

6. Walter Wall is a company which cleans fitted carpets in the home. Walter Wall uses the following scale to estimate the time needed to travel to the house and to clean different numbers of carpets.

Number of carpets	Time in hours
1	2
2	3.25
3	4.5

   (i) Estimate how long will it take Walter Wall to clean
   (a) 4 carpets;
   (b) 5 carpets.

   (ii) Express the number of hours in terms of c, the number of carpets cleaned.

7. When the surrounding temperature is reduced, a person's heart rate in beats per minute decreases. A controlled experiment showed that
   at 37°C heart rate was 98;
   at 36°C heart rate was 95;
   at 35°C heart rate was 92.

   Assuming that this pattern continues, state heart rate at
   (i) 32°C;
   (ii) 27°C;
   (iii) $(37 - n)$°C.

8. Bacteria are being grown in a culture. The time taken for the number of bacteria to double is found to depend on the temperature, as shown below.

Temperature	Time
30°C	4 h 0 min
32°C	4 h 5 min
34°C	4 h 10 min

   State the time taken at
   (i) 36°C
   (ii) 31°C
   (iii) $(30 + n)$°C

9. In Camille's shoe shop, when the price of Footsie shoes is reduced then the number of sales increases. Camille records the following sales figures:
   at £20 per pair, sales are 100 pairs per week;
   at £19 per pair, sales are 120 pairs per week;
   at £18 per pair, sales are 140 pairs per week.

   Assuming this pattern continues, state the number of sales that can be expected if the price is
   (i) £15;
   (ii) £10;
   (iii) £$(20 - n)$.

Quadratic and exponential sequences

Quadratic sequences

The two sequences on page 54 both had constant differences (5 and 300) between their terms. This is not always the case. Look at the sequence of square numbers:

n	1		2		3		4		5		6
n^2	1		4		9		16		25		36
Difference		3		5		7		9		11	

You will notice that these increase by 2 every time. The second differences are constant. Check for yourself that this is always the case for a quadratic expression, like n^2, $2n^2 - 5n + 3$ or $-n^2 + 7$.

Exponential sequences

Look at the sequence 1, 3, 9, 27, 81, 243, … In this case taking the differences between terms is not helpful but you can see that each term is just the previous one multiplied by 3. This sort of sequence is described as *exponential*. There are many situations where exponential sequences are important. Two of these are illustrated in the following examples.

Example

The population of a country starts at 1 000 000 people and increases by 5% each year. Find the population after n years.

Solution

At the end of each year the population is 5% more than it was at the start, or 105% of it. This is the same as multiplying by 1.05.

Year	Start	End
1	1 000 000	1×1.05
2	$1\,000\,000 \times 1.05$	$1\,000\,000 \times 1.05^2$
3	$1\,000\,000 \times 1.05^2$	$1\,000\,000 \times 1.05^3$

and so on.

By the end of the year n, the population is $1\,000\,000 \times 1.05^n$.

Example

Some new office furniture costs £8600. It depreciates each year by 12% of the value at the start of that year. Find its value at the end of the first, second, third and nth year.

Solution

At the end of each year the value is 12% less i.e. 88% of the value at the start of the year.

Year	Start value (£)	End value (£)
1	8600	$8600 \times 0.88 = 7568$
2	$8600 \times 0.88 = 7568$	$8600 \times 0.88^2 = 6659.84$
3	$8600 \times 0.88^2 = 6659.84$	$8600 \times 0.88^3 = 5860.66$

At the end of n years the value is $£8600 \times 0.88^n$.

Exercise

1. A baker icing the tops of circular novelty cakes needs to allow amounts of icing sugar as follows.

Radius of cake (inches)	Icing sugar quantity (g)
1	2
2	8
3	18

   (i) Say why you would expect the amount of icing sugar to be proportional to the square of the radius.

   (ii) Find how much he would need for a cake of radius n inches.

2. Each year Hilary organises a table-tennis league. Every team meets every other team twice. The number of teams entering varies from year to year. Hilary has worked out that
   for 1 team there are 0 matches;
   for 2 teams there are 2 matches;
   for 3 teams there are 6 matches;
   for 4 teams there are 12 matches;

   (i) Explain how she worked out these results.
   (ii) Find how many matches she needs to organise for n teams.

3. Lesley is getting her holiday photos developed. Various sizes of prints are available:
   size 1 is $3'' \times 5''$;
   size 2 is $4'' \times 6''$;
   size 3 is $5'' \times 7''$;
   size 4 is $6'' \times 8''$, and so on.

   (i) Write down the perimeter of each size of photo and then find the perimeter of a size n photo.
   (ii) Use your answer to (i) to find, in terms of n, the lengths of the sides of a photo of size n.
   (iii) Use your answer to (ii) to find, in terms of n, the area of a size n photo.
   (iv) Now find the area of each of the photo sizes 1–4 above, and check that it matches the answer given by your formula for the area.

In questions 4–7 leave all your answers in index form to find the pattern, and then work them out.

4. The value of an oil painting is £15 000. Its value increases at 5% each year. Find the value at the end of
   (i) the 1st year (ii) the 2nd year
   (iii) the 10th year (iv) the nth year.

5. A radioactive substance, radon, has a half-life of 4 days. That means that the mass of radon remaining after 4 days is half the mass at the beginning of the four days. As part of a cancer treatment 10 mg of radon gas in a tiny gold container is implanted in a patient. Find how much remains at the end of
   (i) the 4th day (ii) the 8th day
   (iii) the 12th day (iv) the $4n$th day.

6. The annual depreciation rates (the percentage by which the sale value decreases p.a.) for three types of car are
   25% for an Alpha (price when new: £90 000)
   15% for an Epsilon (price when new: £40 000)
   12.5% for an Omega (price when new: £15 000).

   Find the value of each car at the end of
   (i) the 1st year (ii) the 2nd year
   (iii) the 4th year (iv) the nth year.

7. A chess board has 64 squares. You put 1p on the first square, 2p on the second, 4p on the third and so on, doubling each time.
   (i) Find an expression for the amount on the nth square.
   (ii) How much money is on the last square?

   You now decide to put 1p on the first square, 3p on the second, 9p on the third and so on, trebling each time.
   (iii) Find a formula for the amount on the nth square.
   (iv) How much money is on the last square now?

Investigation

The numbers 1, 3, 6 and 10 are examples of *triangular numbers*. You can see why from the diagram.

Your task is to find the value of the nth triangular number. One way to start is to double each triangular number and draw the dots as a rectangle. This is shown for the number 6 in the diagram below.

Draw the rectangles for the triangular numbers 1, 3, 6 and 10. These correspond to $n = 1, 2, 3$, and 4 in the sequence of triangular numbers.

(i) How many rows of dots are there in each rectangle? Write this in terms of n.
(ii) How many columns of dots are there in each rectangle? Write this in terms of n.
(iii) Assuming this pattern continues, write an expression for the nth term in the sequence of triangular numbers.
(iv) Check that your expression works for some other triangular numbers.
(v) Find the 20th triangular number and find the value of n for the triangular number 55.
(vi) How can you tell whether a number is triangular?

Simultaneous equations

In a situation in which there are two unknown variables, you need two equations in order to find their values. Such equations are described as *simultaneous* equations. You might sometimes meet sets of three or more equations, and if so the approach is similar.

Example

Solve the simultaneous equations

$$3x + 2y = 19$$
$$x - y = 3$$

Solution

$$\begin{aligned} 3x + 2y &= 19 & \longrightarrow\quad 3x + 2y &= 19 \\ x - y &= 3 & \xrightarrow{\times 2}\quad 2x - 2y &= 6 \quad \text{Add} \\ & & 5x &= 25 \\ & & x &= 5 \end{aligned}$$

Substituting for x:

$$15 + 2y = 19$$
$$2y = 4$$
$$y = 2$$

The solution is: $x = 5, y = 2$

To check your answer, substitute into the other equation: $x - y = 5 - 2 = 3$ so this solution is correct.

In order to be able to eliminate one variable by adding (and not subtracting) you sometimes need to multiply one equation by a negative number.

Example

Family Adventure
Mini-breaks
Typical prices
2 adults
2 children
£960
3 adults
1 child
£1060

(i) Write the information given in the advert (left) as two simultaneous equations. Solve them to find the price for one adult and one child.

(ii) Find the price for the Cousins family, which consists of 3 adults and 5 children.

Solution

a is the price for an adult, c the price for a child.

(i)

$$2a + 2c = 960$$
$$3a + c = 1060$$

$$\begin{aligned} 2a + 2c &= 960 & \longrightarrow\quad 2a + 2c &= 960 \\ 3a + c &= 1060 & \xrightarrow{\times -2}\quad -6a - 2c &= -2120 \quad \text{Add} \\ & & -4a &= -1160 \\ & & a &= 290 \end{aligned}$$

Substituting for a:

$$3 \times 290 + c = 1060$$
$$870 + c = 1060$$
$$c = 190.$$

Check: Substituting for a and c in the first equation gives

$$2a + 2c = 580 + 380 = 960.$$

The trip costs £290 for each adult and £190 for each child.

(ii) The total price for the family is:

$$£290 \times 3 + £190 \times 5 = £1820$$

Exercise

1. Solve these pairs of simultaneous equations.

(i) $x + y = 20$, $x - y = 12$

(ii) $x + 2y = 25$, $x - 2y = -11$

(iii) $5x + y = 17$, $3x + y = 11$

(iv) $3x + 5y = 23$, $2x - 5y = 7$

(v) $7x + y = -1$, $x - 6y = -37$

(vi) $6x + y = 26$, $7x - 3y = 22$

(vii) $3x + 5y = 40$, $2x + 3y = 25$

(viii) $x - 4y = 14$, $3x - 5y = 21$

(ix) $5x + 4y = 39$, $6x + 7y = 49$

(x) $2x - 9y = 16$, $7x - 3y = -1$

(xi) $6x + 5y = 30$, $5x + 6y = 25$

(xii) $9x + 7y = 76$, $6x - 5y = 70$

2. A train from Plymouth to London stops 3 times and takes 3 hours for the journey. Another train stops at 10 stations and takes 3 hours 35 minutes.
   (i) Find how long is allowed for stopping at each station.
   (ii) Find the journey time for a non-stop train.

3. Economic theory says that the quantity produced, q tonnes/week, and the price, £p/tonne of a product are governed by the *law of supply and demand*. If production goes up, the product is more widely available and the price goes down.

   For a particular product, the customer's (or demand) equation is
   $$2p - q = 15.$$
   The producer's (or supply) equation is
   $$3q - p = 5.$$
   Find the values of p and q which satisfy both the demand and supply equations.

4. A pharmacist is making up a tonic which is to contain 780 μg (micrograms) of vitamin A and 16 200 μg of vitamin B. To do this, she puts in x ml of Xtravite, which contains 120 μg of vitamin A and 5400 μg of vitamin B in each millilitre and y ml of Yeastavite, which contains 180 μg of vitamin A and 1800 μg of vitamin B in each millilitre. Find the values of x and y.

5. In news reports, this year's profits, £P, for a company are often compared with last year's profits, £L. Actual values of P and L are not always given. Read the headline.

   Profits up by £20 million – a 25% increase!

   Find the values of P and L.

6. A 250 ml bottle of cough mixture has a torn label so the dosages are unknown. Cathy can remember that it holds sufficient for 10 adults and 20 children or for 15 adults and 5 children. What are the dosages for adults and children?

7. Chris makes and delivers buffet meals for office meetings. There are two types: a tray of sandwiches takes $\frac{1}{2}$ hour to make and sells for £6. A pizza takes 1 hour to make and sells for £5. One day he worked for 10 hours to produce lunches to the value of £85. How many trays of sandwiches and how many pizzas did he make?

8. Di is a carpenter. One day, she does 6 hours' work at basic rate and 1 hour of overtime: she is paid £58 gross (i.e. before deductions). On another day, she works for 5 hours at basic pay and 2 hours overtime, for which her gross pay is £60. What is Di's basic hourly rate of pay and what is her overtime rate of pay?

9. For Father's Day, Instagift buys golf balls and bags of tees in bulk. They make up two types of presentation box as shown.

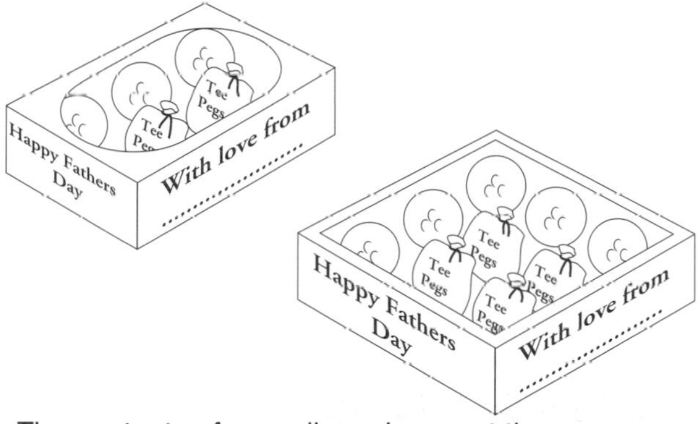

   The contents of an ordinary box cost the company £4.60, and the contents of a de luxe box costs £8.00. Find how much Instagift are paying for one golf ball and for one bag of tee pegs.

10. Flintsky plc makes industrial control units. An old model is to be replaced by two different models, the Alpha and the Beta. The company devoted 760 operator-hours and 500 machine-hours per week to making the old model, and this is now to be used to make the two new models. Each Alpha requires 6 operator-hours and 3 machine-hours; each Beta requires 10 operator-hours and 8 machine hours. Demand is more-or-less unlimited, so the production manager needs to make full use of his resources. How many Alpha units and how many Beta units should be made per week?

11. Quikadd manufactures calculators at two plants, Arborough and Essminster. At Arborough, fixed costs are £7000 per month and the cost of producing each calculator is £7.50. At Essminster the fixed costs are £8800 per month and each calculator costs £6 to produce. Next month Quikadd must produce 1500 calculators. How many should be made at each plant if the total cost at each, in order to comply with company policy, is to be the same?

Quadratic expansions

Look at the expression $(9 + 2) \times (6 + 3)$.

You would normally work this out as $11 \times 9 = 99$ but you could also do the calculation by *expanding* the brackets as follows.

$$(9 + 2) \times (6 + 3) = 9 \times (6 + 3) \quad + \quad 2 \times (6 + 3)$$
$$= 9 \times 6 + 9 \times 3 \quad + \quad 2 \times 6 + 2 \times 3$$

You can check that this comes to 99.

When the same process is applied in algebra, the expansion is written

$$(a + b)(c + d) = a(c + d) + b(c + d)$$
$$= ac + ad + bc + bd$$

and this may be illustrated by the areas of the rectangles as in the diagram below.

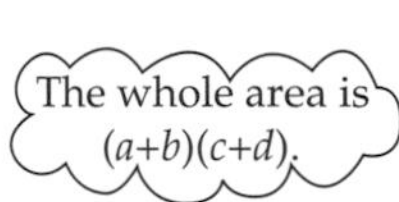
The whole area is
(a+b)(c+d).

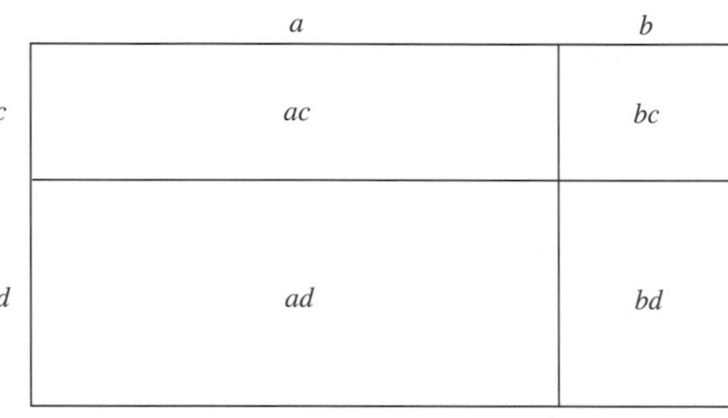
a
b
c
ac
bc
d
ad
bd

This process is particularly useful when the expansion gives you a quadratic expression as in the next example.

Example

	x	7
x	x^2	$7x$
2	$2x$	14

Expand the expressions (i) $(x + 2)(x + 7)$ (ii) $(2x - 1)(3x - 2)$.

Solution

(i) $(x + 2)(x + 7) = x(x + 7) + 2(x + 7)$ Split $(x+2)$ into x and $+2$ but leave $(x+7)$ alone.

$= x^2 + 7x + 2x + 14$ Multiply out each bracket.

$= x^2 + 9x + 14$ Collect like terms.

The expression $(x + 2)(x + 7)$ represents the area of a rectangle with sides $(x + 2)$ and $(x + 7)$.

(ii) $(2x - 1)(3x - 2) = 2x(3x - 2) - 1(3x - 2)$

$= 6x^2 - 4x - 3x + 2$

$= 6x^2 - 7x + 2$

Be careful with
signs here.

Exercise

1. Remove the brackets from the following expressions.

(i)	$(x-3)(x+4)$	(ii)	$(x+2)(x-1)$
(iii)	$(x-5)(x-3)$	(iv)	$(x+1)(x+4)$
(v)	$(x+3)(x+2)$	(vi)	$(x+3)(x-7)$
(vii)	$(x+1)(x-2)$	(viii)	$(x-6)(x-2)$
(ix)	$(x-3)(x-3)$	(x)	$(x+1)(x-1)$
(xi)	$(x-1)(x+6)$	(xii)	$(x-4)(x+4)$
(xiii)	$(x+2)(x+2)$	(xiv)	$(x-1)(x-1)$
(xv)	$(3x+1)(3x+1)$	(xvi)	$(2x+1)(3x+1)$
(xvii)	$(5x+4)(2x+4)$	(xviii)	$(7x+1)(2x-1)$
(xix)	$(6x+1)(3x-4)$	(xx)	$(2x-1)(5x+1)$
(xxi)	$(2x-3)(x+4)$	(xxii)	$(2x-7)(3x+1)$
(xxiii)	$(8x-3)(3x-2)$	(xxiv)	$(5x-1)(6x-1)$
(xxv)	$(5x-4)(3x-5)$		

2. A picture x cm square is mounted on a rectangular piece of card, as shown in the diagram.

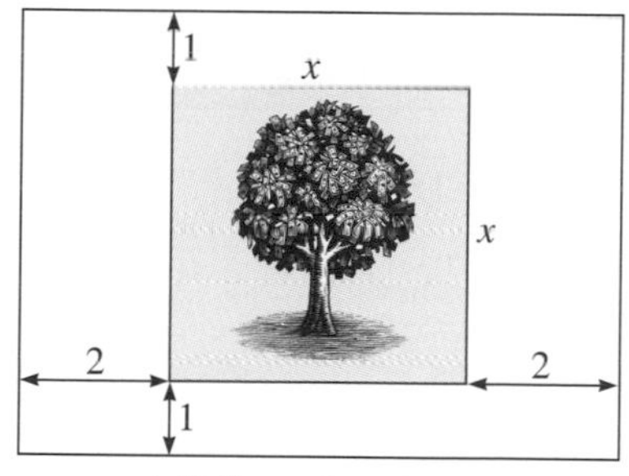

(all dimensions in cm)

   (i) Write down an expression in terms of x for
      (a) the width of the card;
      (b) the height of the card;
      (c) the area of the card.

   (ii) Expand the brackets in (i) (c) to give the area as a quadratic expression.

3. A box is made by cutting squares of side x cm out of the corners of a rectangle 10 cm by 12 cm. The sides are then folded up to make an open box with depth x cm.

   (i) What is the length of the box, in terms of x?
   (ii) What is the width of the box, in terms of x?
   (iii) What is the area of the base of the box?
   (iv) Expand the brackets in the expression you found in part (iii) to give the base area as a quadratic expression.

4. A circular pond is surrounded by a path. The pond has radius r and the path has width x.

   (i) Find expressions for
      (a) the area of the pond;
      (b) the radius of the circle round the outside of the path;
      (c) the area of this circle.

   (ii) By multiplying out and subtracting, find the area of the path.

5. The diagram shows a buffer plate from the inside of a petrol tank. It is made by cutting out the small rectangle (PQRS) from a rectangular plate (ABCD).

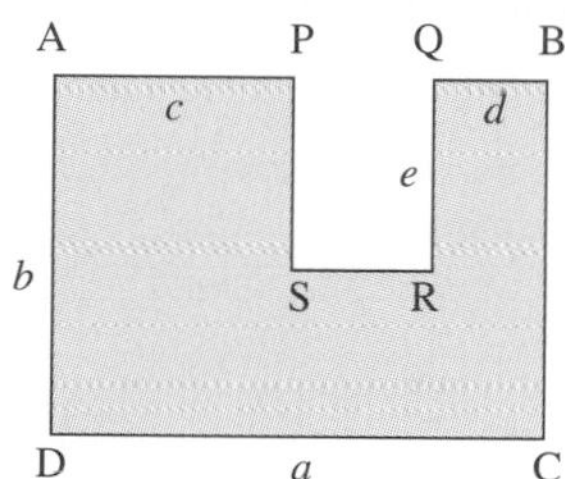

   (i) Find, in terms of a, b, c, d and e, the area of
      (a) the rectangular plate ABCD;
      (b) the cut-out rectangle PQRS;
      (c) the resulting buffer plate.

   (ii) Draw a diagram of the buffer plate and divide it into three rectangles. Add together the areas of these rectangles to find a different expression for the area of the buffer plate.

Activity

The diagram on the right shows a cardboard box. All of its faces are rectangles. Its length, width and height are l, w, and h.

(i) Draw a diagram to show how the box could be made from a rectangular sheet of cardboard of size $(l + 2h)$ by $2(w + h)$.
(ii) Show that, ignoring any pieces used as flaps, the area of offcuts is $2h(w + 2h)$.

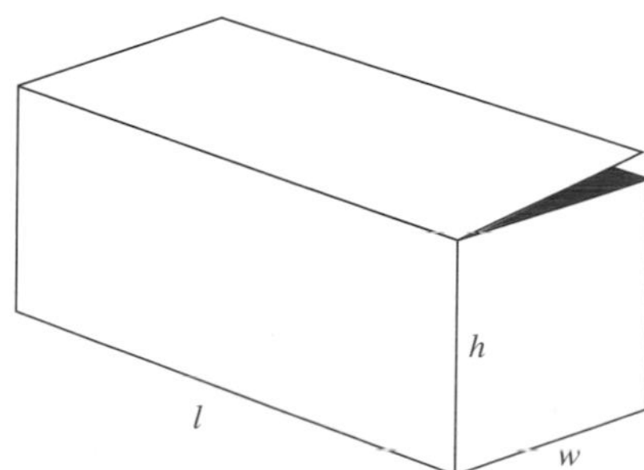

Factors

You have seen how to multiply out brackets. The reverse process, putting in brackets, is called *factorisation*. You may do this just to write an expression more neatly, but it may also reveal important features of the expression.

In each of the four expressions below, you can take out a *common factor*, which might be a letter or a number.

Common factor 3.

Common factor a.

$$4x + 8y = 4(x + 2y)$$

$$6a + 9b + 12c = 3(2a + 3b + 4c)$$

$$a^2 + ab = a(a + b)$$

$$3ab + 9acd = 3a \times b + 3a \times 3cd$$

$$= 3a(b + 3cd)$$

Common factor $3a$. You should always take out the highest possible common factor.

In each of the examples below, there are 4 terms in the expression and you can write the expression as the product of two factors, each with 2 terms.

Example

Factorise the expression $ac + 9a + 2c + 18$.

Solution

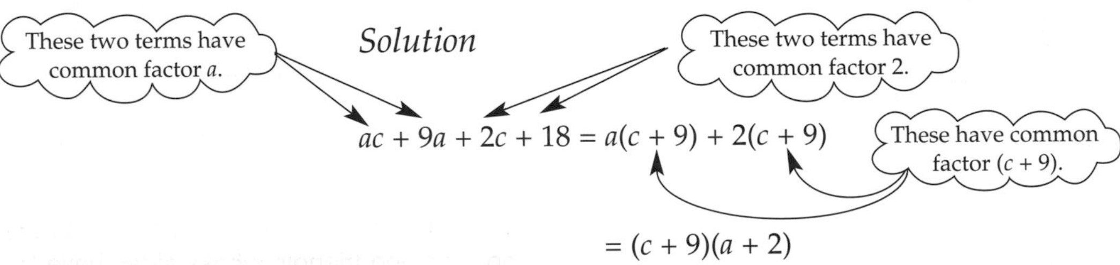

These two terms have common factor a.
These two terms have common factor 2.
ac + 9a + 2c + 18 = a(c + 9) + 2(c + 9)
These have common factor (c + 9).

$$= (c + 9)(a + 2)$$

Example

Factorise the expression $x^2 + 9x - 5x - 45$.

Solution

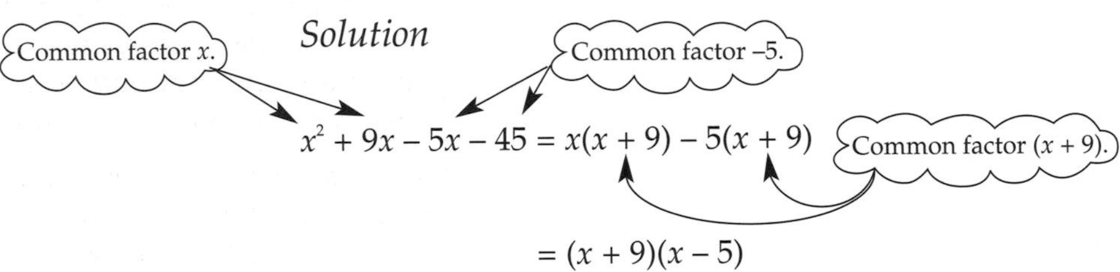

$$= (x + 9)(x - 5)$$

Example

Factorise the expression $x^2 + yz - xy - xz$.

Solution

These two terms have no common factor: you need to change the order of the terms before you can factorise the expression.

$$x^2 + yz - xy - xz = x^2 - xy - xz + yz$$

$$= x(x - y) - z(x - y)$$

$$= (x - y)(x - z)$$

Exercise

Factorise these expressions, each of which has a common factor.

1. $5x + 10$
2. $3x - 9$
3. $5x + 25$
4. $7x - 14$
5. $6x + 3$
6. $15x + 5y$
7. $20x - 16y$
8. $24x - 18y$
9. $ab + 2a$
10. $ab + 6ac$
11. $ac + ad$
12. $bc + bd$
13. $6ac + 3bc$
14. $14x^2 + 21xy$
15. $ab + ac - ad$
16. $3ac + 6ad - 3af$
17. $ap + aq + ar$
18. $ea - ed - ef$
19. $a^2 - ab - ae$
20. $12a^2 - 8ab + 20ac$

Factorise these expressions.

21. $x(a + 3) + y(a + 3)$
22. $a(a + b) + (a + b)c$
23. $a(x + y) + b(x + y)$
24. $ea - ed - bd + ba$
25. $a^2 - ab + ae - be$
26. $ap + aq + bq + bp$
27. $a^2 + ac - 5a - 5c$
28. $ab - 2b - a + 2$
29. $3x^2 - 3xy + 7x - 7y$
30. $ax + ay + az - bx - by - bz$
31. $ab - cd + bc - ad$
32. $am - 2an - bm + 2bn$
33. $y^2 - yz + 2y - 2z$
34. $2ab + 2ac + b + c$
35. $a^2 - ab + 2a - 2b$
36. $am + an + ap + bm + bn + bp$
37. $3a^2 - a - 6a + 2$
38. $5a^2 + 10a - a - 2$
39. $2x^2 + x + 14x + 7$
40. $5x^2 - 3x - 5x + 3$
41. $3x^2 - 6x + 4x - 8$
42. $6x^2 + 10x - 3x - 5$
43. $3y^2 + 5y + 3y + 5$
44. $7x^2 + 14x + 2x + 4$
45. $1 - 8t - 3t + 24t^2$
46. $10d^2 - 5d + 2d - 1$
47. $12y^2 - 3y - 4y + 1$
48. $t^2 - 3t - 4t + 12$
49. $6x^2 - 3x + 8x - 4$
50. $8a^2 - 8a + 6a - 6$
51. $4x^2 + 8x + 6x + 12$
52. $18x^2 - 27x + 12x - 18$
53. $24x^2 + 24x - 15x - 15$
54. $6x^2 + 12x + 10x + 20$
55. $40x^2 + 10x + 180x + 45$
56. $12x^2 + 5 - 10x - 6x$
57. $14a^2 + 6 + 28a + 3a$
58. $6x^2 - 15x - 10 + 4x$
59. $-11x + 33 + 3x^2 - 9x$
60. $4a^2 + 10b^2 - 8ab - 5ab$

Activities

1.

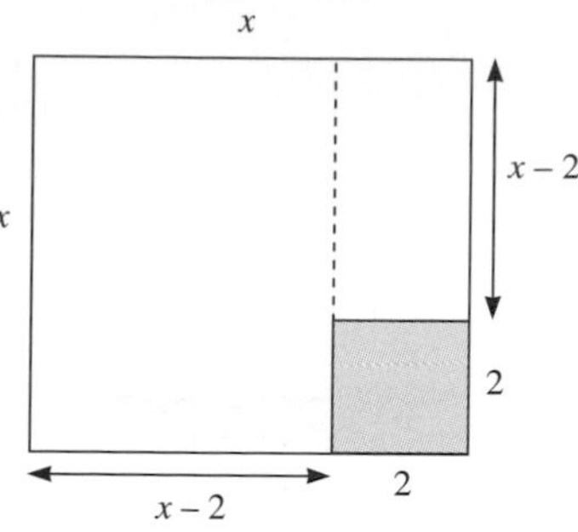

Use the diagram to show that

$$x^2 - 2^2 = (x - 2)(x + 2).$$

To do this, look at the two unshaded regions, and find an expression for the area of each. Add the expressions together, then factorise the result.

2.

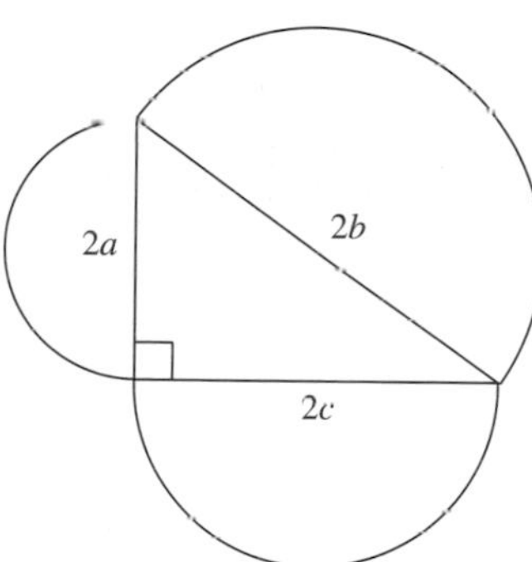

These three semi-circles are drawn on the sides of a right-angled triangle whose sides have lengths $2a$, $2b$ and $2c$.

(i) Use Pythagoras' Theorem to find the relationship between a, b and c.

(ii) Find the area of each semi-circle using the fact that the area of a circle is πr^2.

(iii) Is the area of the largest semi-circle equal to the sum of the areas of the other two?

3. Lip balm is packed in small cylindrical tubs with external height 20 mm and external radius 15 mm. The base and walls of the tubs are t mm thick; the lid is 1 mm thick and there is an air space of depth x mm between the lid and lip balm.

(i) Write down in factor form the volume of lip balm in each tub, in terms of t and x.

(ii) Given that the value of t is 3 and the volume of lip balm in each tub is $5430\,\text{mm}^3$, find the depth, x mm of the air space (to the nearest 0.5 mm).

2 Quadratic expressions

One of the examples on page 62 involved factorising the expression $x^2 + 9x - 5x - 45$. You would normally meet this as $x^2 + 4x - 45$, a quadratic expression with three terms. You will often meet expressions like this, and need to be able to factorise them. The next two examples show you how to do this.

Example

Factorise the quadratic expression $x^2 + 5x + 6$.

Solution

The given expression is the same as $1x^2 + 5x + 6$

Always start with the terms in this order: x^2 term first, then the x term, then the number.

Multiplying the outside numbers gives $1 \times 6 = +6$

The middle number is +5.

Now you need two numbers whose product is +6 and whose sum is +5.

The numbers +2 and +3 satisfy these requirements.

The quadratic expression can now be written as

$$x^2 + 2x + 3x + 6.$$

The middle term has been split into two parts. Now you look for common factors in pairs, as on page 62.

The first two terms have common factor x, and the last two terms have common factor 3, so the expression can be written as

$$x^2 + 2x + 3x + 6 = x(x + 2) + 3(x + 2)$$
$$= (x + 2)(x + 3).$$

The two expressions in brackets should be the same. If not, something has gone wrong.

NOTE

Putting the middle terms the other way round makes no difference to the final answer.

$$x^2 + 3x + 2x + 6 = x(x + 3) + 2(x + 3)$$
$$= (x + 3)(x + 2)$$

Example

Factorise the quadratic expression $4x^2 - 5x - 6$

Solution

The expression is $4x^2 - 5x - 6$.

Multiplying the outside numbers gives –24.

The middle number is –5.

You want two numbers whose product is –24 and whose sum is –5. The numbers –8 and +3 satisfy these requirements:

$$(-8) \times (+3) = -24, \text{ and } -8 + 3 = -5.$$

The quadratic expression can now be written as

$$4x^2 - 8x + 3x - 6 = 4x\,(x - 2) + 3\,(x - 2)$$
$$= (x - 2)(4x + 3)$$

Alternatively you could write:
$4x^2 + 3x - 8x - 6 = x(4x + 3) - 2(4x + 3)$
$= (4x + 3)(x - 2)$

Exercise

Factorise the following expressions.

1. $x^2 + 4x - 21$
2. $x^2 + 3x + 2$
3. $a^2 + a - 72$
4. $a^2 - 11a + 24$
5. $a^2 - 4a - 21$
6. $x^2 + 11x + 18$
7. $a^2 - 8a + 7$
8. $b^2 - 5b + 4$
9. $x^2 + 3x - 4$
10. $y^2 + y - 12$
11. $x^2 - 11x + 10$
12. $x^2 + 5x - 14$
13. $x^2 + 23x + 60$
14. $x^2 + 13x - 30$
15. $x^2 - 12x + 36$
16. $2x^2 + 3x + 1$
17. $4y^2 + 5y + 1$
18. $3p^2 + 5p - 2$
19. $2n^2 + n - 6$
20. $4t^2 + 12t - 72$
21. $3a^2 - 7a + 2$
22. $5a^2 + 9a - 2$
23. $2x^2 + 15x + 7$
24. $5x^2 - 8x + 3$
25. $3x^2 - 2x - 8$
26. $6x^2 + 7x - 5$
27. $3y^2 + 8y + 5$
28. $12c^2 - c - 6$
29. $12x^2 - 16x - 3$
30. $24t^2 - 11t + 1$

Activities

1. Friend or fiend?

Write down ten expressions involving brackets. Multiply out the brackets, and write the answers on a separate sheet of paper.

Exchange answer sheets with a friend. See how many of your friend's expressions you can factorise.

2. Playing-cards

There are 52 cards in an ordinary pack and the diagram shows them sorted into piles, one for each 'suit' of 13 cards.

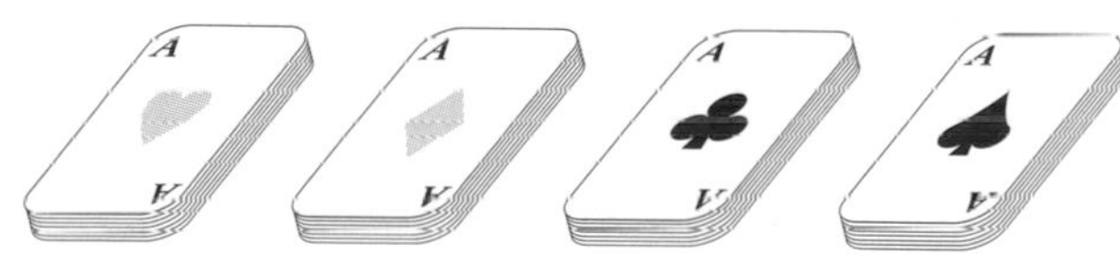

This arrangement corresponds to the factorisation of 52 as

$$(1 + 1 + 1 + 1) \times 13.$$

Describe (or show in a diagram) the arrangements of cards corresponding to

(i) $4 \times (1 + 1 + 1 + \ldots + 1)$
(ii) $(1 + 1 + 1 + 1) \times (1 + 1 + 1 + \ldots + 1)$
(iii) $(2 + 2) \times 13$

Investigations

1. Difference of squares

Write $9x^2 - 4$ as $9x^2 + 0x - 4$. Now factorise this and look very carefully at your answer. Remembering that $(3x)^2 = 9x^2$ and that $2^2 = 4$, can you see a pattern in your answer?

Do the same with $x^2 - 49$.

Now try these:

(i) $x^2 - 81$
(ii) $x^2 - 100$
(iii) $x^2 - 64$
(iv) $x^2 - 25$
(v) $x^2 - 121$
(vi) $x^2 - 144$
(vii) $4a^2 - 25$
(viii) $16a^2 - 49$
(ix) $81 - 9a^2$
(x) $36b^2 - 64$
(xi) $x^2 - y^2$
(xii) $4x^2 - 9y^2$

2. Pythagorean triplets

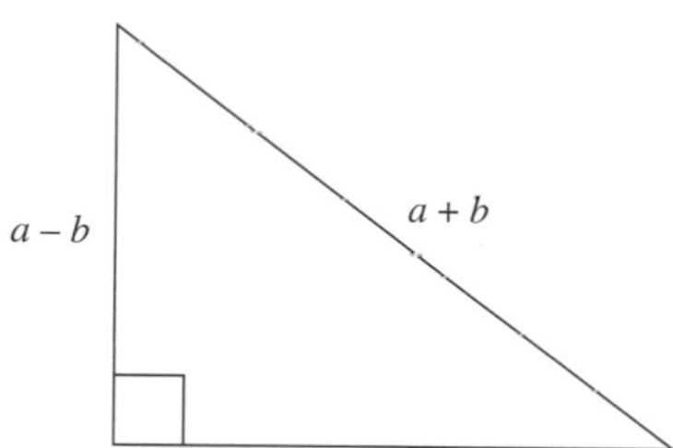

Here is a right-angled triangle. By Pythagoras' Theorem the square of the length of the third side will be

$$(a + b)^2 - (a - b)^2.$$

(i) Use the difference of squares or multiply out the brackets and collect the terms to check that the third side has a length of $2\sqrt{ab}$.

If ab is a perfect square, such as 4, you can form a set of integers like (3, 4, 5) which satisfy Pythagoras' Theorem. Each set is called a *Pythagorean triplet*.

(ii) Draw up a table like the one below, and complete it at least as far as $ab = 144$.

ab	a	b	$a - b$	$2\sqrt{ab}$	$a + b$
4	4	1	3	4	5
9	9	1	8	6	10
16	16	1			
	8	2			
25					

Quadratic equations

Equations like $3x^2 - 5x - 2 = 0$, $x^2 + 7x + 12 = 0$ and $x^2 - 4 = 0$ are called *quadratic equations*. There are several ways of solving them and some of these are shown on this page and page 68.

Example

Solve the equation $3x^2 - 5x - 2 = 0$ by factorisation.

Solution

(i) Factorise the quadratic expression:

$$\begin{aligned} 3x^2 - 5x - 2 &= 0 \\ 3x^2 - 6x + 1x - 2 &= 0 \\ 3x(x-2) + 1(x-2) &= 0 \\ (x-2)(3x+1) &= 0. \end{aligned}$$

Either $x - 2 = 0 \Rightarrow x = 2$ or $3x + 1 = 0 \Rightarrow x = -\frac{1}{3}$

The product is zero, so one of the factors must be zero.

The solution is $x = 2$ or $-\frac{1}{3}$.

Check: substitute $x = 2$ in $3x^2 - 5x - 2$: $3 \times 4 - 5 \times 2 - 2 = 0$;

substitute $x = -\frac{1}{3}$ in $3x^2 - 5x - 2$: $3 \times \frac{1}{9} - 5 \times (-\frac{1}{3}) - 2 = 0$.

There are two important points to check before you start to factorise.

- Always check to see if you can divide through the equation by a whole number. For example both sides of the equation

$$2x^2 - 4x - 6 = 0$$

can be divided by 2 to give $x^2 - 2x - 3 = 0$.

- If the x^2 term is negative it is usual to multiply through by (–1) which changes all three signs:

$$-x^2 + 11x - 18 = 0 \quad \text{becomes} \quad x^2 - 11x + 18 = 0.$$

Example

Solve the equation $3x^2 - 5x - 2 = 0$ graphically.

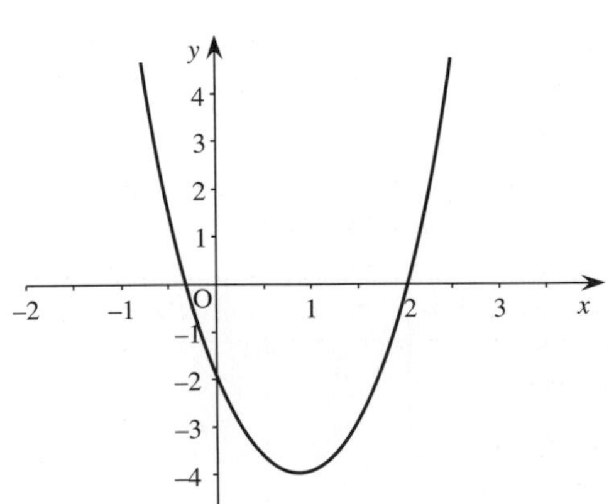

Solution

The graph of $y = 3x^2 - 5x - 2$ cuts the x axis at $x = 2$ and $x = -\frac{1}{3}$ as shown, and these are the *roots* of the equation.

When you draw the graph (by hand or using a calculator) it will be clear that $x = 2$ is an exact root. You can find the other root as accurately as you wish, say between – 0.35 and – 0.30 or between – 0.334 and – 0.333, but the graph will not tell you that it is exactly $-\frac{1}{3}$.

Exercise

Solve the following quadratic equations.

1. $x(x - 3) = 0$
2. $(x - 4)x = 0$
3. $(2x - 3)(x - 5) = 0$
4. $(5x + 1)(5x - 2) = 0$
5. $(x + 1)(x - 2) = 0$
6. $(x - 4)(x + 4) = 0$
7. $(4 + x)(1 - 4x) = 0$
8. $(2x - 5)(x + 3) = 0$
9. $t^2 - 5t + 6 = 0$
10. $t^2 + 5t + 6 = 0$
11. $x^2 - 3x + 2 = 0$
12. $2t^2 - 2t - 12 = 0$
13. $x^2 - 8x + 15 = 0$
14. $x^2 + 8x + 15 = 0$
15. $x^2 + 11x + 28 = 0$
16. $x^2 - 11x + 28 = 0$
17. $-x^2 + 9x = 20$
18. $x^2 - x = 12$
19. $2y^2 + 2 - 5y$
20. $9x^2 = 6 - 3x$
21. $6x^2 = 7x + 3$
22. $4x^2 + 13x + 3 = 0$
23. $-2x^2 + 7x - 6 = 0$
24. $9x^2 - 30x + 25 = 0$
25. A rectangular garden measures 16 m by 10 m. It has a central lawn surrounded by a path of width x metres.

    (i) Find the dimensions of the lawn in terms of x.
    (ii) If the area of the lawn is 72 m², find x and hence the dimensions of the lawn.

26. A ball is thrown up into the air and its height, h metres, above the ground after t seconds is given by

    $$h = 9.8t - 4.9t^2.$$

    The ball returns to the ground when $h = 0$. For how many seconds is it in the air?

27. A plumber needs to bend a length of sheet lead 6 cm wide into the form of an open rectangular duct with a cross-section area of 4 cm². The diagram shows his sketch of the duct. He calls the depth of the duct x cm.

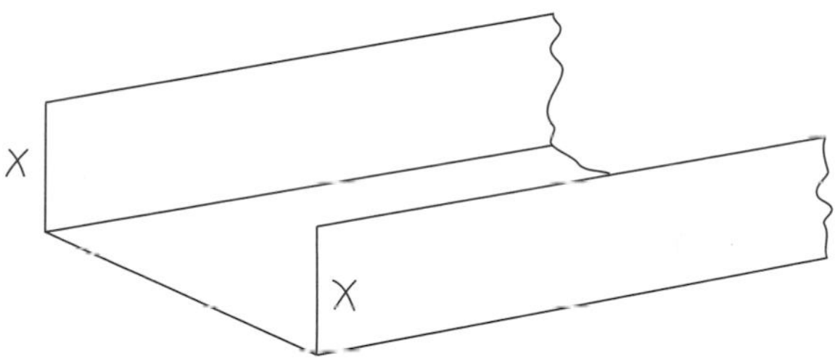

Find the two possible values of x.

Activity

Arrow Airlines charges £200 for a return ticket to Paracano. Its planes have a capacity of 350 passengers but the average load is 300 passengers.

You are a market research consultant advising Arrow Airlines about ticket pricing. You have collected evidence to suggest that increasing the fare by £20 will reduce the average load by 10 passengers, and conversely that reducing the fare by £20 will increase the average load by 10 passengers. What would you recommend as the best ticket price?

To tackle this, you assume that the airline wants to maximise revenue. Note that the revenue per flight is given by

$$\text{revenue} = \text{number of passengers} \times \text{fare}$$

(i) Start by finding the fare that would produce 350 passengers, and calculate the revenue in that case. Now try increasing that fare in steps of £20. You may find it helpful to use a spreadsheet to do the calculations.

(ii) Once you have done the calculations, draw a graph of revenue against number of passengers. What do you notice about its shape?

(iii) Concentrate on the region near the peak of the graph and find the maximum possible revenue, and the number of passengers on the flight in that case (you will need to consider fare increases in smaller steps, say £2, to do this). What price should Arrow Airlines charge for a ticket?

(iv) Taking x to be the number of passengers, write down an expression for the fare in terms of x. Now write down an expression for the revenue in terms of x.

(v) Use your expression from (iv) to explain the shape of your graph of revenue against number of passengers.

The quadratic formula

Sometimes a quadratic equation does not factorise, so you cannot solve it by factorising. Fortunately, there is a formula that will give you the solution of any equation, whether or not it can be factorised.

All quadratic equations can be written in the form $ax^2 + bx + c = 0$, where a, b, and c are constants. For example, the equation $2x^2 + 8x + 3 = 0$ has $a = 2$, $b = 8$ and $c = 3$. To solve an equation using the formula, you first work out the values of a, b and c in your equation. The quadratic formula then gives the solution of the equation as

a is the coefficient of x^2, b is the coefficient of x, c is the constant term.

$$x = \frac{-b \pm \sqrt{b^2 - 4ac}}{2a}$$

Substituting the values of a, b and c for your equation into this formula will give you the two values of x which satisfy the equation.

Example

Use the quadratic formula to solve the equation $3x^2 - 5x - 2 = 0$.

Solution

In this equation, $a = 3$, $b = -5$ and $c = -2$.

$$\Rightarrow \quad b^2 - 4ac = 25 - 4 \times 3 \times (-2) = 25 + 24 = 49 \quad \Rightarrow \quad \sqrt{b^2 - 4ac} = \sqrt{49} = 7$$

The formula therefore gives $x = \dfrac{-(-5) \pm 7}{2(3)} = \dfrac{5 \pm 7}{6} = \dfrac{12}{6}$ or $\dfrac{-2}{6}$

Cancelling these gives $x = 2$ or $-\frac{1}{3}$.

Notice that you have to find the square root of **all** of $(b^2 - 4ac)$.

You can check this against the answer on page 66.

When solving problems you can often reject a negative answer straight away, and sometimes an answer is too big or too small to fit the problem.

Example

A drinks manufacturer has developed a new drink. If the development costs are to be recouped, the manufacturer needs the drink to be bought by 30 000 households within the first 6 months of sales. The marketing manager estimates that n months ($n < 6$) after the introduction of the drink the number H of households buying it will be given by

$$H = 1000n(12 - n).$$

Assuming that this equation is a good guide, when will the target be reached?

Solution

This information may be written as $\quad 30\,000 = 1000n(12 - n)$.

Dividing both sides by 1000, $\quad 30 = n(12 - n)$

$30 = 12n - n^2$.

This is a quadratic equation: $\quad n^2 - 12n + 30 = 0$.

The left-hand side cannot be factorised so you use the formula with $a = 1$, $b = -12$ and $c = 30$. This gives

$$n = \frac{+12 \pm \sqrt{12^2 - 4 \times 1 \times 30}}{2 \times 1} = 3.55 \text{ or } 8.45.$$

So it takes just over $3\frac{1}{2}$ months for the number of households to reach the target level. Notice that the second answer, 8.45, is outside the period for which the equation is valid.

Exercise

Use the formula to solve the quadratic equations in questions 1–21. Give your answers correct to 2 decimal places. If you have access to a graphical calculator or computer package with a 'zoom' facility, use it to check your answers

1. $x^2 - 5x - 14 = 0$
2. $x^2 - 12x + 11 = 0$
3. $x^2 - 4x - 7 = 0$
4. $x^2 - 10x - 3 = 0$
5. $x^2 - 6x + 4 = 0$
6. $x(x + 8) + 14 = 0$
7. $2x^2 + 10x + 11 = 0$
8. $4x^2 + 8x - 7 = 0$
9. $3x^2 - 18x - 1 = 0$
10. $2x^2 - 10x + 11 = 0$
11. $5x^2 + 8x = 10$
12. $4x^2 + 3x - 5 = 0$
13. $2x^2 + 7x + 4 = 0$
14. $6x^2 + 8x - 15 = 0$
15. $3x^2 + 7x - 11 = 0$
16. $6x^2 - 10x - 3 = 0$
17. $9x^2 - 30x + 13 = 0$
18. $2x^2 - 4x = 5$
19. $2x^2 - 6x - 3 = 0$
20. $4x(x - 7) + 3 = 0$
21. $4 + 2x - x^2 = 0$

22. The production department estimates that the weekly cost, £C, of producing x tonnes of product per week is

$$C = 100 + 48x \quad (\text{for } x \leq 100).$$

Selling this product creates revenue of £R for the company. The sales department estimates the revenue to be

$$R = 100x - x^2.$$

Calculate the break-even points, at which $R = C$.

23. A specialist packaging company packs hand-made chocolates in decorated card boxes with clear plastic lids. Each box is 5.25 cm high, and has a square base. The area of card used for each box is $100\,\text{cm}^2$. Find the length of the sides of the base.

24. When the nozzle of a fire-hose is held at a certain angle at ground level, the height, h metres, of the water jet at a horizontal distance x metres from the nozzle is given by $200h = x(200 - x)$.

If the jet can be directed onto a tower block window 48 metres above ground level, find the horizontal distance of the window from the nozzle. Explain the two solutions.

25. A group of biologists studied the nutritional effect of a diet of yeast and cornflour containing 10% protein. The percentage P of yeast in the protein mix and the gain g in weight were found to be connected by the equation

$$g = -200P^2 + 200P + 20.$$

What percentage of yeast gives a weight gain, g, of 70?

Investigation

(i) What happens when you try to use the quadratic formula to solve the equation

$$x^2 - 4x + 5 = 0?$$

(ii) What happens when you try to solve the equation graphically?

(iii) How many roots does the quadratic equation

$$ax^2 + bx + c = 0$$

have in the cases when the value of $b^2 - 4ac$ is

(a) greater than 0?
(b) equal to 0?
(c) less than 0?

Activity

A motor manufacturer is investigating the fuel consumption, f, (in miles per gallon), at different speeds v (in miles per hour) for a new car. Following a set of experiments this graph of f against v is drawn.

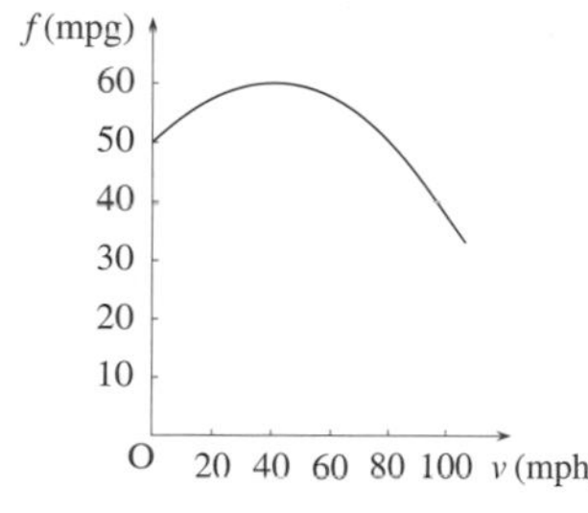

(i) Show that the relationship

$$f = 60 - \frac{1}{160}(v - 40)^2$$

fits the graph exactly for $v = 0$, 40 and 80.

(ii) Show that when $f = 20$ the equation can be rearranged as

$$(v - 40)^2 = 160 \times 40$$

and that this gives

$$v = +40 + \sqrt{6400} \quad \text{or} \quad v = +40 - \sqrt{6400}.$$

Explain why you would only consider one of these two values of v.

Fractions

You handle fractions in algebra just as you handle them in arithmetic.

Example

Simplify (i) $\frac{1}{2} + \frac{2}{3} - \frac{3}{4}$ (ii) $\frac{w}{2} + \frac{2w}{3} - \frac{3w}{4}$

Solution

(i) $\frac{1}{2} + \frac{2}{3} - \frac{3}{4} = \frac{6}{12} + \frac{8}{12} - \frac{9}{12}$

$= \frac{6+8-9}{12} = \frac{5}{12}$.

The bottom numbers, 2, 3 and 4, are all factors of 12.

(ii) The method is the same: start by putting the fractions over a common denominator:

$$\frac{w}{2} + \frac{2w}{3} - \frac{3w}{4} = \frac{6w}{12} + \frac{8w}{12} - \frac{9w}{12}$$

$$= \frac{6w+8w-9w}{12} = \frac{5w}{12}.$$

All of the fractions have the same denominator so you can write them as a single fraction.

Remember that you can only add or subtract *like* terms. For example, the expression $\frac{6x+8y-9z}{12}$ cannot be simplified any further.

When you have an equation, you can often get rid of any fractions by multiplying both sides of the equation by the same number, as shown in the next example. (You cannot do this, though, when you are just simplifying an expression as in the example above.)

Example

Solve the equation $\frac{2x}{5} + 7 = \frac{3x}{4}$.

Solution

Write the equation as it was given: $\frac{2x}{5} + 7 = \frac{3x}{4}$

Notice that 5 and 4 are both factors of 20, so you can get rid of the fractions by multiplying both sides by 20:

$$20 \times \frac{2x}{5} + 20 \times 7 = 20 \times \frac{3x}{4}.$$

Remember to multiply *each term* by 20.

Simplifying each term, you obtain

$$4 \times 2x + 140 = 5 \times 3x$$

$$8x + 140 = 15x$$

Collecting all x terms on one side, $140 = 7x$,

and dividing by 7, $20 = x$.

Substituting $x = 20$ into the original equation gives

LHS $= \frac{40}{5} + 7 = 8 + 7 = 15$;

RHS $= \frac{60}{4} = 15$.

Always remember to check your answer.

So $x = 20$ does indeed satisfy the equation.

Exercise

Simplify the fractions in questions 1–18.

1. $\frac{x}{2}+\frac{x}{3}$

2. $\frac{x}{2}+\frac{2x}{5}$

3. $\frac{2x}{3}-\frac{x}{6}$

4. $\frac{3x}{4}-\frac{2x}{3}$

5. $\frac{x}{5}+\frac{x}{3}$

6. $\frac{x}{3}-\frac{x}{5}$

7. $\frac{3x}{5}-\frac{2x}{3}$

8. $\frac{3x}{4}+\frac{4x}{5}$

9. $\frac{3x}{4}-\frac{5x}{7}$

10. $\frac{2x}{3}-\frac{x}{6}+\frac{x}{2}$

11. $\frac{2a}{15}-\frac{a}{3}+\frac{2a}{5}$

12. $\frac{3b}{4}-\frac{5b}{6}+\frac{7b}{9}$

13. $\frac{x+2}{2}-\frac{x}{6}$

14. $\frac{2(x+3)}{3}-\frac{(x-2)}{6}$

15. $\frac{3x+7}{5}-\frac{5x+2}{7}$

16. $\frac{3x-1}{2}-\frac{4x+1}{3}$

17. $\frac{x+1}{3}-\frac{x-1}{4}$

18. $\frac{x+6}{5}-\frac{x-4}{10}$

Solve the equations in questions 19 – 36.

19. $\frac{x}{2}+\frac{x}{3} = 5$

20. $\frac{x}{2}+\frac{2x}{5} = 9$

21. $\frac{2x}{3}-\frac{x}{6} = 4$

22. $\frac{3x}{4}-\frac{2x}{3} = 2$

23. $\frac{x}{5}+\frac{x}{3} = 16$

24. $\frac{x}{3}-\frac{x}{5} = 2$

25. $\frac{3x}{5}-\frac{2x}{3} = 1$

26. $\frac{3x}{4}+\frac{4x}{5} + 62 = 0$

27. $\frac{5x}{7}-\frac{3x}{4} = \frac{1}{4}$

28. $\frac{2x}{3}+\frac{x}{6}+\frac{x}{2} = 1$

29. $\frac{2a}{15}-\frac{a}{3}+\frac{2a}{5} = 1$

30. $\frac{3b}{4}-\frac{5b}{6}+\frac{7b}{9} = 25$

31. $\frac{x+2}{2}-\frac{x}{6} = \frac{7}{3}$

32. $\frac{2(x+3)}{3}-\frac{(x-2)}{6} = \frac{17}{6}$

33. $\frac{3x+7}{5}-\frac{(5x+2)}{7} = 1$

34. $\frac{(4x+1)}{3}-\frac{3x-1}{2} = 1$

35. $\frac{x-2}{5}-\frac{(x-1)}{3} = 1$

36. $\frac{4(x+2)}{5}-\frac{3(x-2)}{10} = 2$

37.

8 mph

12 mph

Peter cycles x miles uphill to work at 8 miles per hour. He can cycle back at 12 miles per hour. The two-way journey takes him 1 hour 15 mins. State, in terms of x, how long it takes Peter to cycle

(i) to work;
(ii) home;
(iii) both ways.
(iv) Form an equation in x and solve it to find how far Peter has to travel to work.

38. A river excursion travels x miles upstream at 3 miles per hour and back to the same place at 5 miles per hour. The round trip takes 48 minutes. State in terms of x how long the journey takes

(i) on the upstream leg;
(ii) on the downstream leg;
(iii) in total.
(iv) Form an equation in x and solve it to find how far upstream the excursion goes.

Activity

In question **1** you found that

$$\frac{x}{2}+\frac{x}{3} = \frac{5x}{6}.$$

(i) Solve the equation

$$\frac{5x}{6} = 10.$$

(ii) Multiply both sides of the equation

$$\frac{x}{2}+\frac{x}{3} = 10$$

by 6. Hence solve the equation

$$\frac{x}{2}+\frac{x}{3} = 10.$$

(iii) Explain why your answers to parts (i) and (ii) are the same. Which method is quicker?

Rearranging formulae

Look at the formula relating voltage, V, current, I and resistance, R in an electric circuit; $V = IR$. (You may recognise this as Ohm's Law.)

As this is written, V is the subject of the formula. It may be, however, that you know V and R and want to find the current, I. In that case, you want to make I the subject of the formula. To do this you divide both sides by R:

$$\frac{V}{R} = I \text{ i.e. } I = \frac{V}{R}.$$

You will often need to change the subject of a formula. The examples on this page illustrate the techniques involved.

Example

When a quantity £P is invested at a R% simple interest, the interest due, I, is calculated using the formula

$$I = \frac{PRT}{100}.$$

Rearrange the formula to make T the subject.

Solution

Write the formula as it was given: $I = \dfrac{PRT}{100}$

Multiply by 100: $100\,I = PRT$

Divide by P and R: $\dfrac{100I}{PR} = T$ i.e. $T = \dfrac{100I}{PR}$.

Example

The distance, s, travelled by a car accelerating steadily from speed u to speed v in time t is given by

$$s = (u + v)\,\frac{t}{2}.$$

Rearrange this equation to find t in terms of u, v and s.

Solution

Write the equation as it was given: $s = (u + v)\,\dfrac{t}{2}$

Multiply by 2: $2s = (u + v)t$

Divide by $(u + v)$: $\dfrac{2s}{(u+v)} = t$ i.e. $t = \dfrac{2s}{(u+v)}$.

Example

The volume, V, of a spherical ball is given by $V = \frac{4}{3}\pi r^3$. Find an expression for the radius in terms of the volume.

Solution

Write the formula as it was given: $V = \dfrac{4}{3}\pi r^3$

Multiply by 3: $3V = 4\pi r^3$

Divide by 4π: $\dfrac{3V}{4\pi} = r^3$

Take the cube root of each side: $\sqrt[3]{\dfrac{3V}{4\pi}} = r$ i.e. $r = \sqrt[3]{\dfrac{3V}{4\pi}}$.

Exercise

1. Rearrange $T = \frac{2\pi}{\omega}$, making ω the subject.

2. Rearrange $A = \frac{bh}{2}$, making (i) b (ii) h the subject.

3. Rearrange $v^2 = u^2 + 2as$, making (i) a (ii) s the subject.

4. Rearrange $P = 2(L + B)$, making (i) L (ii) B the subject.

5. Rearrange $5hR = 12dV^2$, making V the subject.

6. Rearrange $V = \pi r^2 h$, making (i) h (ii) r the subject.

7. Rearrange $v = u + at$ making (i) u (ii) a the subject.

8. Rearrange $T = \frac{\lambda x}{l}$, making (i) l (ii) x the subject.

9. Rearrange $v = 8\sqrt{H}$, making H the subject.

10. Given that $T = 2\pi\sqrt{\frac{l}{g}}$, rearrange to find l.

11. A leatherworker makes jewellery by setting beads into leather. The beads are surrounded by a coloured strand whose length, L, is the circumference of the bead, C, plus $1\frac{1}{2}$ times the thickness of the leather, T:

$$L = C + 1.5T$$

(i) Find the length of the strand needed for a bead of circumference 24 mm using leather of thickness 1 mm.
(ii) Rearrange the equation to give T in terms of L and C.
(iii) An oval bead has a circumference of 52 mm and the corresponding strand length is 55 mm. What is the thickness of the leather in this case?

12. Every Munchito bar I eat contains 320 calories. If I then walk briskly, I will lose 8 calories per minute so the number of calories remaining to be used up is given by

$$c = 320m - 8t,$$

where c is the number of calories, m is the number of Munchito bars, and t is the walking time in minutes.
(i) Calculate the calories remaining to be used up if I eat two Munchito bars and then walk briskly for 40 minutes.
(ii) Rearrange the formula to give t in terms of c and m.
(iii) If I eat three Munchito bars, for how long will I need to walk to use up all the calories they contain?

13. The number of 1 metre square non-slip paving slabs needed to surround a rectangular swimming pool with dimensions x metres by y metres is given by

$$N = 2(x + y) + 4.$$

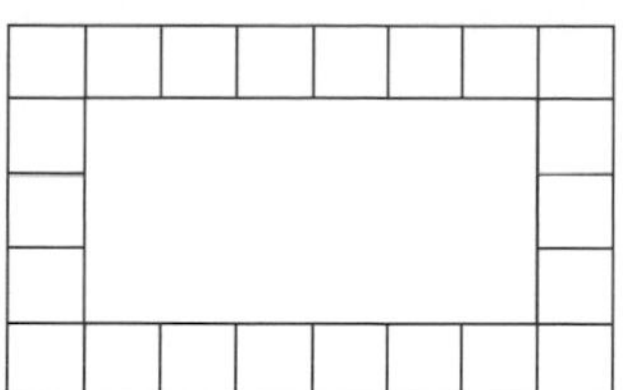

(i) How many slabs are needed for a pool measuring 12 metres by 10 metres?
(ii) Rearrange the formula to give y in terms of N and x.
(iii) Use your new formula to find the longest possible pool of 25 metres width given that 114 slabs are available.
(iv) Adapt the original formula to find the number of slabs needed for a pool which is x metres square.
(v) Rearrange this formula to give x in terms of N.
(vi) What should the dimensions of a square pool be if there are 164 slabs available for the surround?

14. The money invested in a company is called capital. The *return on capital employed* (ROCE) is calculated using the formula

$$\text{ROCE} = \frac{\text{net profit}}{\text{capital}} \times 100 \quad \text{or} \quad R = \frac{P \times 100}{C}$$

where R is a percentage and P and C are in £.
(i) What is the return if $P = 3520$ and $C = 115\,000$?
(ii) Rearrange the formula to give P in terms of R and C.
(iii) What is the net profit if $C = 7500$ and $R = 21$?

15. The cardiac output C, of a person is given by

$$C = HS$$

where H is the heartbeat rate in beats per minute and S the volume of each stroke in cm^3.
Ben has a heartbeat of 50 beats per minute and a stroke volume of 91 cm^3.
(i) Find Ben's cardiac output.

Ben's friend, Simon, has the same cardiac output as Ben, but his stroke volume is 56 cm^3.
(ii) Rearrange the formula to find an expression for heartbeat rate in terms of C and S.
(iii) Find Simon's heartbeat rate.

Chapter 3

Graphs

Introduction

A graph is a way of displaying the relationship between two or more variables. The diagram below shows the grid for a Cartesian graph. The point where the axes meet is called the origin, O.

Point A has a displacement from the origin of +4 in the x direction and +3 in the y direction; its co-ordinates are (4, 3). Point B has a displacement from the origin of –3 in the x direction and +2 in the y direction; its co-ordinates are (–3, 2).

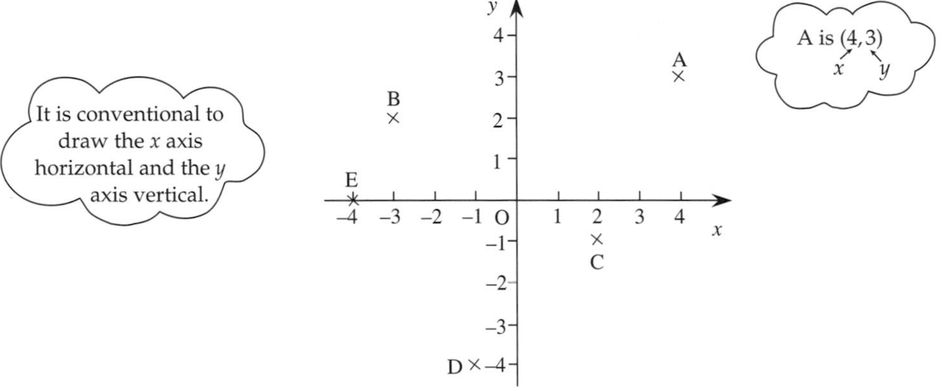

Before reading on, check that you agree that C, D and E are the points (2, –1), (–1, –4) and (–4, 0) respectively.

Scales

Each axis has a scale. The scale for x need not be the same as that for y. Sometimes the quantities being plotted may be very large (measured in tens, hundreds or even thousands of units) and sometimes they may be very small (measured in tenths, hundredths or thousandths of a unit). You need to be able to plot and read values on any scale, particularly the values that come between the main scale marks. Look carefully at the following examples to make sure you agree with the indicated readings.

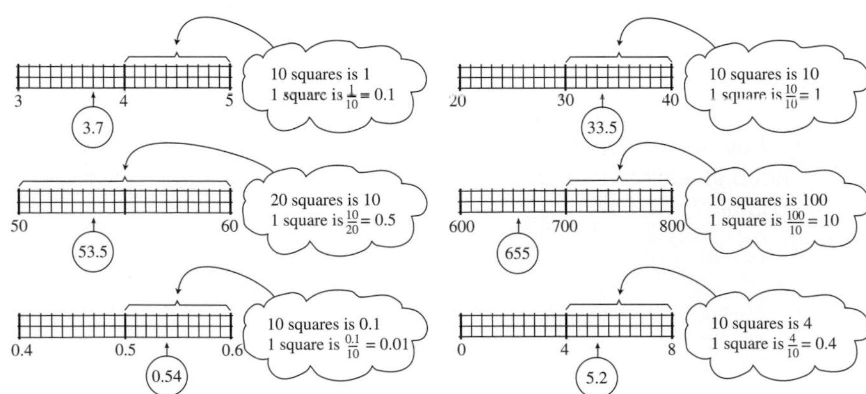

Exercise

1. (i) Draw co-ordinate axes for values of x and y from −5 to +5.
   (ii) Plot and label each of the following points.
   A (2, 3), B (3, −4), C (−5, 2),
   D (5, −3), E (−1, −5), F (−2, 1),
   G (4, 0), H (0, −2).

2. Write down the co-ordinates of the points labelled A-J in the diagram below.

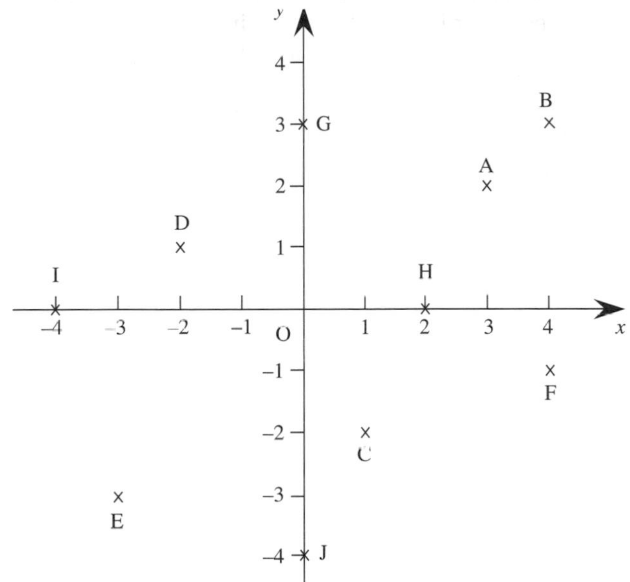

3. (i) Draw co-ordinate axes for values of x from 0 to 700 (scale 2 cm = 100 units) and values of y from 0 to 160 (scale 2 cm = 20 units).
   (ii) Plot and label the following points.
   P (600, 80), Q (150, 70), R (200, 25),
   S (340, 110), T (660, 38), U (465, 145).

4. In each of the following, write down the reading.

(a)
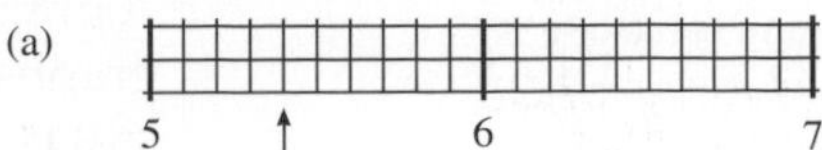

(b)
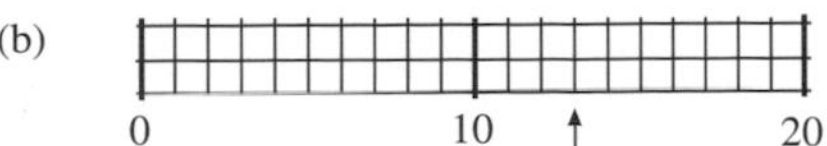

(c)

(d)
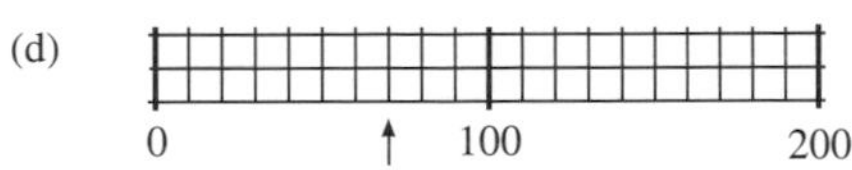

(e)
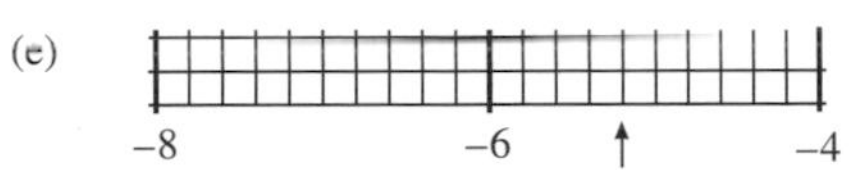

(f)
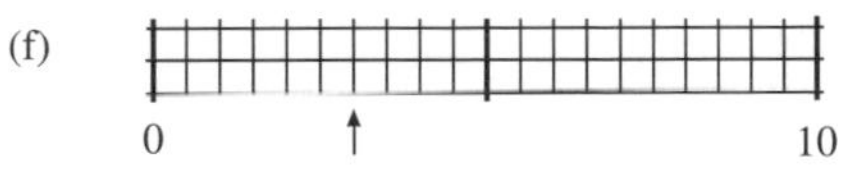

(g)
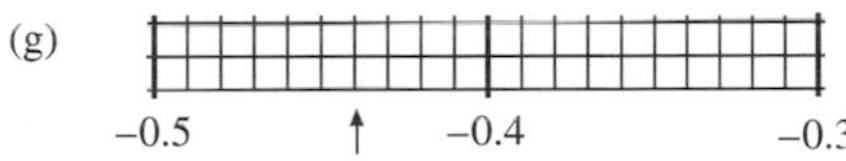

(h)
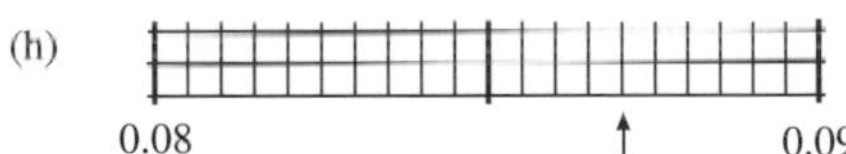

Activity

The Cartesian co-ordinate system can be extended to three dimensions with axes x, y and z as shown on the right.

A rectangular box measuring 9 units by 8 units by 6 units is placed with one corner at the origin. The point A has co-ordinates (8, 0, 0). Write down the co-ordinates of the other corners of the box.

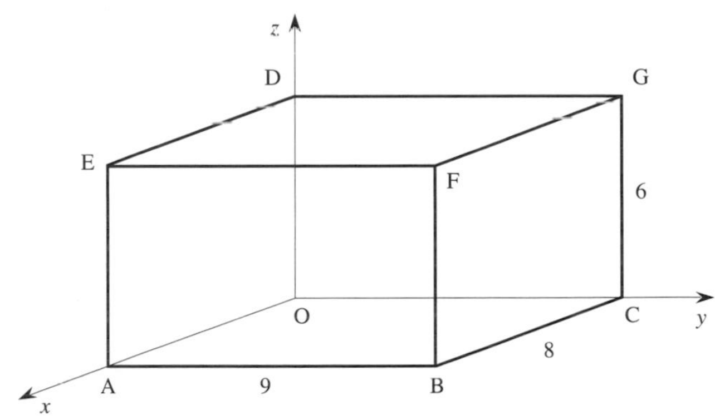

Information from graphs

The maximum value of y is 5; it occurs when $x = 1$.

A graph can give you a great deal of information. Look at the three graphs on this page, and notice how much they tell you about the situations they represent.

Where the curve crosses the x axis, $y = 0$.

1. A curve from an equation

Part of the graph of $y = 4 + 2x - x^2$ is shown on the left. By taking suitable readings you can find the approximate solution of the equation

$$4 + 2x - x^2 = 0.$$

You will see how to use graphs to solve equations on page 88. You can also read off maximum (or minimum) values, as shown.

2. Experiment with a spring

In an experiment, the length of a spring was measured when objects with different mass were suspended from it. The results are shown in the graph.

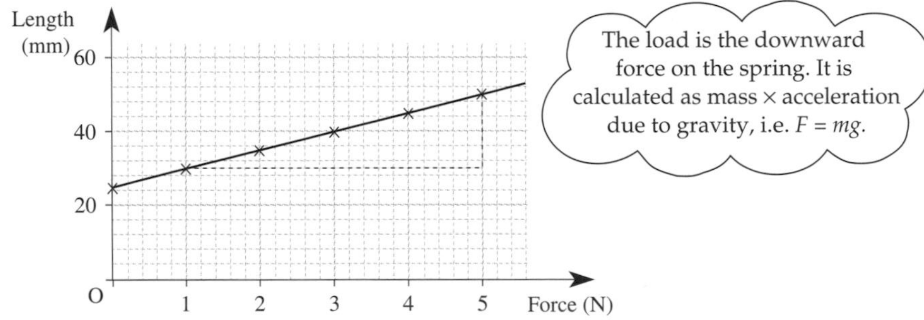
Length (mm)
60
40
20
O
1
2
3
4
5
Force (N)
The load is the downward force on the spring. It is calculated as mass × acceleration due to gravity, i.e. $F = mg$.

It appears that the length against load graph is a straight line (if allowances are made for small experimental errors). This means that the length and the load can be related by a simple equation: you will see how to find the equation on page 86.

3. Cyclist

A cyclist accelerates from rest, travels at constant speed for some time, and then decelerates. The graph shows speed v, plotted against time t.

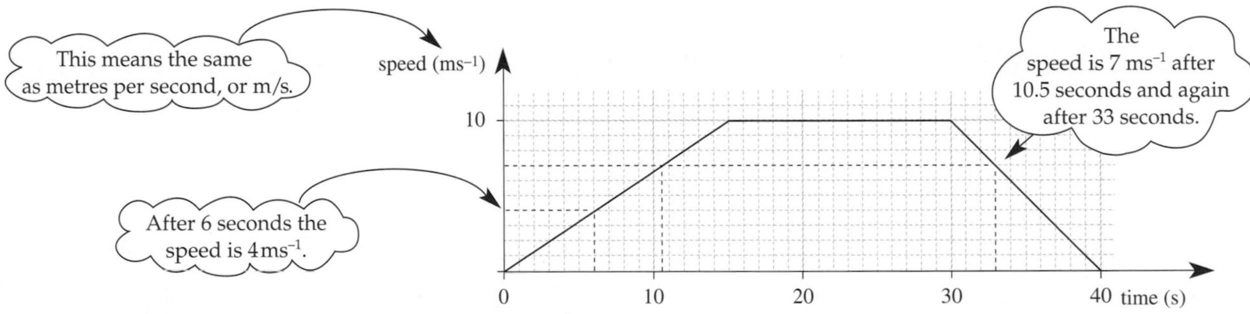
This means the same as metres per second, or m/s.
speed (ms^{-1})
10
After 6 seconds the speed is $4 ms^{-1}$.
The speed is $7 ms^{-1}$ after 10.5 seconds and again after 33 seconds.
0
10
20
30
40
time (s)

The slope of the graph (page 92) and the area under the graph (page 94) also give useful information about the cyclist's progress.

Choice of axes

When plotting a graph you usually choose values of one variable at even intervals, and plot the corresponding values of the other variable. The horizontal axis is generally used for the first variable, which is often called the *independent* variable.

Exercise

1. The graph shows how the cost of hiring a van varies with the number of miles travelled.

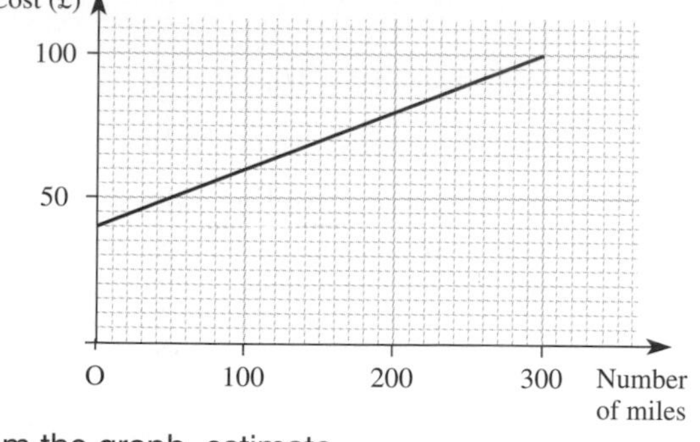

From the graph, estimate
   (i) the cost of travelling 100 miles;
   (ii) the cost of travelling 240 miles;
   (iii) the maximum mileage if the cost is not to exceed £75.

2. Helen's height was monitored from birth until she reached the age of 20 years and the results are as shown.

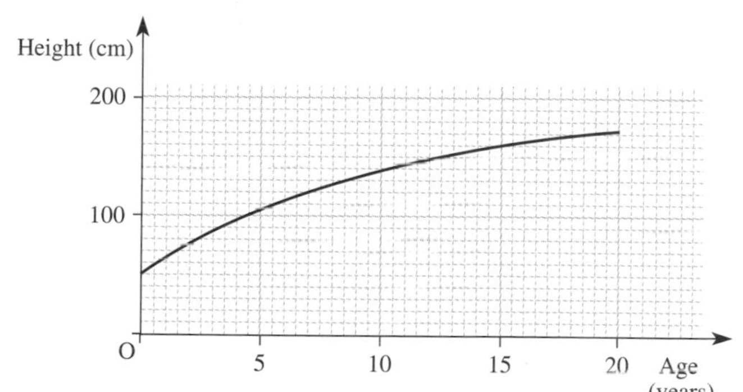

Use the graph to estimate
   (i) Helen's height on her fourth birthday;
   (ii) Helen's height on her eighth birthday;
   (iii) the age at which Helen reached a height of 1.5 m.

3. The graph shows how the petrol consumption of Jason's car varies with its speed.

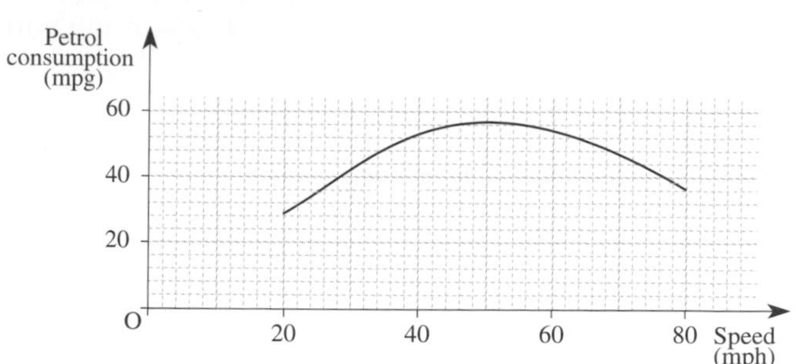

Use the graph to estimate
   (i) the petrol consumption (in miles per gallon) when Jason drives at 30 mph;
   (ii) the speed at which the number of miles per gallon is maximised;
   (iii) the range of speeds for which the petrol consumption is at least 40 miles per gallon.

4. John and Sue take out a mortgage of £50 000. The repayments are spread over a 20 year period. The graph shows how the outstanding loan varies during the term of the mortgage.

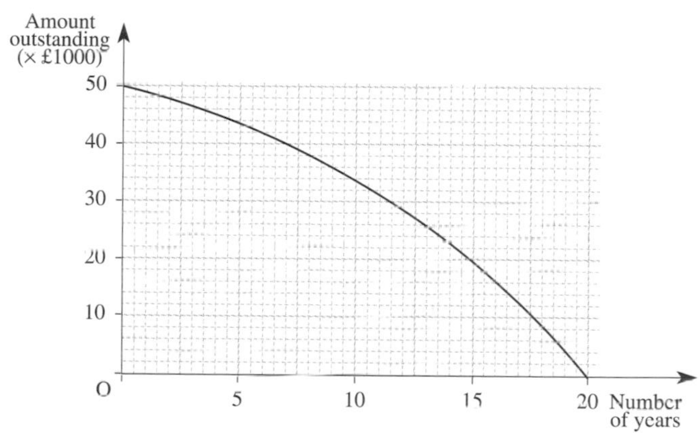

From the graph estimate
   (i) the loan outstanding after 7 years;
   (ii) the loan outstanding after 13 years;
   (iii) when the outstanding loan first falls below £20 000.

Activity

Speed *(mph)*	20	30	40	50	60	70
Total stopping distance *(feet)*	40	75	120	175	240	315

The table shows the total stopping distances that drivers should allow at different speeds. The information is from the Highway Code.

(i) Draw a graph of stopping distance (vertical axis) against speed (horizontal axis).
(ii) A graph often helps you to understand a situation more clearly. What extra information can you see from your graph that was not clear from the table?

3 Graphs from equations

Straight line graphs

Look at the equations

$$y = 2x + 1, \quad 2y = x - 3, \quad 5x - 3y = 16, \quad y = -2.$$

When you plot y against x, all of these produce straight line graphs.

To plot the graph in each case, you choose a range of values of x, and substitute each x value into the equation to obtain the corresponding y value. It is often helpful to draw up a table of x and y values, as in part (i) of the following example.

Example

Draw the graphs of

(i) $y = 2x + 1$ for values of x from -2 to 2;

(ii) $2y = x - 3$ for values of x from 0 to 4.

Solution

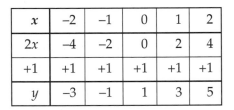

x	-2	-1	0	1	2
$2x$	-4	-2	0	2	4
$+1$	$+1$	$+1$	$+1$	$+1$	$+1$
y	-3	-1	1	3	5

(i) First we set up a table (left), for values of x from -2 to $+2$.

The points are $(-2, -3)$, $(-1, -1)$, $(0, 1)$, $(1, 3)$ and $(2, 5)$, and these are plotted below.

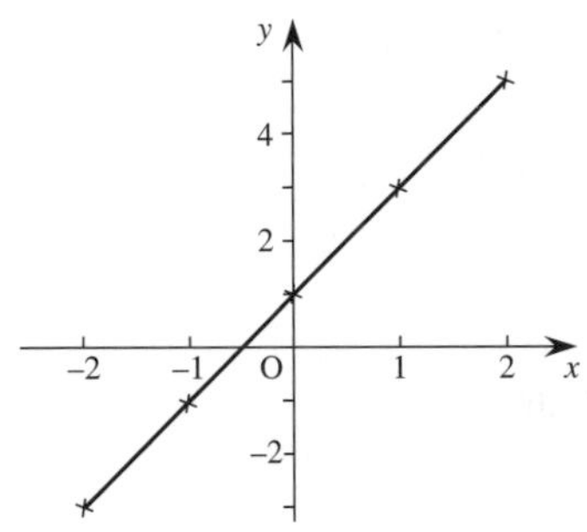

(ii) Rearranging the equation in the form $y = \frac{1}{2}(x - 3)$, we can then substitute $x = 0, 1, 2, 3$ and 4 into this in turn.

When $x = 0$, $y = \frac{1}{2}(0-3) = \frac{1}{2}(-3) = -1\frac{1}{2} \Rightarrow (0, -1\frac{1}{2})$ is on the line.

When $x = 1$, $y = \frac{1}{2}(1-3) = \frac{1}{2}(-2) = -1 \Rightarrow (1, -1)$ is on the line.

Continuing like this, we get the points $(0, -1\frac{1}{2})$, $(1,-1)$, $(2,-\frac{1}{2})$, $(3,0)$, and $(4, \frac{1}{2})$. These points can then be plotted.

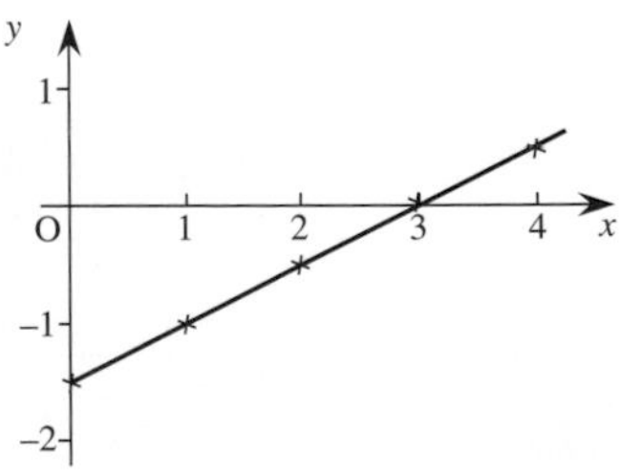

Quick method: If you know that a graph is going to be a straight line you only need to take two points, but it is wise to take a third to act as a check.

Exercise

1. Draw the graphs of the following straight lines over the intervals given.

(i) $y = 3x + 2$	$x = -3$ to 3
(ii) $4y = x + 6$	$x = -2$ to 6
(iii) $y = 15 - 2x$	$x = 0$ to 6
(iv) $x + y = 3$	$x = -1$ to 4
(v) $y = 0.2x + 1.3$	$x = -2$ to 2
(vi) $y = 20x - 600$	$x = 0$ to 100
(vii) $3y - 2x = 9$	$x = 0$ to 6
(viii) $y = 6 - 0.1x$	$x = 0$ to 60

2. The total cost, £C, of manufacturing n items is given by

$$C = F + nV,$$

where £F is a fixed cost and £V is the cost per item. Given that $F = 200$ and $V = 2$, draw the graph for $0 \le n \le 100$. From your graph find
(i) the total cost when 35 items are produced;
(ii) the number of items produced when the total cost is £336.

3. An athletics coach believes that the optimum heart-rate, H beats per minute, during strenuous exercise is given by

$$H = 214 - 0.8A$$

where A is the athlete's age in years.
Draw the graph for values of A from 20 to 60.
From your graph find
(i) the optimum heart-rate for a 43-year-old athlete;
(ii) the age of an athlete whose optimum heart-rate is 189 beats per minute.

4. The cooking time, t minutes, for a piece of meat with mass m kg is given by the formula

$$t = 15 + 40m$$

Draw the graph for the interval $0.5 \le m \le 2$. From your graph estimate
(i) the cooking time for a piece of meat with mass 800 grams;
(ii) the mass of a piece of meat that will require a cooking time of one and a quarter hours.

5. The conversion of temperature from degrees Celsius (°C) to degrees Fahrenheit (°F) is given by

$$F = 1.8C + 32$$

where F is the temperature in °F and C is the temperature in °C. Draw the graph of F against C for values of C from 0 to 100, and use it to estimate
(i) the Fahrenheit temperature that corresponds to 20°C;
(ii) the Celsius temperature that corresponds to 98°F.

6. An accountant decides to write off the cost of an asset using the straight-line method of depreciation. She uses the formula

$$V = 12\,000 - 200N$$

where V is the valuation of the asset (in pounds) after N months.
(i) Draw the graph of V against N for N from 0 to 60.
(ii) After how many months has the valuation of the asset fallen to £8000?
(iii) After how many months has the valuation of the asset fallen by £9000?
(iv) By how much has the valuation fallen after 21 months?

Activities

1. Find out the exchange rate between the euro (€) and the pound Sterling. Draw a conversion graph with pounds (from £0 to £10) on the horizontal axis and euros on the vertical axis. Use your graph to convert the following prices.

(i) Sherry: 12 euros	(into pounds)
(ii) Tuna fish: £0.60	(into euros)
(iii) Calamares: 3.50 euros	(into pounds)
(iv) Prawns: £5.40	(into euros)

2. To convert between inches and centimetres you can use the fact that 39.37 inches = 100 centimetres. Draw a conversion graph with centimetres (from 0 to 100) on the horizontal axis and inches (from 0 to 40) on the vertical axis.

Use your graph to convert the following measurements.
(i) A waist measurement of 32 inches (into centimetres).
(ii) The length (29.7 cm) and width (21.0 cm) of A4 paper (into inches).
(iii) A young child's height, 2 feet 4 inches (into centimetres).
(iv) The width (600 mm) and length (1200 mm) of a desk top (into inches).

3 Curved graphs

Equations such as

$$y = 2x^2 - 10x + 7, \quad y = \frac{30}{x}, \quad y = x^3 - 4x + 6$$

produce curved graphs. These result from the presence of terms such as x^2, x^3 and $\frac{1}{x}$ (i.e. terms other than x, y, or numbers).

Example

Plot the graphs of

(i) $y = 2x^2 - 10x + 7$ for values of x from 0 to 5;

(ii) $y = \frac{30}{x}$ for values of x from 1 to 5.

Solution

(i) First we set up a table of values.

x	0	1	2	3	4	5
$2x^2$	0	2	8	18	32	50
$-10x$	0	–10	–20	–30	–40	–50
+7	+7	+7	+7	+7	+7	+7
y	7	–1	–5	–5	–1	7

The points are (0, 7), (1, –1), (2, –5), (3, –5), (4, –1) and (5, 7). They are plotted below.

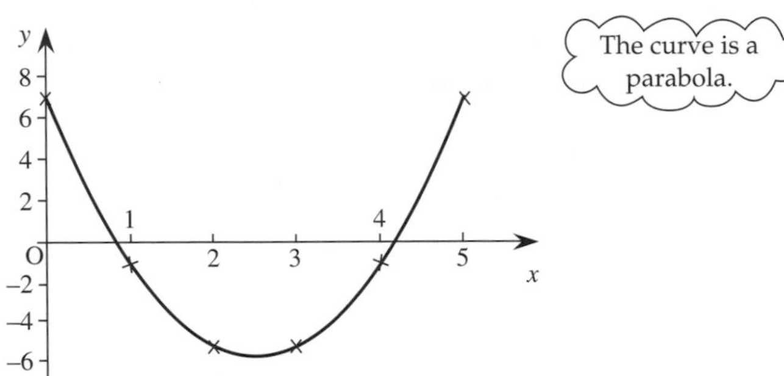

(ii) Substituting the values $x = 1, 2, 3, 4$ and 5 into $y = \frac{30}{x}$ gives the points (1, 30), (2, 15), (3, 10), (4, $7\frac{1}{2}$) and (5, 6).

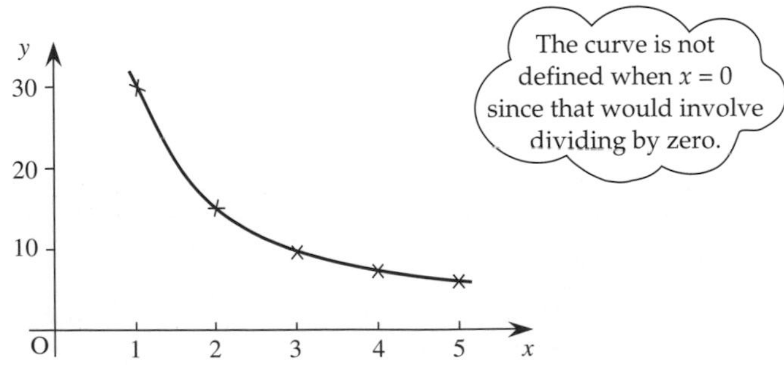

The curve is not defined when x = 0 since that would involve dividing by zero.

NOTE *There is a separate branch of the curve for negative x values.*

Exercise

1. Draw the graphs of the following curves over the given intervals.
   (i) $y = x^2 - 6x + 8$ $\quad x = 0$ to 6
   (ii) $y = 9 + 3x - x^2$ $\quad x = -2$ to 5
   (iii) $y = \dfrac{24}{x}$ $\quad x = 1$ to 6
   (iv) $y = x + \dfrac{4}{x}$ $\quad x = 0.5$ to 4 at intervals of 0.5
   (v) $y = 15 + x - 2x^2$ $\quad x = -2$ to 5
   (vi) $y = 1.2x^2 - 0.8x - 3$ $\quad x = -2$ to 3
   (vii) $y = x^3 - 10x^2 + 30x$ $\quad x = 0$ to 7
   (viii) $y = 2x - 5 + \dfrac{10}{x}$ $\quad x = 1$ to 7

2. The surface area, $A\,\text{cm}^2$, of a cube of edge $x\,\text{cm}$, is given by $A = 6x^2$. Copy and complete the following table.

x	0	1	2	3
A	0			

   Draw the graph of $A = 6x^2$.

   From your graph estimate
   (i) the surface area of a cube of edge $2.6\,\text{cm}$;
   (ii) the edge length of a cube whose surface area is $18\,\text{cm}^2$.

3. A ball is thrown vertically upwards. Its height h metres above the ground after time t seconds is given by $h = 15t - 5t^2 + 1.5$. Draw up a table to calculate the values of h corresponding to $t = 0, 0.5, 1, 1.5, 2, 2.5$ and 3.
   Draw the graph of $h = 15t - 5t^2 + 1.5$. From your graph, estimate
   (i) the height of the ball above the ground when $t = 0.8$;
   (ii) the values of t when the ball is 8.4 metres above the ground;
   (iii) the value of t when the ball is at its greatest height;
   (iv) the greatest height that the ball reaches.

4. A business analyst calculates that the average unit cost (AUC), in pounds, of Q hundred items is

$$\text{AUC} = Q + 3 + \frac{80}{Q}.$$

   Copy and complete the following table.

Q	2	4	6	8	10	12	14
AUC			22.3				

   Using scales of $1\,\text{cm}$ to 1 unit on the Q axis and $2\,\text{cm}$ to 5 units on the AUC axis, draw the graph. From your graph, estimate
   (i) the AUC corresponding to 450 items;
   (ii) the minimum AUC and the number of items when this occurs;
   (iii) the values of Q between which the AUC is less than £22.

5. Squares of side $x\,\text{cm}$ are cut from each corner of a sheet of cardboard $20\,\text{cm}$ long and $12\,\text{cm}$ wide.
   The cardboard is then folded along the dotted lines to form an open rectangular box.

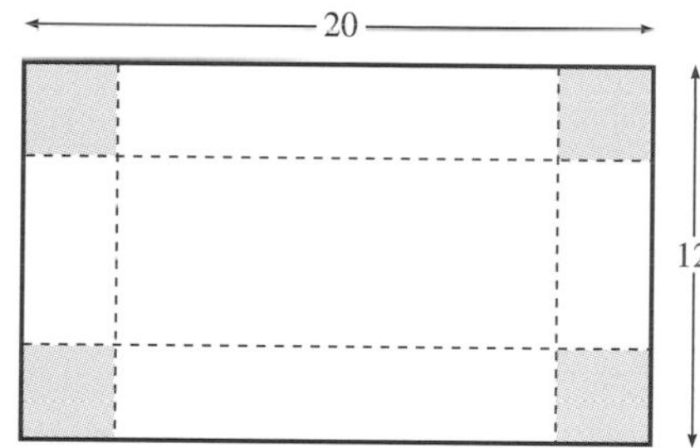

   (i) Show that the volume of the box, $V\,\text{cm}^3$, can be written in the form

$$V = 4x(10 - x)(6 - x).$$

   (ii) Calculate the values of V when $x = 0, 1, 2, 3, 4, 5$ and 6.
   (iii) Using a scale of $2\,\text{cm}$ to 1 unit on the x axis and $2\,\text{cm}$ to 50 units on the V axis draw the graph of

$$V = 4x(10 - x)(6 - x)$$

   for values of x from 0 to 6.

   Use your graph to estimate
   (iv) the maximum volume of the box and the value of x for which it occurs;
   (v) the values of x between which the volume of the box exceeds $140\,\text{cm}^3$.

Investigation

Graphs of equations such as $y = 2x^2 - 10x + 7$, which contain only x^2, x and number terms, are called *quadratic* graphs. You have met several of these in this section, and in all cases the graphs have been either ‘u’-shaped or ‘n’-shaped curves.

How can you tell by looking at an equation whether it will produce a ‘u’ shape or an ‘n’ shape? Use a graphics calculator or a suitable computer package to test your ideas.

Gradient and y intercept of straight lines

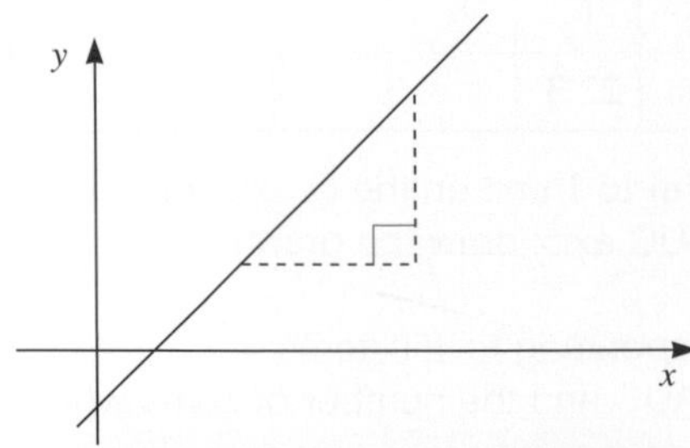

Look at the straight line graph on the left. The gradient (or slope) of a straight line is the rate of change of y relative to x, and is calculated as

$$\text{gradient} = \frac{\text{change in } y}{\text{change in } x}.$$

The y intercept is the value of y when $x = 0$, i.e. the value of y when the graph crosses the y axis.

Example

Draw the graphs of (i) $y = 2x - 1$ (ii) $y = 4 - \frac{1}{2}x$.

For each graph, write down its gradient and its y intercept.

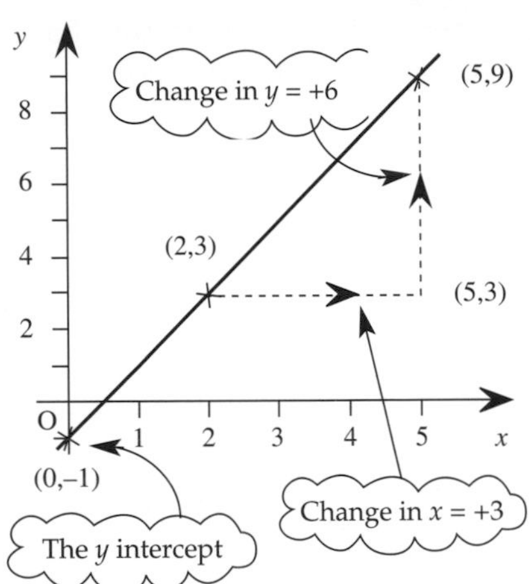

Solution

(i) To draw the straight line $y = 2x - 1$ we need two points, but we plot three so the third can act as a check. In this case we choose (0, –1), (2, 3) and (5, 9). The resulting graph is shown on the left.

To find the gradient we start by drawing a triangle like the one shown, then

$$\text{gradient} = \frac{\text{change in } y}{\text{change in } x} = \frac{+6}{+3} = +2$$

The line crosses the y axis at (0,–1) so the y intercept is –1.

(ii) First we draw the line $y = 4 - \frac{1}{2}x$ by choosing 3 points e.g. (–2, 5), (4, 2) and (6, 1). Then we use the graph to find the changes in y and x between two of these, e.g. (4, 2) and (6, 1).

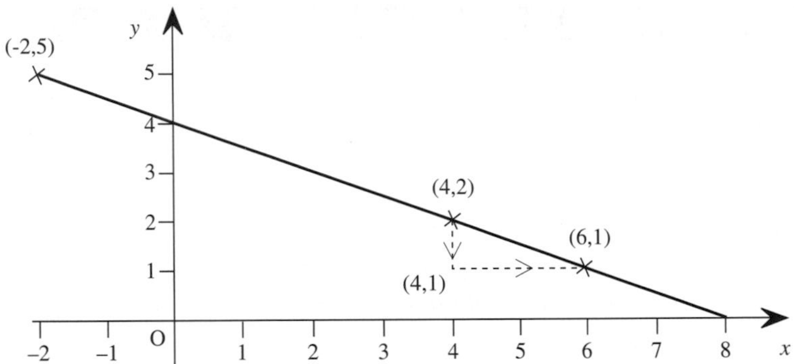

As before, $\text{gradient} = \dfrac{\text{change in } y}{\text{change in } x}$.

In this case, the change in y is –1 and the change in x is +2, so the gradient is $-\frac{1}{2}$. The y intercept can again be read from the graph directly: it is +4.

Question

Does it matter which two points you use to find the gradient? You can find this out for yourself in question 1, opposite.

Exercise

1. (i) For the graph of $y = 2x - 1$ (opposite page), calculate the gradient using the points
   (a) $(0, -1)$ and $(2, 3)$, (b) $(5, 9)$ and $(0, -1)$.
   (ii) For the graph of $y = 4 - \frac{1}{2}x$ (opposite page), calculate the gradient using the points
   (a) $(-2, 5)$ and $(4, 2)$, (b) $(-2, 5)$ and $(6, 1)$.
   (iii) Compare the gradients you found in (i) and (ii) with those found in the example. What do you conclude?

2. In each of the following
   (i) find the y co-ordinates for the three points given;
   (ii) draw a graph of the straight line;
   (iii) find the gradient and y intercept.
   (a) $y = x + 1$ $(1,\ldots), (3,\ldots), (6,\ldots)$
   (b) $y = 5 - 2x$ $(-1,\ldots), (0,\ldots), (4,\ldots)$
   (c) $y = -x + 4$ $(-4,\ldots), (-2,\ldots), (3,\ldots)$
   (d) $y = 4$ $(-1,\ldots), (2,\ldots), (4,\ldots)$
   (e) $y = \dfrac{3x}{4} - 3$ $(0,\ldots), (4,\ldots), (8,\ldots)$
   (f) $y = \dfrac{x}{3}$ $(-3,\ldots), (0,\ldots), (3,\ldots)$

3. Draw the graph of each of the following lines and find the gradient and y intercept.
   (i) $2y = x + 3$ (ii) $x + y = 6$
   (iii) $3y + x = 12$ (iv) $5y = 2x - 7$
   (v) $y + 4 = 2x$ (vi) $3x + 4y = 24$

4. Find the gradient of each of the lines shown.
   (a)

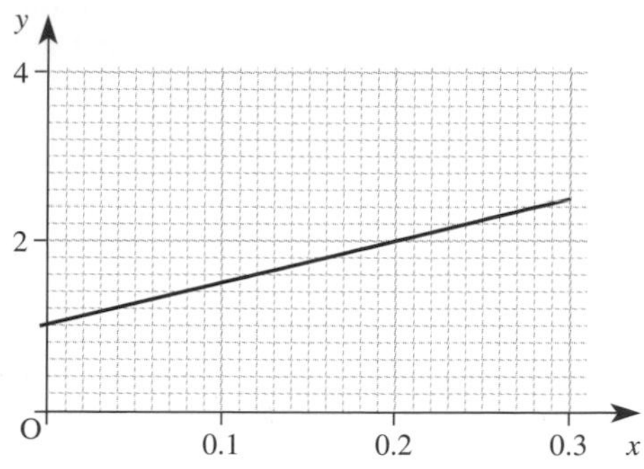

(b)

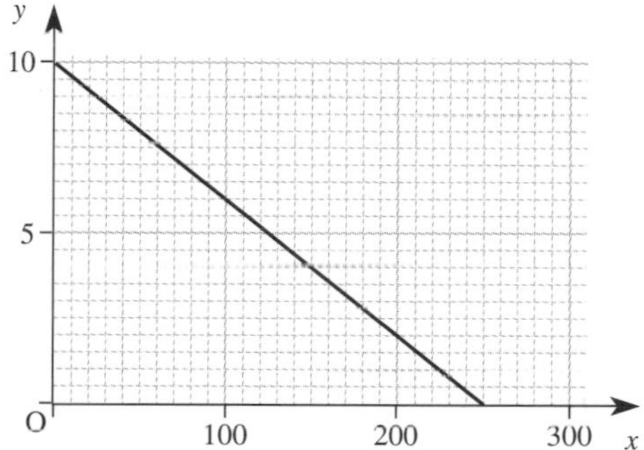

Investigation (A very important one!)

How can you find the gradient and y intercept of a straight line without drawing it?

You can find out by using some of your results from question 2 to complete the following table.

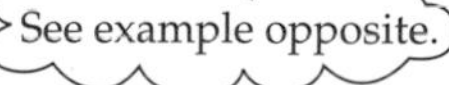

It is convenient to label these in this way for future use.

Equation of line in form $y = \ldots\ldots$	Coefficient (m) of x	Number term (c) (constant)	Gradient	y intercept
$y = 2x - 1$	2	-1	2	-1
$y = 4 - \frac{1}{2}x$	$-\frac{1}{2}$	4	$-\frac{1}{2}$	4
$y = x + 1$				
$y = 5 - 2x$				
$y = -x + 4$				
$y = 4$				

Can you see the pattern in the numbers in the table?

Once you can see a pattern, check that it works for the straight lines in question 3. (Remember to write each one in the form $y = \ldots\ldots$ first.)

Equation of a straight line

The equation of a straight line contains only a y term, an x term and a constant. This means that the equation of a straight line can be written in the form

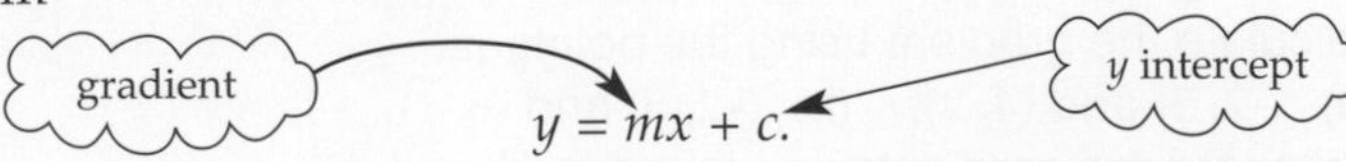

Lines such as $x = 1$, $x = 2$ etc., which are parallel to the y axis, are the only exception to this.

In the last investigation you found that m the coefficient of x, gives the gradient of the line, and the constant, c, gives the y intercept.

So if you are given the equation $y = 2 - 3x$, you can find its gradient and y intercept by rearranging it:

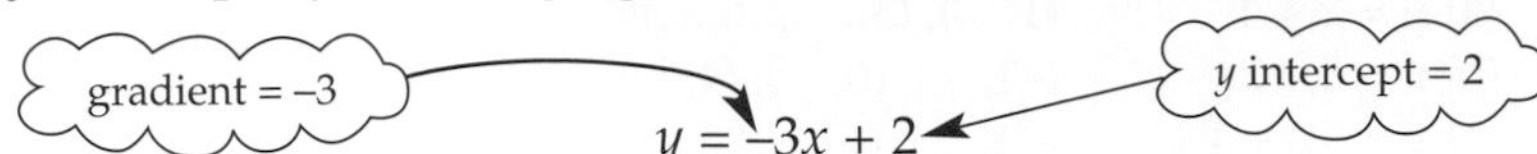

If you have the equation $2y - x + 9 = 0$, you can rearrange it to get

$$2y = x - 9$$

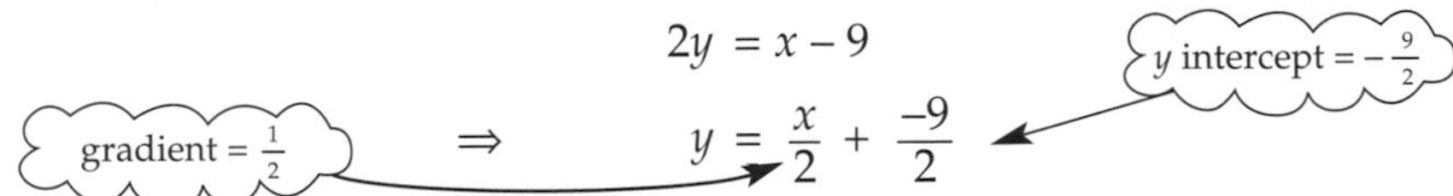

Finding the equation of a straight line from its graph

If you have a straight line graph and need to find its equation, you can work out the gradient (m) and the y intercept (c) and then substitute these values into $y = mx + c$.

Example

Find the equation of the straight line in the graph below.

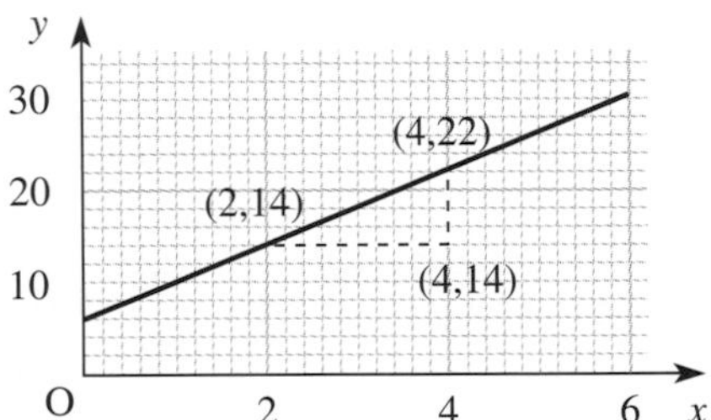

Solution

$$\text{Gradient} = \frac{\text{change in } y}{\text{change in } x} = \frac{+8}{+2} = +4 \quad \Rightarrow \quad m = 4.$$

The line cuts the y axis at $y = 6$, so the y intercept is 6 $\Rightarrow$ $c = 6$.

Substituting these values of m and c into $y = mx + c$ gives the equation of the line:

$$y = 4x + 6$$

NOTE

You will use this method later to establish straight line relationships from experimental data (page 86).

Exercise

1. For each of the following straight line equations, identify the gradient and the y intercept.
   (i) $y = 4x - 1$
   (ii) $y = 7 - 3x$
   (iii) $x + y = 8$
   (iv) $2y = 3x - 12$
   (v) $3x = y + 5$
   (vi) $4y - 1 = 10x$
   (vii) $7x = 4y$
   (viii) $2x + 5y - 20 = 0$

2. For each of these straight line graphs, find the gradient and the y intercept and hence write down the equation of the line.

(i)

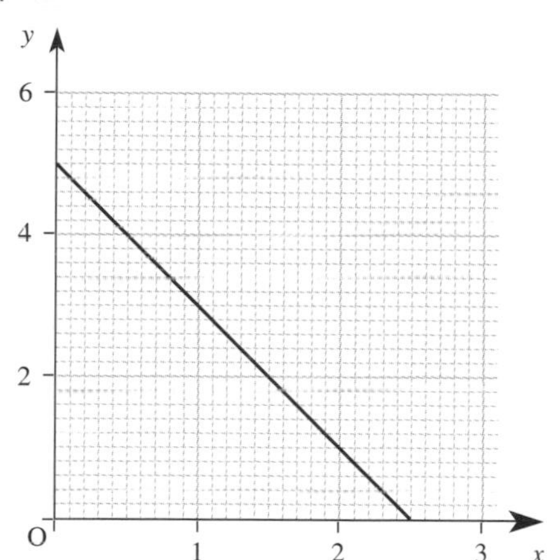

(ii)

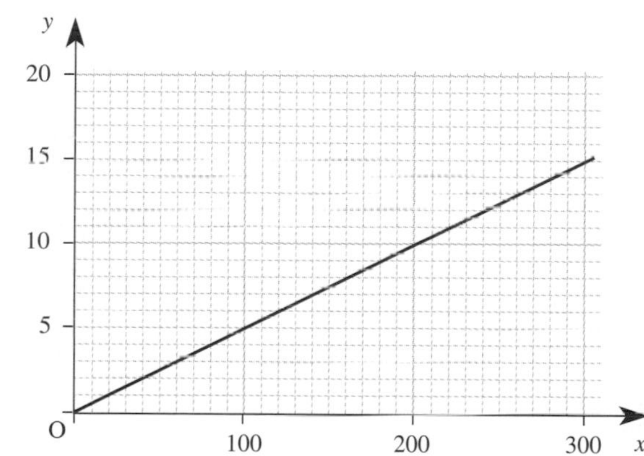

(iii)

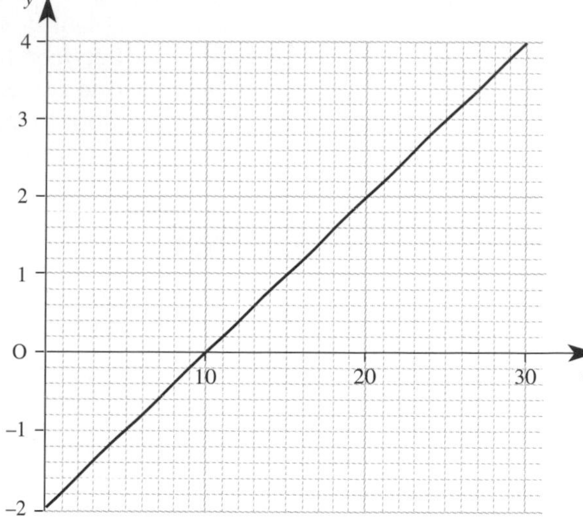

3. (i) Find the gradient of the straight line shown below.

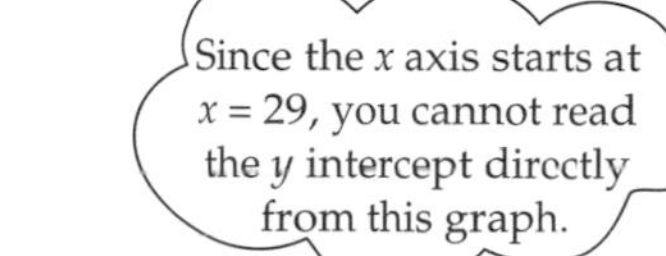
Since the x axis starts at $x = 29$, you cannot read the y intercept directly from this graph.

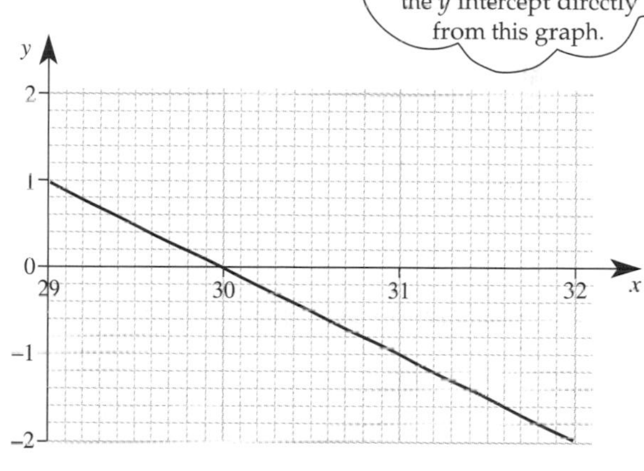

   (ii) The point $(29, 1)$ lies on the line. Substitute these values of x and y, and the value of m from (i), into $y = mx + c$ to find the value of c.
   (iii) Hence write down the equation of the line.

Investigations

1. (i) Construct a set of axes for values of x from –3 to +3 and values of y from –6 to +6. (Use the same scale for both axes.)
   (ii) Draw the lines $y = 3x$ and $y = 3x - 4$. What do you notice?
   (iii) Construct a second set of axes as in (i) and draw the lines $y = 2x$ and $y = 2x + 3$. What do you notice?
   (iv) How can you recognise parallel lines from their equations?

2. (i) On your graph of $y = 3x$, draw the line $y = -\frac{1}{3}x + 1$. What is the angle between these two lines?
   (ii) On your graph of $y = 2x$ construct as accurately as you can a line perpendicular to it. Measure its gradient.
   (iii) Find a relationship between the gradients of lines that are perpendicular to one another. Test your ideas on some more lines.

Experimental data

Scientific experiments are often carried out to investigate the relationship between two variable quantities. In such cases it is usual to plot a graph of one variable against the other. If it is possible to draw a straight line that is close to the points, the variables can be related by an equation of form $y = mx + c$, where m is the gradient of the graph and c is the y intercept.

Example

In an experiment to find how the length of a spring varies with the force applied to it, you hang objects with different mass from it and obtain the following results.

Force (newtons)	0	1	2	3	4	5
Length (mm)	25	31	34	40	46	49

(i) On a pair of suitably labelled axes, plot the length of the spring against the force.

(ii) Draw a straight line of best fit.

(iii) Find the gradient of your line, and its vertical (y) intercept.

(iv) Write an equation relating length (call this y) to force (x).

Solution

(i) and (ii)

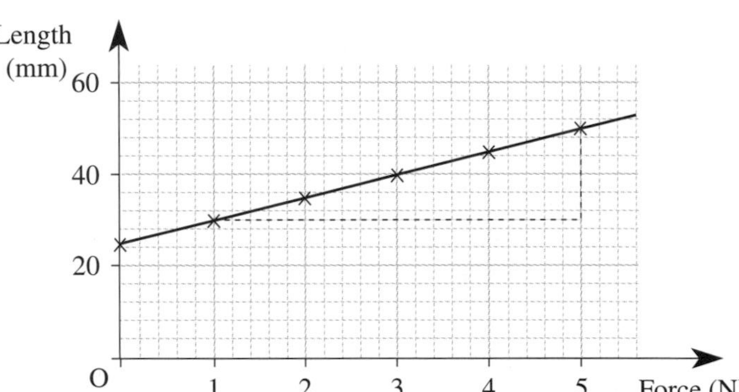

(iii) From the graph,

$$\text{gradient} = \frac{\text{change in length}}{\text{change in force}} = \frac{20}{4} = 5.$$

Vertical intercept $= 25$.

(iv) Substituting the gradient and intercept from (iii) into the straight line equation $y = mx + c$ gives

$$y = 5x + 25$$

where x is the force in newtons and y is the length in millimetres.

NOTE

(i) *Equations obtained in this way often enable you to predict intermediate values quite accurately, but you should be cautious about making predictions outside the range of values used in the experiment.*

(ii) *In some situations it is inappropriate to start the x axis at $x = 0$, so you cannot read the y intercept directly from the graph. Once m has been found it is necessary to use the co-ordinates of a point on the line to find c.* *(See question 3 on page 85.)*

Exercise

1. The table below shows the effort required to raise different loads using a simple machine. Both the load and the effort are in newtons.

Load, L	10	20	30	40	50	60
Effort, E	6.4	7.9	9.6	10.9	12.6	13.8

Using scales of 2 cm to 10 units on the L axis and 1 cm to 1 unit on the E axis, plot the points and explain why it is reasonable to assume that the load and effort are related by an equation of the form

$$E = mL + c$$

where m and c are constants. From your graph, estimate the values of m and c.

2. In an experiment to find the internal resistance of a cell, the current, I amperes, was varied and the potential difference, V volts, across the cell terminals was measured. The results are shown below.

I (amps)	0.2	0.4	0.6	0.8	1.0	1.2
V (volts)	1.35	1.15	1.01	0.83	0.69	0.52

Using a scale of 1 cm to 0.1 units on each axis, plot the points and draw a line of best fit.
Given that $V = E - Ir$, where E is the electromotive force of the cell in volts, and r is the internal resistance in ohms, use your graph to estimate the values of the constants E and r.

3. An experiment is carried out to measure the volume, $V\,\text{m}^3$ of a quantity of gas for different values of temperature, T °C. The results are shown in the table.

T (°C)	20	25	30	35	40	45
V (m^3)	20.83	21.01	21.20	21.39	21.62	21.81

Using scales of 2 cm to 5 units on the T axis and 1 cm to 0.1 units on the V axis, plot the points and show that the relationship between V and T has the form $V = mT + c$ where m and c are constants. From your graph, find the values of m and c.

4. The experimental results below show the solubility, $S\,\text{g}/100\,\text{g}$, of potassium chloride in water at different values of the temperature, T °C.

T(°C)	20	40	60	80	100
S(g/100 g)	33.5	39.0	46.0	52.5	57.0

Plot these points and show that for potassium chloride, the relationship between S and T is linear, with the form $S = aT + b$, where a and b are constants. From your graph, estimate the values of a and b.

Activity

The examples above illustrate situations in which two variables are related by a law of the form $y = mx + c$ where m and c are constants. You will meet many situations in which the relationship is not so simple. For example, it may be that y is related to a power of x. In this activity you will see how to establish such rules.

(i) Two variables x and y are observed to take the following values.

x	0	1	2	3	4	5	6
y	6	7.8	13.6	23.4	36.6	54.2	75.6

Plot these on a graph using scales of 2 cm to 1 unit on the x axis and 2 cm to 10 units on the y axis. What do you find?

(ii) Now try plotting y against x^2 using scales of 2 cm to 5 units on the x axis and 2 cm to 10 units on the y axis. The table of values is on the right.

x^2	0	1	4	9	16	25	36
y	6	7.8	13.6	23.4	36.6	54.2	75.6

You should find that the points lie close to a straight line, showing that a relationship of the form $y = ax^2 + b$ exists.

(iii) Use your graph to find the values of a and b in this relationship.

(iv) What graph would you plot if you suspected the relationship to be of the form (a) $y = ax^3 + b$ (b) $y = a\sqrt{x} + b$?

Graphs to solve equations

In Chapter 2 you solved quadratic equations such as $4 + 2x - x^2 = 0$ by factorisation (where possible) or by using the formula. But there are many types of equation that you cannot solve using algebra.

For example, you cannot at present solve the cubic equation

$$x^3 + 6x^2 + 11x + 7 = 0$$

or the trigonometric equation

$$\sin 5\theta = 1 - \cos 3\theta.$$

Fortunately, you can find approximate solutions of these equations, and of many others, using a graphical approach.

Example

By drawing the graph of $y = 4 + 2x - x^2$ for values of x from –2 to 4, solve the equation $4 + 2x - x^2 = 0$.

Solution

The table of values is as follows.

x	−2	−1	0	1	2	3	4
4	4	4	4	4	4	4	4
$+2x$	−4	−2	0	2	4	6	8
$-x^2$	−4	−1	0	−1	−4	−9	−16
y	−4	1	4	5	4	1	−4

Plotting the values on a graph gives the curve shown below.

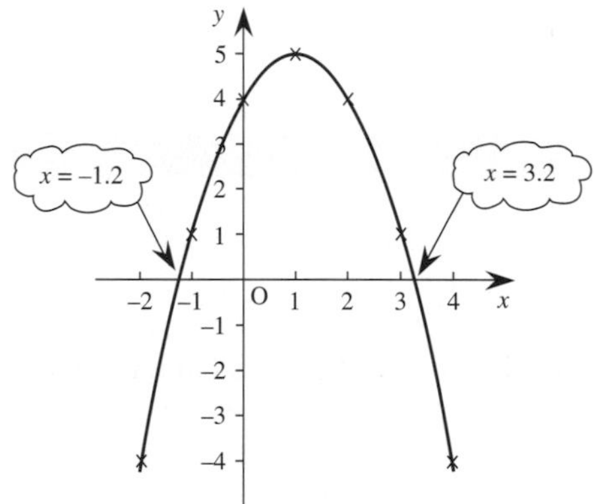

Where the curve crosses the x axis, $y = 0$, so the x value at each of these points is a solution or a *root* of the equation $4 + 2x - x^2 = 0$.

From the graph, the solution is $x = -1.2$ or $x = 3.2$.

NOTE

You have already solved this equation using algebra, on page 69. You obtained the solution $x = -1.24$ or $x = 3.24$. The graphical approach described above is less accurate, since you cannot draw curves or take readings from scales without introducing small errors. Another way of solving this equation graphically is to use the zoom facility on a graphics calculator. You should find out how to do this and then compare your results with those above.

Exercise

1. Solve the following equations by drawing suitable graphs over the suggested intervals.

(i) $2x^2 - 9x + 5 = 0$	$x = 0$ to $x = 5$	(v) $5x - x^2 - 2 = 0$	$x = 0$ to $x = 5$
(ii) $1 - x - x^2 = 0$	$x = -3$ to $x = 2$	(vi) $3x - 14 + \dfrac{12}{x} = 0$	$x = 1$ to $x = 5$
(iii) $x^2 + 4x + 1 = 0$	$x = -5$ to $x = 1$	(vii) $x^3 - 6x^2 + 9x - 3 = 0$	$x = 0$ to $x = 4$
(iv) $6 - x^2 = 0$	$x = -3$ to $x = 3$	(viii) $9 - 14x + 7x^2 - x^3 = 0$	$x = 0$ to $x = 5$

Investigation

How many roots does a quadratic equation have?

A juggler throws a ball straight up from $1\,\text{m}$ above ground level with a speed of $20\,\text{ms}^{-1}$. The height, h metres, of the ball t seconds later is given to reasonable accuracy by

$$h = 1 + 20t - 5t^2$$

and this is shown on the graph on the right.

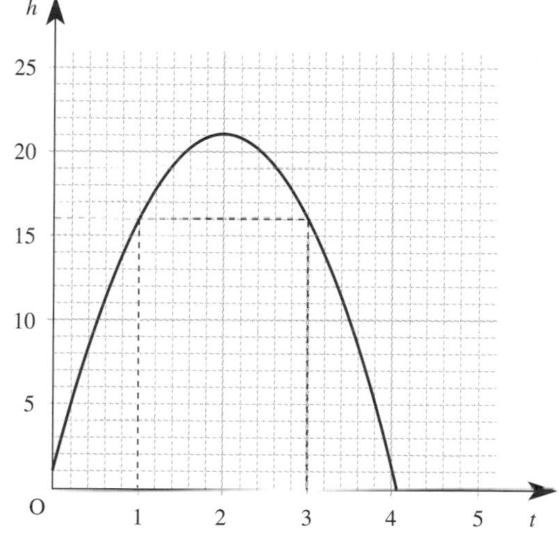

You can see from the graph that the ball is at height $16\,\text{m}$ when $t = 1$ and $t = 3$. When $t = 1$ the ball is going up and when $t = 3$ it is coming down.

(i) Use the graph to find the values of t when the height of the ball is

(a) $19\,\text{m}$ (b) $21\,\text{m}$ (c) $23\,\text{m}$.

Give brief explanations of your answers.

(ii) Show that the times asked for in part (i) are the roots of the equations

(a) $5t^2 - 20t + 18 = 0$

(b) $5t^2 - 20t + 20 = 0$

(c) $5t^2 - 20t + 22 = 0$.

(iii) Use the quadratic formula, $t = \dfrac{-b \pm \sqrt{b^2 - 4ac}}{2a}$, (see page 68) to solve equations (a) and (b). What happens when you try to use the formula on equation (c)?

(iv) For equation (a) the expression $(b^2 - 4ac)$ is positive and there are two roots. For the others it is zero or negative. How many roots are there in each of these cases?

Activity

(i) For each of the following equations, work out the value of $(b^2 - 4ac)$ and use it to predict how many roots the equation has.

(a) $x^2 - 4x + 3 = 0$

(b) $x^2 - 4x + 4 = 0$

(c) $x^2 - 4x + 5 = 0$

(ii) Draw the graphs of

(a) $y = x^2 - 4x + 3$

(b) $y = x^2 - 4x + 4$

(c) $y = x^2 - 4x + 5$.

Do they confirm your predictions in part (i)?

3 Graphical solution of simultaneous equations

In Chapter 2 you used algebra to solve simultaneous equations. You can also use graphs to solve such equations, as shown below.

Example

Use a graphical method for values of x from –1 to 4 to solve the simultaneous equations $y = x^2 - 3x + 4$ and $y = x + 2$.

Solution

The tables of values are as follows.

x	–1	0	1	2	3	4
x^2	1	0	1	4	9	16
$-3x$	3	0	–3	–6	–9	–12
+4	+4	+4	+4	+4	+4	+4
$y = x^2 - 3x + 4$	8	4	2	2	4	8

x	–1	0	1	2	3	4
$y = x + 2$	1	2	3	4	5	6

Plotting the two sets of points on the same axes gives the graph below.

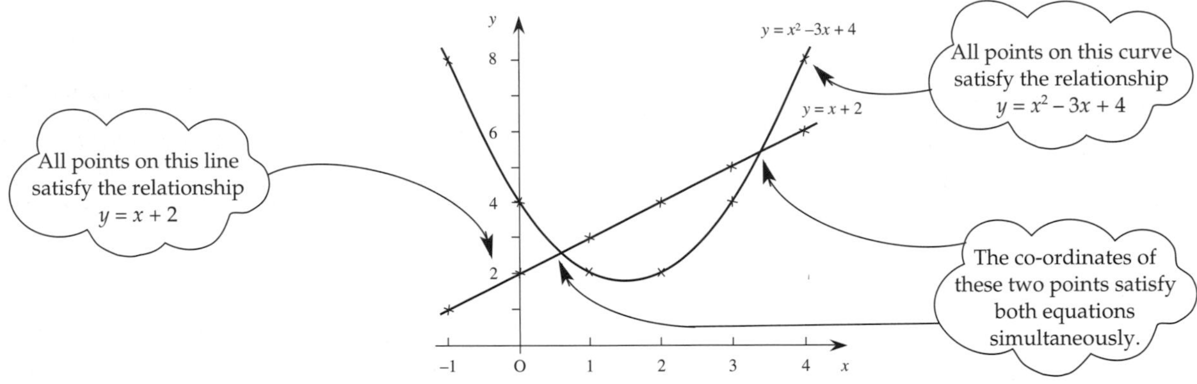

From the graph, the co-ordinates of the points of intersection of the two curves are approximately (0.6, 2.6) and (3.4, 5.4). The solution of the simultaneous equations $y = x^2 - 3x + 4$ and $y = x + 2$ is therefore

$$x = 0.6, y = 2.6 \text{ or } x = 3.4, y = 5.4.$$

NOTE

You can solve these same equations algebraically by writing

$$x^2 - 3x + 4 = x + 2,$$

which can be simplified to give the quadratic equation

$$x^2 - 4x + 2 = 0.$$

You can then solve this using the quadratic formula. The graphical approach is of course less accurate, since you cannot draw the curves perfectly or take readings from them without introducing small errors.

Exercise

1. For each pair of equations below, draw a graph for the given x intervals, and find the point(s) of intersection.
   (i) $y = \frac{1}{2}x + 3$ and $y = 2x - 1$; $x = 0$ to 4.
   (ii) $x + y = 6$ and $3y = 2x + 4$; $x = 0$ to 6.
   (iii) $y = x^2 - 4x + 6$ and $y = x + 1$; $x = 0$ to 4.
   (iv) $y = 9 - x^2$ and $2y = x + 5$; $x = -3$ to 3.
   (v) $y = \dfrac{5}{x}$ and $y = 5 - x$; $x = 1$ to 5.
   (vi) $y = x^3 - 3x^2 + 7$ and $y = 2x$; $x = -2$ to 4.

2. The total cost of production of a set of wedding invitations depends on the process used and the number printed. The total cost, £C, for N invitations can be calculated as follows.
   Process A: $C = 120 + 0.2N$
   Process B: $C = 50 + 0.5N$

   Draw the graph for each process for values of N from 0 to 500. From your graph,
   (i) which process would you choose, and what would be the total cost, if you needed 400 invitations?
   (ii) for what number of invitations would both processes cost the same, and what is that cost?

3. A ball is thrown vertically upwards from ground level with an initial speed of $10\,\text{ms}^{-1}$. Its height, s metres, at time t seconds, is given by

$$s = 10t - 5t^2 \quad (0 \leqslant t \leqslant 2).$$

   One second later another ball is thrown vertically upwards with an initial speed of $12.5\,\text{ms}^{-1}$ and its height, s metres, t seconds after the first ball was thrown, is given by

$$s = 22.5t - 5t^2 - 17.5 \quad (1 \leqslant t \leqslant 3.5).$$

   Draw a set of axes, using a scale of 2 cm to 1 metre on the vertical axis and 4 cm to 1 second on the horizontal axis. Starting at $t = 0$ (when the first ball is thrown), plot the height of each ball every 0.5 seconds up to 3.5 seconds.

   From your graph, estimate the time at which the two balls are the same height above the ground, and state this height.

4. In a large recreation ground it is decided to fence off a rectangular play area of $1000\,\text{m}^2$ using 100 m of fencing. One side of the play area is bounded by an existing fence of the recreation ground as shown. The play area is x metres by y metres.

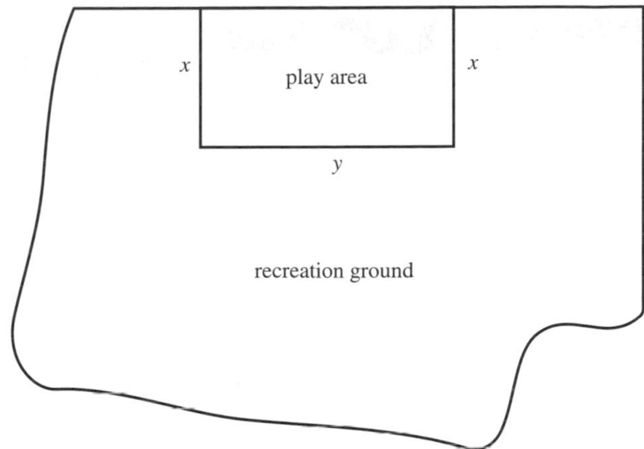

   (i) Show that $y = 100 - 2x$ and $y = \dfrac{1000}{x}$.
   (ii) Using scales of 2 cm to 10 metres on the x axis and 1 cm to 10 metres on the y axis, draw the graphs of
   (a) $y = 100 - 2x$ for $x = 0$ to 50, and
   (b) $y = \dfrac{1000}{x}$ for $x = 10$ to 50.

   Hence find the possible dimensions of the play area.

5. Ben needs to borrow £1000. Sunil offers to lend him the money, but will charge simple interest at 0.5% per month. After M months the interest payable, £I, on the loan is given by

$$I = 5M.$$

   Diana also offers to lend him the money, but she will charge compound interest at 0.45% per month. After M months the interest payable, £I, on the loan is given by

$$I = 1000 \times (1.0045)^M - 1000.$$

   Draw the graph of interest I against time M for each offer, for $M = 0$ to $M = 60$.
   (i) After how many months do the two loans incur equal interest?
   (ii) If Ben knows he can pay off the loan and interest within 6 months, which offer should he accept?

Activity

Use the zoom facility on a graphics calculator to check your results from question 1 (i) - (vi).

Rate of change

The gradient of a graph can give you valuable information. If you plot a graph of distance against time for a journey, the gradient of the graph at any point gives you the *rate of change* of distance at that instant, in other words the *speed*. If you plot the volume of water in a reservoir against time, the gradient of the graph tells you the rate of change of the volume.

For a straight line graph, $\text{gradient} = \dfrac{\text{change in } y}{\text{change in } x}$.

You know how to find the gradient of a straight line graph - either from its equation, or by drawing right-angled triangles on the graph. Now you want to find the gradient (i.e. the rate of change) when the graph is not a straight line.

Look at the table below, which shows the temperature of a cup of tea as it cools down.

Time (minutes)	0	2	4	6	8	10	12	14
Temperature (°C)	100	77	61	51	44	40	37	35

Suppose you want to know the rate of cooling of the tea. First, you draw the graph of temperature against time, which looks like this.

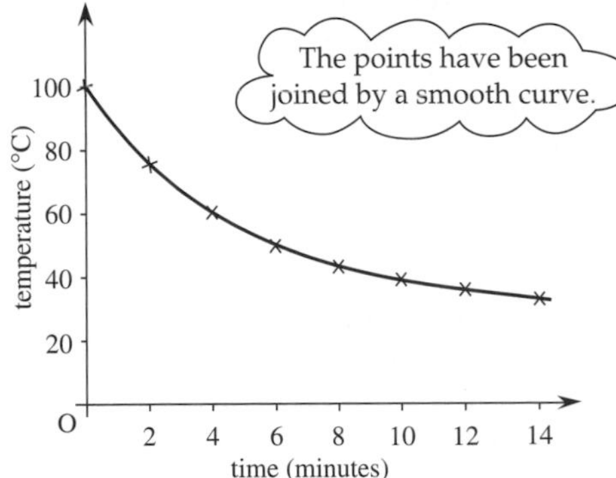

You can see that the graph is steeper at the beginning than at the end, so the gradient (i.e. the rate of change of temperature) is not constant.

Look at a particular point such as $t = 6$. How quickly is the tea cooling down after 6 minutes? Here you have a problem: the gradient is changing continuously, so even when you look at a particular point it isn't obvious how to draw the triangles. How do you decide which points to use?

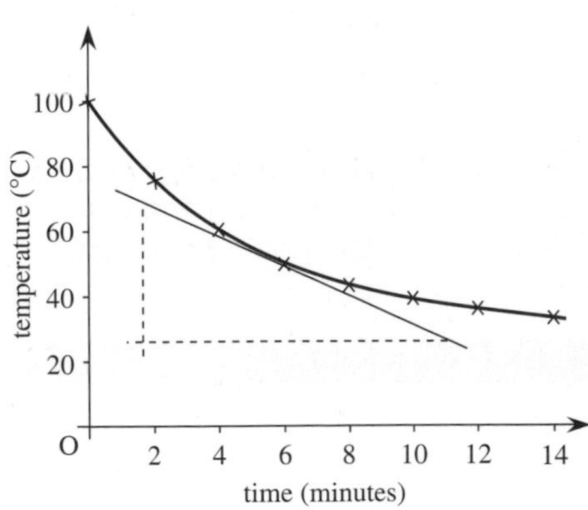

The gradient of the curve at $t = 6$ is actually the gradient of the *tangent* to the curve at this point. To find the gradient of a curve from its graph you first have to draw a tangent at the relevant point, as shown on the left. You then find the gradient of the tangent as you would for any other straight line.

In this case, $\text{gradient of tangent} = \dfrac{-43°\text{C}}{9.5 \text{ minutes}} = -4.526...°\text{C min}^{-1}$,

$\Rightarrow$ rate of change of temperature after 6 minutes = $-4.5°\text{C min}^{-1}$ (to 1 decimal place).

The negative sign indicates that the temperature is decreasing: another way of expressing the result is to say the tea is *cooling* at 4.5°C per minute.

Exercise

1. A cyclist accelerates from rest, travels at constant speed and then decelerates. The graph below shows speed plotted against time (you may recall seeing the same graph on page 76). The gradient at any point on the graph gives the acceleration at that time. Find the acceleration after
   (i) 10 seconds; (ii) 20 seconds; (iii) 35 seconds.

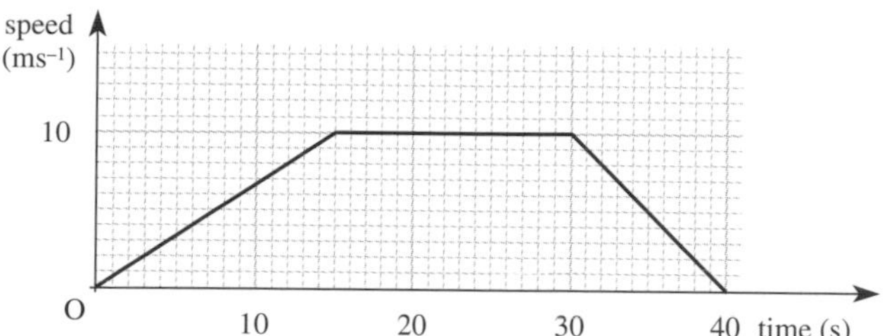

2. In 1798 Thomas Malthus put forward his theory on population growth. One interpretation of his theory today would be that if a country has a population of one million in the year 2000 then it will increase as follows.

Year	Population (millions)
2000	1
2025	2
2050	4
2075	8
2100	16
2125	32
2150	64
2175	128

Plot the points on a graph and join them with a smooth curve. From your graph estimate
(i) the population expected in the year 2140;
(ii) the year in which the population is predicted to reach 100 million;
(iii) the rate at which the population is predicted to increase in 2125.

3. A firm's total revenue from a particular item depends on the volume of sales as shown.

Volume of sales	Total revenue (£)
0	0
100	2000
200	6000
300	12 000
400	15 000
500	16 000
600	15 000

(i) Plot these points on a graph and join them with a smooth curve.

The marginal revenue per unit sold is obtained by finding the gradient of the tangent. From your graph estimate:
(ii) the marginal revenue when 350 items are sold;
(iii) the marginal revenue when 550 items are sold.

4. Samantha goes out jogging. The graph shows the distance she covers, plotted against time. The slope of the graph represents her speed in metres per minute. Use the graph to estimate
   (i) Samantha's speed 2 minutes after she sets off;
   (ii) Samantha's speed 10 minutes after she sets off;
   (iii) after how many minutes Samantha reaches her maximum speed;
   (iv) after how many minutes Samantha has to stop to cross a road.

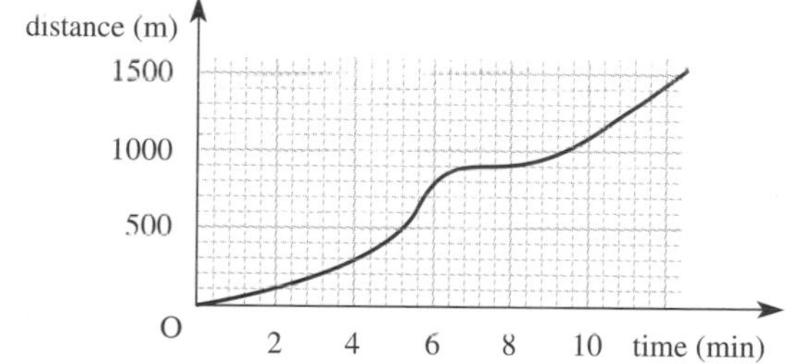

Activity

Oliver is investigating the tide patterns at his local port. On 1 August high water occurred at 0200 hours and Oliver monitored the water level every hour after that for a 15-hour period. His results are shown in the table. Plot these data on a graph with time on the horizontal axis.

From your graph estimate
(i) when low water occurred and the water level at this time;
(ii) the time interval from one high water to the next;
(iii) when the water level was rising most quickly;
(iv) the rate at which the water level was changing at 0300 hours.

Time	Level (m)	Time	Level (m)
0200	8.8	1000	2.2
0300	8.3	1100	4.0
0400	6.9	1200	6.1
0500	5.1	1300	7.7
0600	3.1	1400	8.7
0700	1.6	1500	8.7
0800	0.9	1600	7.6
0900	1.1	1700	6.0

Area under a graph

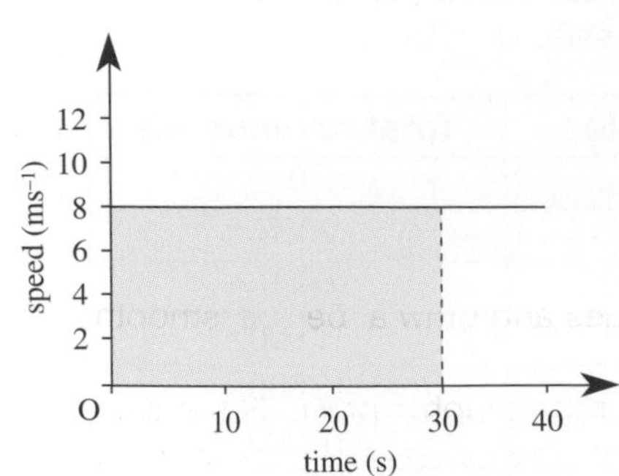

The graph on the left shows speed plotted against time for a car travelling at a constant speed of $8\,\text{ms}^{-1}$. The car passed a set of traffic lights at $t = 0$, and another set at $t = 30$, (both showing green). How far apart are the sets of lights?

You know that for constant speed, distance = speed × time. In this case, the speed is $8\,\text{ms}^{-1}$ and the time is 30 s, so the distance is

$$8\text{ ms}^{-1} \times 30\text{ s} = 240\text{ m}.$$

The area under the graph (shaded) is a rectangle. For a rectangle, area = base × height, and this also gives $8\,\text{ms}^{-1} \times 30\,\text{s}$. The area under the speed–time graph is the same as the distance travelled and this is true for any speed–time graph.

Not all areas are as straightforward to find, as shown in the following examples. They refer to the same traffic lights but a different car which had to stop at both lights.

1 Red lights – simplified graph

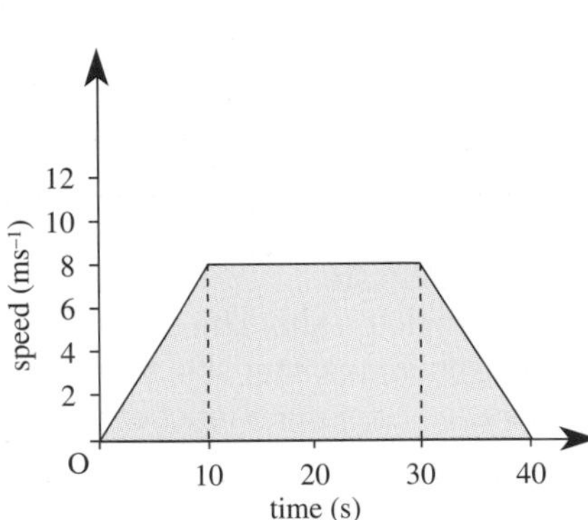

The speed–time graph on the left is simplified. The car is assumed to accelerate at a constant rate from the first light, then to travel at constant speed for a while, then to slow down at a constant rate for the next red light.

The graph is a trapezium, so you can calculate the area directly using

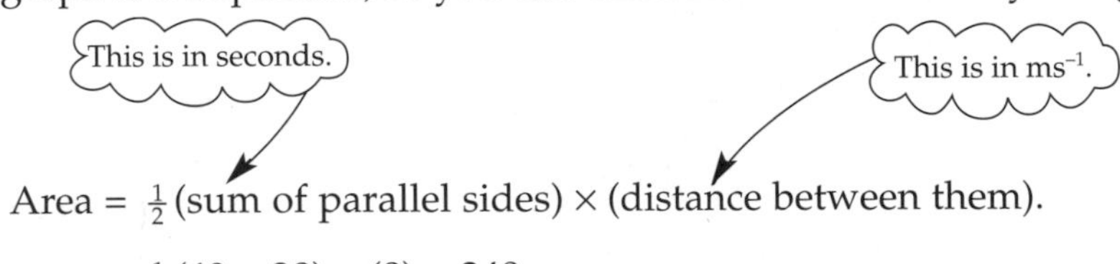

$$\text{Area} = \tfrac{1}{2}(\text{sum of parallel sides}) \times (\text{distance between them}).$$
$$= \tfrac{1}{2}(40 + 20) \times (8) = 240.$$

The total distance between the lights is therefore 240 m, as before. (Alternatively, you can split the area into three parts – two triangles and a rectangle – and add together the separate areas.)

2 Red lights – realistic graph

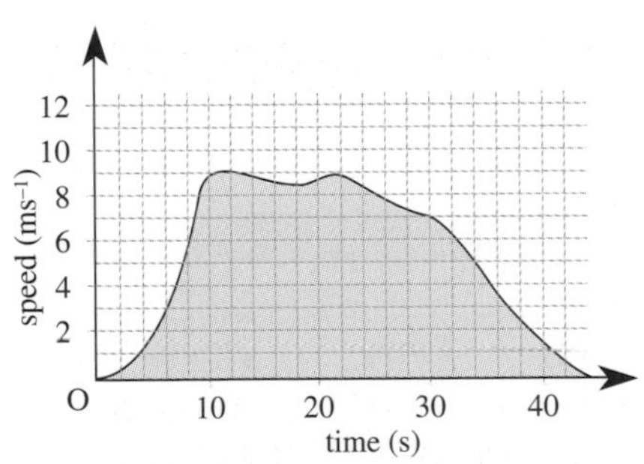

A more realistic graph of the car's speed between the two red lights is shown on the left. The car does not accelerate and decelerate at a constant rate, and the presence of other traffic means the car cannot travel at a steady speed. The resulting area is not a simple shape.

One approximate method of finding areas like this is to count the squares under the curve. In this graph, each square represents

$$1\,\text{ms}^{-1} \times 2\,\text{s} = 2\,\text{m}.$$

You can check, if you like, that the traffic lights are *still* 240 m apart, by counting squares in the graph on the left! You can also practise this method in question 2, opposite.

More sophisticated methods of finding complicated areas involve dividing the area into strips and finding the approximate area of each strip (see Activity opposite).

Exercise

1. A cyclist accelerates from rest, travels at constant speed and then decelerates. The graph shows his speed plotted against time.
The area under the graph represents the distance travelled. Calculate the distance he travels
   (i) in the first 15 seconds;
   (ii) in the last 10 seconds;
   (iii) in the whole 40-second period.

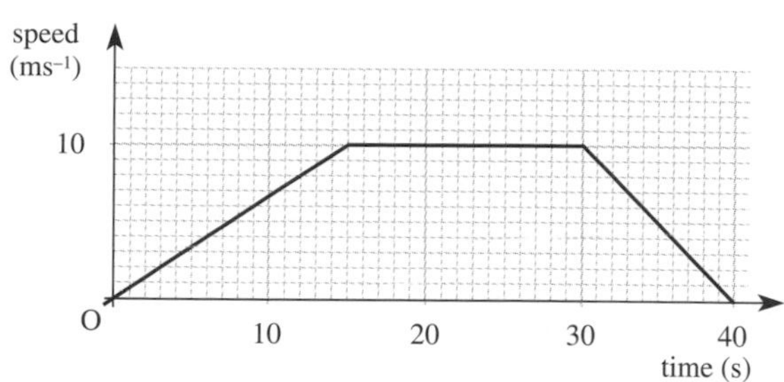

2. The aerial photograph below shows the outline of an oil slick following a leak from a supertanker. The photo has a 100 km grid superimposed on it.

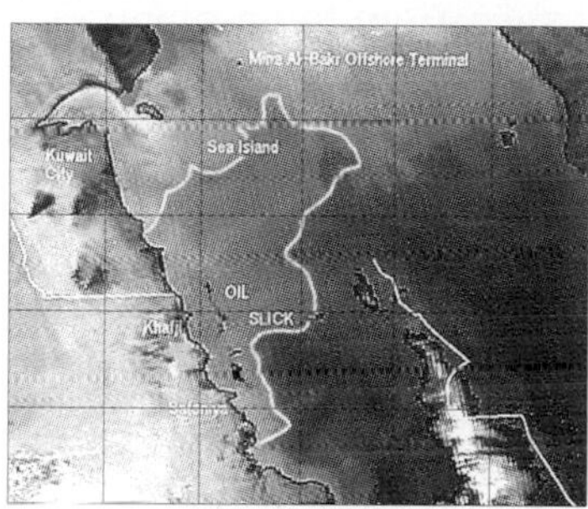

   (i) Use the method of counting squares to estimate the total area of the slick.
   (ii) Do you think your estimate is an underestimate, or an overestimate? Justify your answer.

3. The electric current in a charging circuit is monitored against time.

Time (s)	0	10	20	30	40	50
Current (μA)	80	62	48	37	28	21

   (i) Plot these values and draw a 'best-fit' smooth curve.
   (ii) The area under the graph represents the charge accumulated (measured in microcoulombs). Use an appropriate technique to estimate the area and hence find the charge.

4. During a fitness test, Julia's rate of air intake is recorded at one minute intervals as follows.

Time (minutes)	**Air intake ($dm^3\ min^{-1}$)**
0	6.0
1	15.3
2	19.8
3	19.4
4	15.9
5	13.2
6	11.3
7	9.8
8	8.6

   (i) Plot these values and join them with a smooth curve.
   The area under the graph gives the total volume of air that Julia has inhaled in the eight minute period.
   (ii) Estimate this volume.

Activity

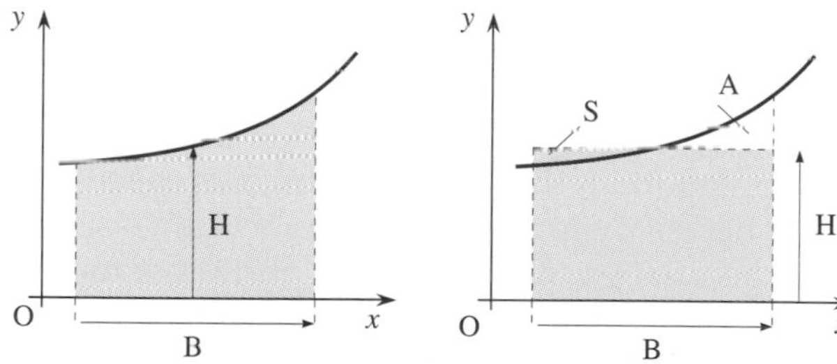

One method for approximating the area under a graph is to take the y value (H) at the midpoint of the base, and multiply it by the length of the base (B), then use the formula

$$\text{area under graph} \approx H \times B.$$

This assumes that the area marked A is approximately equal to the area marked S.

(i) Using a scale of 2 cm to 1 unit on each axis draw the graph of $y = 1 + 1.2x - 0.1x^2$ for $0 \leqslant x \leqslant 12$.

(ii) Divide the area into six strips of equal width, and read off the values of y when $x=1, 3, 5, 7, 9$ and 11. Use these as your values of H in the formula to obtain estimates of the area of each section. Hence estimate the total area under the curve.

(iii) Explain whether you expect your estimate to be less than or greater than the actual area.

Chapter 

Statistics and Probability

Introduction

Statistics involves the collection, display and analysis of data. This chapter covers each of these in turn and finishes with a section on probability.

Collection of data

There are four main methods of data collection. An example of the use of each is given below.

- **Direct measurement or observation:** town planners make counts of the number of pedestrians crossing a road and of the number of vehicles using it, in order to decide whether a zebra crossing is necessary.
- **Interviews:** chemical companies might employ interviewers to ask people in the street (or over the phone) their opinions of different washing powders. This helps companies to develop products that meet people's needs.
- **Questionnaires:** package-tour operators often hand out printed lists of questions to holidaymakers on their homeward journey. Their aim is to collect data relating to the success or otherwise of the holiday. (Questionnaires are discussed in more detail on page 100.)
- **Secondary data:** the social services department of a local authority makes use of data already collected, e.g. the latest Census, in order to obtain information about the age profile of the local population. This is much quicker and cheaper than collecting new data.

Sampling

In some situations it is realistic to gather information from (or about) every member of a population. In this case the survey is called a *census.* More often it is necessary for financial or practical reasons to take a *sample* from the population. This is a small portion of the population, usually chosen at random. Information is then collected from (or about) each member of the sample, and conclusions are drawn about the whole population based on the sample data.

Note that in statistics, the term *population* does not necessarily mean a group of people. For example, if you were the quality inspector for a banana importer, the *population* might be a whole container-load of bananas. You would probably check a sample from this population to decide whether to accept the whole container-load.

Activity

A researcher recorded the number of children in each of the 100 households on Century Estate. The data she collected are shown in the table.

Household	No. of children in household
01–10	2 1 0 2 4 0 0 2 1 2
11–20	3 0 2 1 3 4 0 2 1 1
21–30	0 1 0 2 1 1 2 0 3 5
31–40	2 2 1 0 1 0 1 2 0 3
41–50	1 1 0 2 3 2 2 3 1 4
51–60	0 1 0 2 0 2 3 0 0 2
61–70	4 3 1 1 2 4 1 1 0 2
71–80	2 2 0 3 1 0 3 0 1 1
81–90	1 0 3 5 2 1 0 0 2 1
91–100	3 1 3 0 0 2 2 0 1 1

These figures can be summarised in a line diagram as shown below.

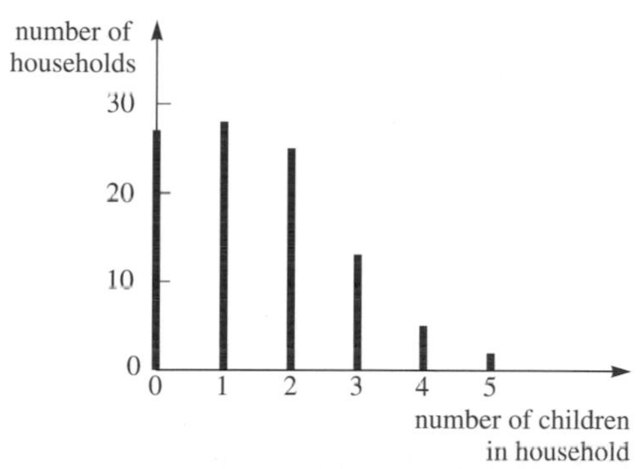

This was a census because all the households were included in the survey. Suppose instead the researcher had collected data only from a random sample of 15 households. She would know the number of households in the sample containing 0, 1, 2, 3, 4 and 5 children, and could draw a line diagram for this. Could she use these figures to predict with reasonable accuracy the values for the line diagram for the population?

The purpose of this activity is to find out how well the data from a sample match the data from the whole population.

Your task

(i) Select a random sample of size 15 as follows. Use your calculator to generate a 2-digit random number which will represent the household number. (Consult the instruction manual or ask a friend if you are not sure how to do this.) For each random number, refer to the table and write down the number of children in the corresponding household. For example, household number 08 contains 2 children, and household number 62 contains 3 children. Continue in this way until you have a sample of size 15.

Note: treat 00 as 100 and ignore repeats.

(ii) Summarise your sample data in a table like the one below.

No. of children	0	1	2	3	4	5
No. of households						

(iii) Draw a line diagram for the sample and see if it resembles the line diagram for the whole population.

(iv) Repeat this for further samples of 15. Do your results support the principle of sampling?

Investigation

Your college, school or local library should have copies of various statistical publications in book or CD ROM form. You may also have access to data on the Internet. Find out what is available to you and make some notes on data of particular interest to you. For example,

(i) locate the latest Census report and find the populations of the districts in your locality;

(ii) locate the latest Annual Abstract of Statistics and find the ages at which people get married;

(iii) locate the latest edition of Social Trends and find out which environmental issues most concern people;

(iv) locate the latest edition of Regional Trends and find out the percentage of 16 year olds in your region who achieve 5 or more GCSE grades A to C;

(v) find out the titles of publications relevant to your course of study and make notes about the sorts of data they contain.

Sampling

When taking a sample you need to ensure that it is representative of the whole population. A *biased* sample may lead you to conclusions that do not reflect the whole population. If, in the last activity, your sample of 15 consisted of the smallest 15 houses, this would probably show rather fewer children per household than the estate actually has.

The 15 houses could have been chosen by writing the address of every house on Century Estate on a piece of paper, putting them all in a box, mixing them and drawing out 15 of them. Each household would have an equal chance of being drawn. This illustrates *simple random sampling*. In the activity you used random numbers to achieve the same effect, and this method is far more convenient for large populations.

The accuracy of any results can often be improved by taking a *stratified sample*. Suppose that there are 100 houses on Century Estate: 20 two-bedroom, 40 three-bedroom and 40 four-bedroom. These are in the ratio of 1:2:2. The researcher may have chosen to make up her sample of 15 in the same ratio and chosen 3 two-bedroom, 6 three-bedroom and 6 four-bedroom houses. Sampling like this is called *stratified sampling*.

Sometimes you cannot make a list of the population. A scientist wishing to study the parasite infestation of guillemots around our coasts would need to capture a sample of birds, but there is no list of guillemots! The scientist would probably go to a number of nesting sites and capture a few birds at each; this is called *cluster sampling*.

Market research companies employ interviewers who are given a list of the sort of people to interview in one day: so many men, so many women etc. This is called *quota sampling*. It is definitely not random: interviewers are more likely to choose people with smiles on their faces and not too much to do.

How large a sample do you take? There is no simple answer: it depends on the situation.

- A new road is proposed, but it is opposed by 90% of the residents in the area. A small sample, say 20, would be enough to tell you the proposal is deeply unpopular, but you would need a larger sample if you wanted to be able to say "About 90% of the residents oppose it".
- A forthcoming election is very close. The last opinion poll suggested that 51% support one candidate and 49% the other. A very large sample would be needed to predict the outcome, and any bias in the sample may invalidate your forecast. A sample of several thousands would be needed.

A sample needs to be large enough for you to be confident of your result. It is common practice to start with a pilot survey to help you decide what size of sample to take. The calculation of the minimum sample size for any level of confidence requires deeper knowledge of statistics than that covered in this book, but you will find the activity on the opposite page will help you to understand the ideas involved.

For discussion

For each of the situations on the right, answer these questions.

(i) What is the population from which a sample should be taken?
(ii) Is it possible to obtain a list of the whole population?
(iii) What method of sampling would you recommend?
(iv) Who would be a suitable person to collect the data? Who would be unsuitable?
(v) What data would you collect? What questions would you ask?
(vi) How large a sample do you think would be reasonable?

(a) An education authority wants to decide how much bullying is taking place in its schools.
(b) The police want to find out why they are viewed with hostility on a large housing estate.
(c) A member of Parliament wants to know what proportion of her constituents favour the death penalty for drug dealers.
(d) A scientist wishes to determine the proportion of pepper moths in a particular region which are dark coloured.
(e) A railway company wishes to know what proportion of its customers travel First Class.
(f) A geneticist wishes to discover the proportion of people in a remote jungle tribe who are red-green colour blind. (Note: colour blindness is usually more common in men than women.)

Activity

This activity will help you to get a feel for what is a reasonable sample size.

Take 50 counters, about 20 red and the rest yellow, and place them in a bag. Shake the bag, select one at random and record its colour. You have now taken a sample of size 1. If the counter is red, the proportion of reds in your sample is 1, otherwise it is zero.

Repeat the procedure. You now have a sample of size 2. What is the proportion of reds?

Continue in this way, recording the proportion of reds at each stage.

Draw a graph of your results like the one on the right.

How large a sample do you need to be fairly confident that you know the proportion of red counters in the bag to the nearest 0.1? To the nearest 0.05?

What happens if you change the proportion of red counters in the bag?

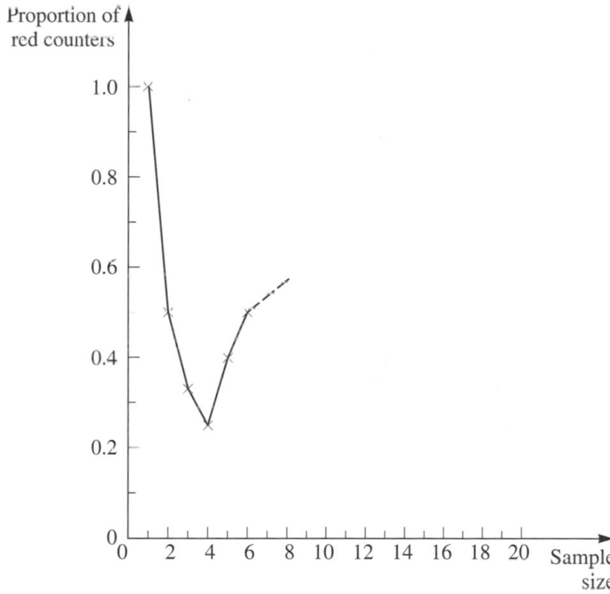

Questionnaires

A questionnaire is often the best way of collecting data about people and their opinions and preferences. You might expect it also to be a quick way, but it actually requires careful planning and thought.

There are five stages in the process:

1 designing and piloting the questionnaire;
2 selecting random samples of people to fill it in;
3 processing the data;
4 collating and presenting the results;
5 decision making.

This section concentrates on the first stage.

Designing the questionnaire

When writing the questions the following guidelines are important.

- Be clear as to the purpose of the questionnaire before you start.
- Be sure that the respondent is clear as to the purpose before he/she responds.
- Make sure that each question is designed to provide relevant information.
- Make all questions clear and concise.
- Avoid questions of a personal nature if possible.
- Avoid asking *leading questions* (i.e. questions which try to influence the response), such as 'Do you agree that the new road layout is better?'
- Avoid biased questions, such as 'Why do you think car X is better than car Y?' (This assumes that car X is better.)
- Don't ask too many questions.

It is also important to think about the form of response you want for each question. It is helpful for later analysis if responses are in a common format. This can be encouraged by offering a range of appropriate answers, and asking respondents to indicate (by circling, ticking or deleting) which answer is the most appropriate. Here are some commonly used sets of alternatives.

- Yes / No
- Yes / No / Don't know.
- Always / Often / Sometimes / Rarely / Never
- Strongly agree / Agree / No opinion / Disagree / Strongly disagree.
- 0 / 1 / 2 / 3 / 4 or more. (This avoids responses like '1 or 2' or 'less than 3'.)
- 0 to 4 / 5 to 9 / 10 to 14. (This aims to group responses.)
- Headache / backache / stress / other (please specify) .... (This form is useful when your range of alternatives cannot possibly cover all possibilities.)

Pilot questionnaire

Before circulating questionnaires to the target audience it is a good idea to try them on a small scale. What seemed a clear and unambiguous question to the writer may produce all sorts of unexpected responses when trialled. These difficulties can then be addressed before the main batch is distributed.

Exercise

Gerry has just written a questionnaire to help him find out about people's television viewing habits. He shows it to you and asks you to suggest ways in which he can improve it.

Consider each of Gerry's questions in turn and either accept it, re-write it or delete it. You should explain the reasons for your actions.

QUESTIONNAIRE

As part of my college course I am investigating people's television-viewing habits. I would be grateful if you could spare a few minutes to complete my questionnaire.

1. How much television do you watch?

2. What is your favourite television programme?

3. Which channel do you watch most often?

4. Do you have a television licence?

5. Do you receive Satellite T.V.? Yes/No

6. Rank the following types of programme according to your preference

(a) News / current affairs	**1.**	.... (like most)
(b) Sport	**2.**	....
(c) Films	**3.**	....
(d) Documentaries	**4.**	....
(e) Drama	**5.**	....
(f) Light entertainment	**6.**	.... (like least)

7. How many T.V. sets are there in your home? 1 2 3 4 (Please circle)

8. How frequently do you watch T.V. alone? Always / Often / Sometimes / Rarely / Never (Please circle)

9. How frequently do you watch T.V. with others? Always / Often / Sometimes / Rarely / Never (Please circle)

10. Why do you think weekend T.V. is better than weeknight T.V.?

11. Do you think that there is too much sport on T.V.?

12. The amount of violence on T.V. should be

(a) increased (b) maintained at its present level (c) reduced (Please circle)

Thank you for completing my questionnaire. *Gerry Cotterill*

Organisation of data

Mel, a statistical assistant, has three sets of data.

Set 1

Set 1	
1 3 2 1 2	0 1 3 4 1
2 0 3 2 1	2 2 0 1 3
2 1 0 0 3	1 2 1 4 0

Mel is overwhelmed by so many figures and decides to present them in a way that is easier to read. She does this by setting up a *tally chart.*

Outcome	**Tally**	**Frequency**
0	~~1111~~ 1	6
1	~~1111~~ 1111	9
2	~~1111~~ 111	8
3	~~1111~~	5
4	11	2

She now presents the data in the *frequency table* shown below.

Outcome	0	1	2	3	4
Frequency	6	9	8	5	2

Set 2

Set 2				
10	14	8	11	7
13	12	9	9	10
11	16	19	3	7
22	19	8	2	4
7	15	11	21	13

Mel again produces a frequency table, but with over 20 columns it is unwieldy. So she combines the numbers into equal classes as shown.

Outcome	1–4	5–8	9–12	13–16	17–20	21–24
Frequency	3	5	8	5	2	2

This is more compact but some information is lost. For example, from this table alone you cannot tell the individual values of the 3 entries in the 1–4 class: they might be three 1s, three 4s, or any other combination of values within the interval 1–4.

When organising data into classes, you do not want the number of classes to be so small that a lot of information is lost. Neither do you want the number of classes to be unmanageably large. In many cases, five to eight classes is about right.

Set 3

Set 3		
£2.45	£4.16	£5.99
£2.05	£2.00	£1.63
£4.82	£4.91	£9.38
£0.98	£4.25	£7.81
£7.69	£6.03	£8.52
£3.58		

Mel decides to group the figures as £0–£2, £2–£4, etc. This is fine until she finds the figure £2.00. Which class should it go in? Mel can solve this by redefining the classes as £0.00–£1.99, £2.00–£3.99, and so on.

Amount (£)	0.00–1.99	2.00–3.99	4.00–5.99	6.00–7.99	8.00–9.99
Frequency	2	4	5	3	2

This works well, but the mid-points, which are used to find the mean, are now £0.99$\frac{1}{2}$, £2.99$\frac{1}{2}$, and so on – much messier than the £1, £3 etc. from Mel's original groups. This problem is discussed on page 118.

Exercise

1. Marcus is a quality control inspector at a pottery. He has recorded the number of defective teacups in each of 30 production batches as shown below.

0 2 1 3 0   1 2 1 0 3   3 1 1 0 2
1 1 0 4 2   1 2 1 0 2   1 2 3 0 1

Draw up a tally chart and from it construct a frequency table to summarise the data.

2. Sophie records the number of spelling errors made by each of her students in their latest essay.

5	7	12	2	8	16	10
3	19	5	15	4	2	8
22	10	8	12	6	1	17

Using classes of 1–5, 6–10, etc. construct a frequency table to summarise the data.

3. The attendance at football matches in the Carling Premier and Endsleigh Leagues on 3rd February 1996 are shown below. (Source: *The Sunday Telegraph*.)

Premier League	Endsleigh League		
	Division 1	Division 2	Division 3
35 623	6 139	4 713	1 674
35 982	7 818	4 447	4 114
30 419	12 041	5 195	2 313
40 628	10 796	4 617	1 880
27 509	14 821	4 948	1 927
36 567	10 956	4 050	2 531
15 136	7 302	5 067	2 981
21 257	26 537	2 842	5 714
25 380		3 605	1 307
		8 242	2 594
			5 567

Present these data in a frequency table.

4. Dave did a survey to find out the amount of time for which drivers were parked by a petrol pump when refuelling. He recorded the following times in minutes and seconds.

2 min 43 s	4 min 35 s	1 min 56 s	3 min 29 s
3 min 21 s	2 min 56 s	1 min 53 s	4 min 39 s
3 min 12 s	3 min 41 s	2 min 58 s	4 min 04 s
2 min 06 s	5 min 17 s	3 min 22 s	1 min 22 s
2 min 49 s	5 min 46 s	6 min 23 s	2 min 38 s

(i) Re-write the data, showing each time to the nearest minute.
(ii) Hence complete the frequency table below.

Time parked (nearest min)	1	2	3	4	5	6
Number of drivers						

5. Twenty-four students work part-time at a local superstore. Their weekly wages for last week are listed below.

£40.50	£50.72	£44.96
£44.96	£38.64	£36.84
£37.68	£42.72	£35.88
£46.82	£45.22	£37.84
£35.88	£46.75	£39.64
£55.20	£53.80	£50.30
£35.88	£44.62	£55.20
£42.72	£38.75	£46.75

Summarise the data in the form of a suitable frequency table. You must decide for yourself how many classes to use and what their boundaries should be.

Activities

1. Collect the following data from the students in your group. In each case, organise the data into suitable categories, and produce a frequency table.
(i) height in centimetres;
(ii) time taken to reach school or college;
(iii) distance of journey from home to school or college;
(iv) time spent doing the last Maths homework.

2. Another way of organising data is a stem and leaf diagram. This one shows the speeds in mph of 30 cars observed by a speed camera on a motorway.

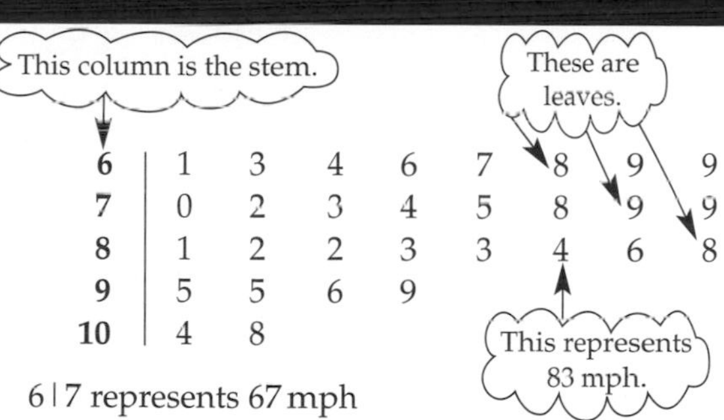

(i) How many cars were travelling between 80 and 89 mph?
(ii) What percentage of cars were exceeding the 70 mph speed limit?
(iii) Redraw the stem and leaf diagram with the numbers in the leaves sorted into order, smallest on the left.

Displaying data

The number of houses built by Quickbuild each year from 1991 to 1995 is shown below.

Year	Number of houses
1991	500
1992	400
1993	275
1994	300
1995	350

It is hard to see from a table like this whether there are any patterns in the data. A clearer picture is given by presenting the data on a diagram or a chart. There are many kinds of diagram and chart, and you will meet some of the most common in the next few pages.

Pictograms

The Quickbuild data are displayed below in the form of a pictogram using one house symbol to represent 100 houses. You can see that since the yearly figures are not all exact multiples of a hundred, the pictogram includes fractions of symbols as well as whole ones.

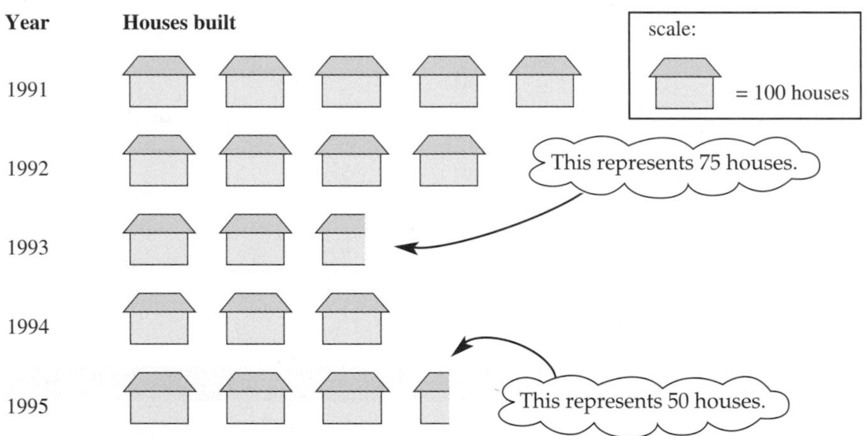

From the pictogram you can see that

- 1991 was the most productive year;
- 1993 was the least productive year;
- the 1993 figure is around half the 1991 figure;
- the figures declined from 1991 to 1993 but increased after that.

NOTE

It is wrong to draw a pictogram containing two or more different sizes of symbol, because it is not clear to the reader whether the frequency is proportional to the height, width or area of the symbol. If you see this done, be suspicious: someone may be trying to mislead you.

Exercise

1. The pictogram below shows the number of incoming flights at each of four airports on a given day.

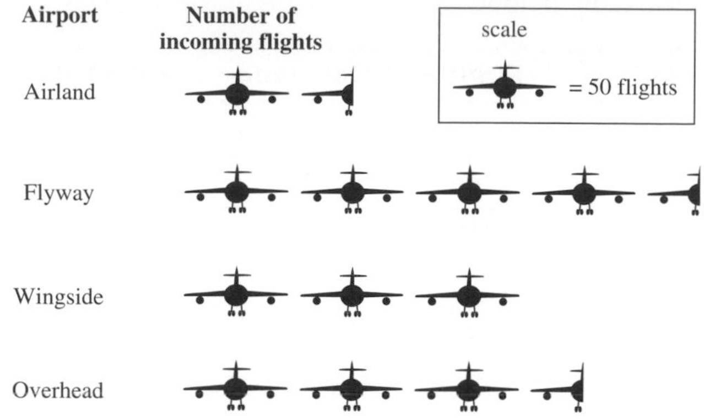

(i) How many flights are there into Wingside?
(ii) Which airport has most incoming flights and approximately how many are there?
(iii) Airport X has twice as many incoming flights as Airport Y. Identify X and Y.

2. The table shows the number of motorcycles, scooters and mopeds licensed in Great Britain from 1984 to 2000. (Source: *Monthly Digest of Statistics*.)

Year	**Number** ('000)
1984	1225
1988	912
1992	686
1994	630
1996	609
1998	684
2000	825

(i) Draw a pictogram to represent this information.
(ii) By Year X the number of licences had fallen by around 25% since 1984. Identify X.
(iii) If the number of licences in 1984 compared with the number ten years later is in the ratio of approximately n:1, state the value of n.
(iv) By year Y the number of licences had fallen by almost one third since 1988. Identify Y.

3. The pictogram shows the number of women candidates in British general elections 1974–1992. (Source: *The Times Magazine*.)

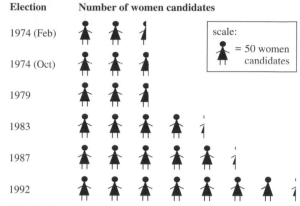

Election — Number of women candidates

scale: = 50 women candidates

1974 (Feb)
1974 (Oct)
1979
1983
1987
1992

(i) Estimate the number of women candidates in
(a) February 1974 (b) 1983.
(ii) In election X there were approximately 50 more women candidates than in election Y. Identify elections X and Y.
(iii) Which election had approximately three times as many women candidates as February 1974?

4. The numbers of unemployed people in four districts are shown in the table below.

District	North	East	South	West
Number unemployed	544	792	232	376

(i) If you were to display these data in a pictogram, which of the following scales would you choose? Explain your answer.

(a) = 8 people (b) = 200 people

(c) = 100 people (d) = 50 people

(ii) Using your chosen scale draw a pictogram to represent the data.

Activity

(i) Look through a variety of daily newspapers for statistical data presented in diagrams and charts.
(ii) Collect 12 examples of diagrams and charts.
(iii) How many of your examples are pictograms?
(iv) Which types of chart are used most often?
(v) Do you think the presentation of data is generally clear and appropriate?
(vi) Can you find any examples of what you consider to be misleading presentation of data? If so, do you think the presentation was deliberately misleading?
(vii) Discuss your findings with those of other students in your group.

Bar charts

The number of male and female employees at each of Treecraft plc's four sites is shown in the table below.

Site	Ashbury	Firside	Oakwood	Pinecroft
Males	155	80	200	145
Females	65	60	65	55
Total	220	140	265	200

These data are used to illustrate three different forms of bar chart.

Vertical and horizontal bar charts

The bar charts below show the number of employees of Treecraft plc at each of its four sites.

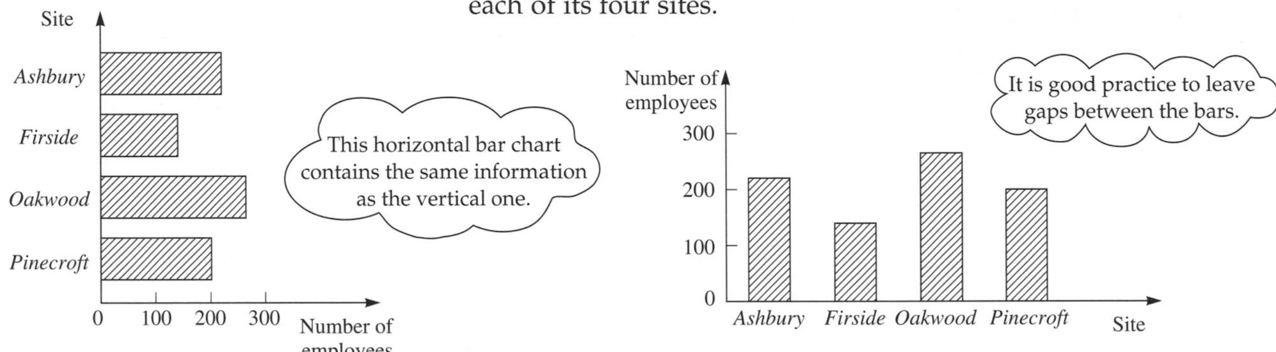

From each of these bar charts you can see that

- Oakwood has most employees and Firside has least;
- Oakwood has nearly twice as many employees as Firside;
- Ashbury has a similar number of employees to Pinecroft.

What you cannot see easily is

- the proportion of the company's employees who work on a given site (although you could if necessary estimate this from the chart);
- the ratio of male to female employees.

Misleading representation

You may sometimes come across bar charts in which part of the vertical axis is omitted. For some data (where the numbers themselves are large but the differences between them are small) this presentation can be useful. However, in many situations it serves only to confuse, since the differences between the numbers look misleadingly large. Here are the Treecraft data presented in this form.

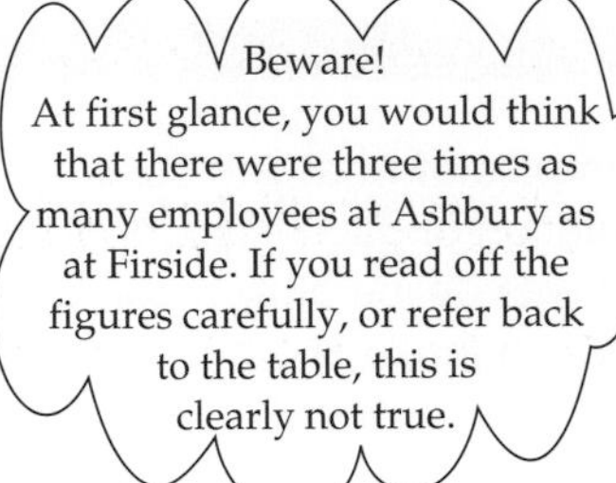

Beware!
At first glance, you would think that there were three times as many employees at Ashbury as at Firside. If you read off the figures carefully, or refer back to the table, this is clearly not true.

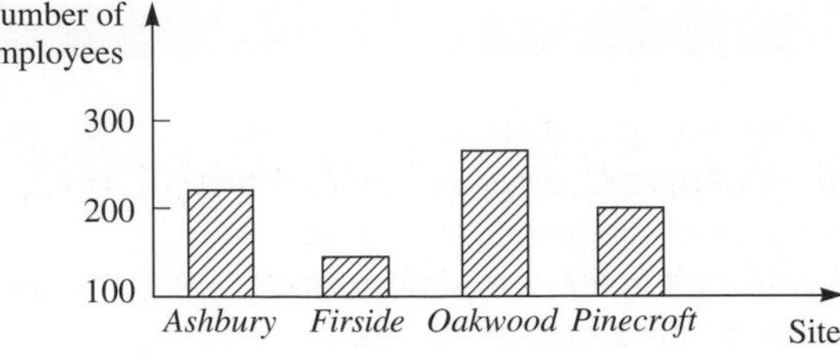

Exercise

1. The percentage of 16–24 year olds unable to spell certain words is shown below.

Word	% unable to spell
because	7
writing	14
sincerely	41
receive	49
necessary	51
accommodation	73

Draw a horizontal bar chart to represent these data. (Source: *Basic Skills Agency*)

2. The horizontal bar chart below shows the number of new AIDS cases reported in the UK, France, Spain and Italy in 1993. (Source: *Social Trends*)

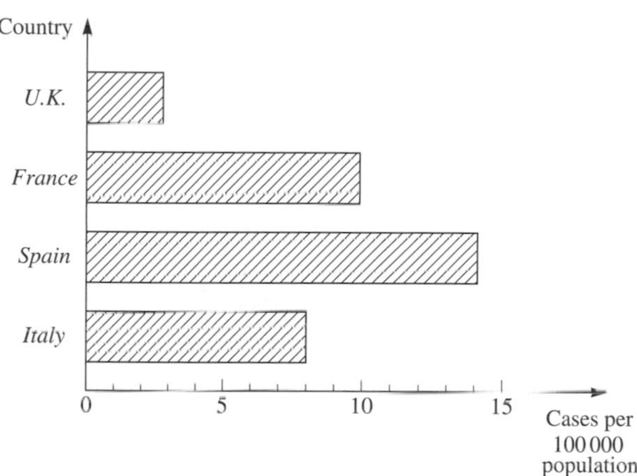

Estimate

(i) the number of new AIDS cases reported per 100 000 population in France;

(ii) how many more new AIDS cases were reported per 100 000 population in Spain than in the UK;

(iii) the ratio of the number of new AIDS cases reported in Italy to the number reported in the UK;

(iv) the actual number of new AIDS cases reported in the UK, given that the population of the UK is 56 million.

3. Using the data below, draw and label a vertical bar chart showing the average cost of a dwelling in different parts of the UK in 2000. (Source: *Regional Trends*, 2001)

Area	Average cost of dwelling
North East	£62 900
South East	£147 300
London	£177 900
South West	£110 100
West Midlands	£87 700
North West	£70 800
Wales	£67 600

4. The populations of six Hertfordshire districts are shown in the bar chart below. (Source: *1991 Census*)

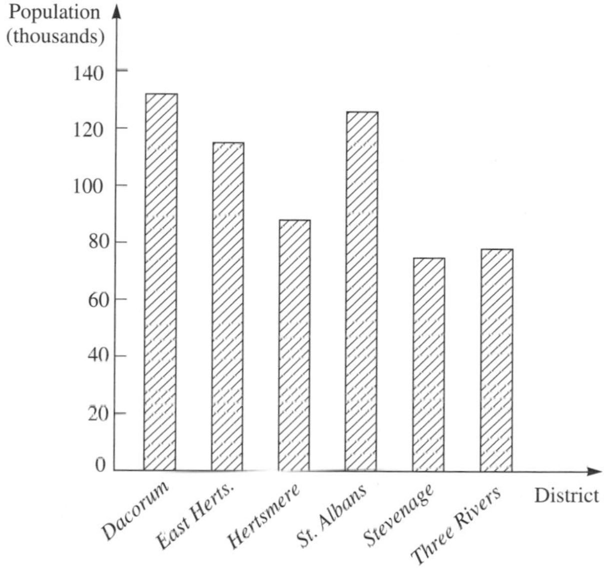

(i) Estimate the population of St. Albans.

(ii) Which district has approximately two-thirds the population of Dacorum?

(iii) Which two districts have roughly the same population?

(iv) Estimate by how many the population of East Herts exceeds the population of Stevenage.

Activity

(i) Find out the names of two computer packages which you can use to create a bar chart.

(ii) Use one of the packages to reproduce your bar chart from question **1** above.

Percentage (or component) bar charts

The diagram below is a percentage bar chart showing the proportion of Treecraft's employees at each of its four sites (see data on page 106).

Ashbury	*Firside*	*Oakwood*	*Pinecroft*

The length of each section is proportional to the number of employees on that site. Once a suitable length for the whole chart has been decided (in this case 10 cm), the section lengths are calculated in the following way. (The results are given correct to 1 decimal place.)

825 is the total number of employees at all four sites.

Ashbury: $\frac{220}{825} \times 10 = 2.7\text{cm}$

Firside: $\frac{140}{825} \times 10 = 1.7\text{cm}$

Oakwood: $\frac{265}{825} \times 10 = 3.2\text{cm}$

Pinecroft: $\frac{200}{825} \times 10 = 2.4\text{cm}$

Total 10.0cm

2.7cm is 27% of 10 cm: Ashbury has 27% of the employees.

This total will match the length of the whole bar (subject t small rounding errors) if the calculations are correct: it is a useful check.

From the percentage bar chart you can see that

- of all Treecraft's employees, just over a quarter work at Ashbury and just less than a quarter work at Pinecroft;
- Oakwood has most employees and Firside least (but this is less obvious here than on the vertical bar chart, page 106).

Dual (and multiple) bar charts

The diagram below is a dual bar chart displaying the numbers of male and female employees at each of Treecraft's four sites.

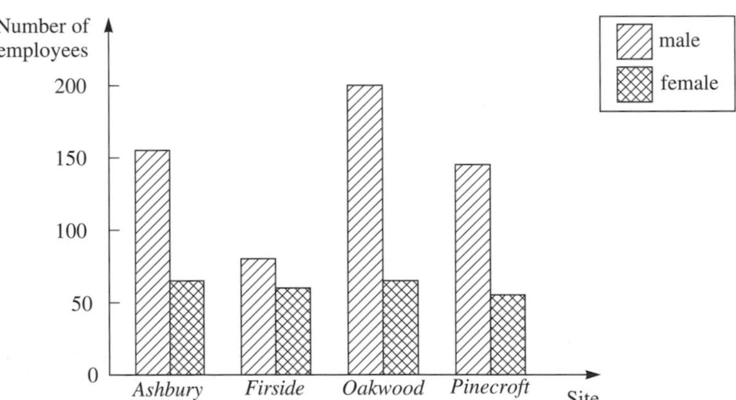

From this chart it is easy to see that

- each site has more male employees than female employees;
- each site has approximately 60 female employees.

A multiple bar chart might have 3 or 4 bars in each main category, rather than just 2 as here. For instance, Treecraft may have some staff who are full-time, some who are part-time, and some who are on temporary contracts at each site. This information could be displayed in a chart with 3 bars for each site.

Exercise

1. Five hundred people were asked how many holidays they had last year and the findings are shown in the table below. Construct an accurate percentage bar chart to represent this. A 10 cm bar should work well.

Number of holidays	Number of people
none	200
1	150
2	100
3 or more	50

2. The percentage bar chart below represents the amounts of fat, starch, sugar and protein in Amy's diet.

fat	*starch*	*sugar*	*protein*

   (i) By inspection, estimate the percentage of
   (a) fat,
   (b) starch,
   (c) sugar,
   (d) protein,
   in Amy's diet.
   (ii) By measuring and calculation, obtain a more accurate estimate of the percentages found in (i).

3. The total cost of manufacturing an item is made up of:
   wages 35%
   materials 45%
   overheads 20%

   Draw a percentage bar chart to represent these data.

4. The percentage of households owning (i) a home computer and (ii) a mobile phone in 1999/2000 is shown in the table below. Draw a fully-labelled dual bar chart to display these data. (Source: *Family Expenditure*)

Country	Home computer	Mobile phone
England	39	46
Scotland	33	39
Wales	27	44
N Ireland	22	25

5. The component bar chart below shows the numbers of men and women at each area office of a company.

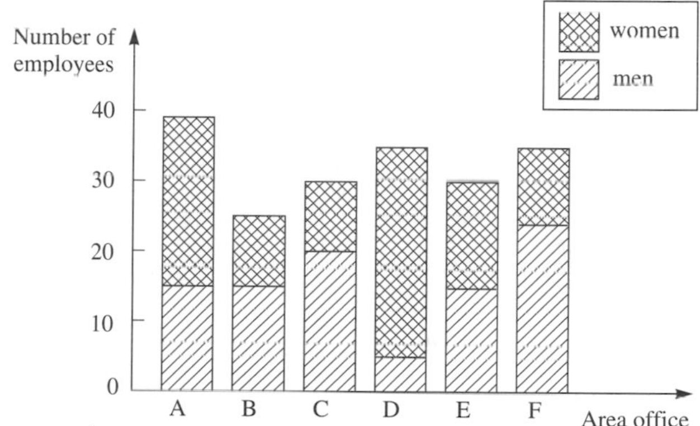

From the chart, estimate
(i) the number of men at A;
(ii) the number of women at B;
(iii) how many more men there are at C than E;
(iv) how many more women there are at D than F;
(v) which area office has most women.

State one advantage and one disadvantage of this type of bar chart compared with the dual bar chart on page 108.

Activity

At Hill Fort School the pupils' choice of sports activities is shown below.

Sport	Boys	Girls
Football	25	0
Hockey	8	15
Badminton	6	7
Cross country	8	2
Basketball	9	3
Aerobics	2	30

Use a computer package to draw
(i) a percentage bar chart showing the sport chosen by
(a) all pupils,
(b) boys only,
(c) girls only;
(ii) a dual bar chart showing each sport and the number of boys and girls choosing it.

For each bar chart make a comment of the form, 'You can see that …'.

Pie charts

The table below shows how the scientific research grant awarded to Proftown University is distributed.

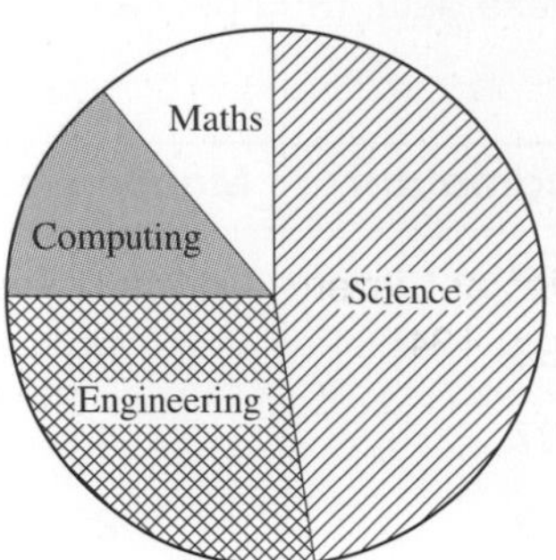

One way of displaying these data is in a pie chart like the one on the left. To produce the pie chart from these numerical data, you would first decide on the radius of the 'pie': in this case 2 cm has been chosen.

Science	£420 000
Engineering	£240 000
Computing	£120 000
Mathematics	£100 000
Total	£880 000

The angle for each sector can be calculated as follows.

Science: $\frac{420\,000}{880\,000} \times 360° = 171.8°$

Engineering: $\frac{240\,000}{880\,000} \times 360° = 98.2°$

Computing: $\frac{120\,000}{880\,000} \times 360° = 49.1°$

Mathematics: $\frac{100\,000}{880\,000} \times 360° = 40.9°$

Total $360.0°$

If the calculations are done correctly, the total will be 360°: a useful check.

Using these angles you can construct the pie chart.

From the pie chart you can see that

- just under half the money is allocated to science and just over a quarter to engineering.

If two sets of data are being compared, they may need to be displayed in different-sized pie charts. For example, if the total amount of research funding at Proftown is different in another year, the total area of the pie chart must be changed accordingly.

Example

Proftown University receives £1 026 400 the next year. If the data for the two years are to be compared, what should be the radius of the second pie chart?

Solution

$$\frac{\text{Area of second chart}}{\text{Area of first chart}} = \frac{1\,026\,400}{880\,000} = 1.166\ldots$$

$$\frac{\pi r_2^2}{\pi r_1^2} = 1.166\ldots$$

$$r_2^2 = (1.166\ldots)r_1^2.$$

Keep the rest of the figures on your calculator and perform rounding at the end.

Taking the square root of both sides gives

$$r_2 = (1.079\ldots)r_1.$$

As r_1 is 2 cm, the new pie chart has radius $(1.079\ldots) \times 2\,\text{cm} = 2.16\,\text{cm}$, correct to 2 decimal places.

Exercise

1. The following data, which show 90 domestic noise complaints analysed by type of source, are to be represented on a pie chart. (Based on *Social Trends 22*)

Type of source	No. of cases
Music	34
Dogs	26
Domestic activities	11
Voices	6
DIY	4
Other	9

Calculate the angle required for each sector and construct an accurate pie chart.

2. The pie chart below represents the radiation exposure of the U.K. population by source in 1993. (Source: *Social Trends 25*)

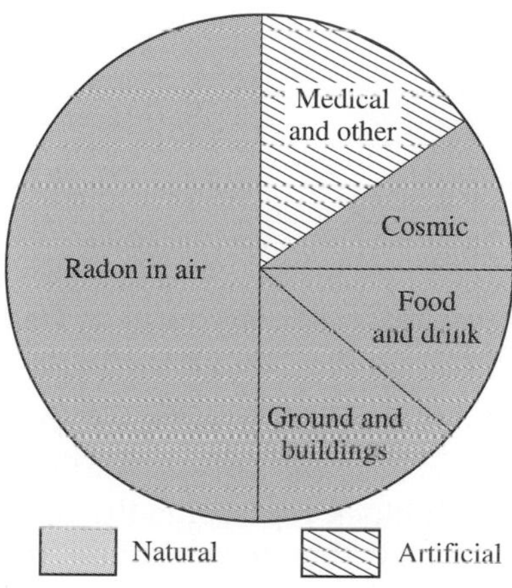

From the diagram make a rough estimate of
(i) the percentage from radon,
(ii) the fraction from food and drink,
(iii) the percentage from ground and buildings,
(iv) the ratio of artificial sources to natural sources.

Now use a protractor to make a more accurate estimate of (ii) and (iii)

3. The air you breathe out consists approximately of 80% nitrogen, 16% oxygen and 4% carbon dioxide. Represent this information on a pie chart.

4. A pie chart of radius 4 cm represents an annual budget of £80 000. What is the correct radius to use when drawing a pie chart for the following year when the budget will be £100 000?

5. The number of enquiries at Citizens' Advice Bureaux in the U.K. in 1980/1 was 4 345 000 and in 1993/4 was 8 253 000 (Source: *Social Trends 25*). The pie charts below show the types of enquiry.

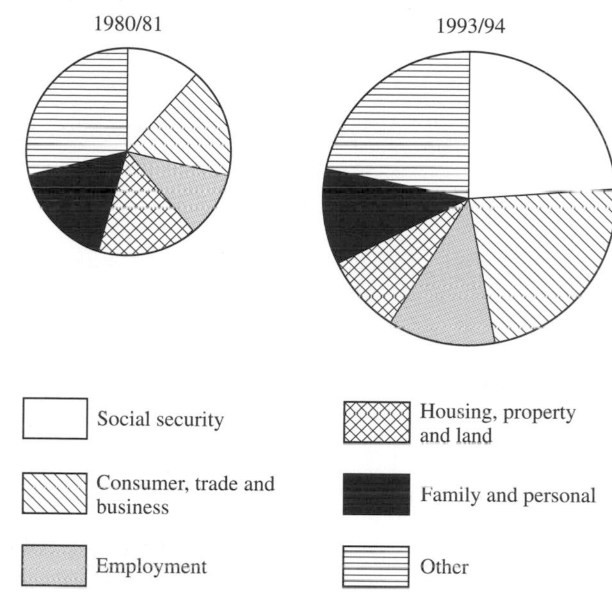

(i) From the diagrams make a rough estimate of
   (a) the percentage of employment enquiries in 1980/1,
   (b) the actual number of social security enquiries in 1993/4.
(ii) Which type of enquiry shows the biggest percentage increase from 1980/1 to 1993/4?
(iii) What has happened to family and personal enquiries between 1980/1 and 1993/4?
(iv) If in 2006/7 there are 12 million enquiries, how will the radius for the 2006/7 pie chart compare with that for 1993/4?

Activities

1. Use a computer package to present the data from question **1** on a pie chart.

2. You will find pie charts useful in other areas of study. Choose a data set that interests you, and use a computer package to present it on a pie chart.

Displaying numerical data

The data on pages 104–111 have been sorted into categories (e.g. employees at different sites). The data in the next two examples are different because the classes are *numerical*.

It is important to decide whether the data are

- *continuous* (which means that the data could take any numerical value within a particular interval), or
- *discrete* (which means that only particular values are possible).

Continuous data are usually displayed as a *histogram*, while discrete data are usually displayed as a *line diagram*.

Histograms

The table shows the weekly wages of a group of workers. The actual categories were £125.00–£149.99 etc. but the form shown here is neater.

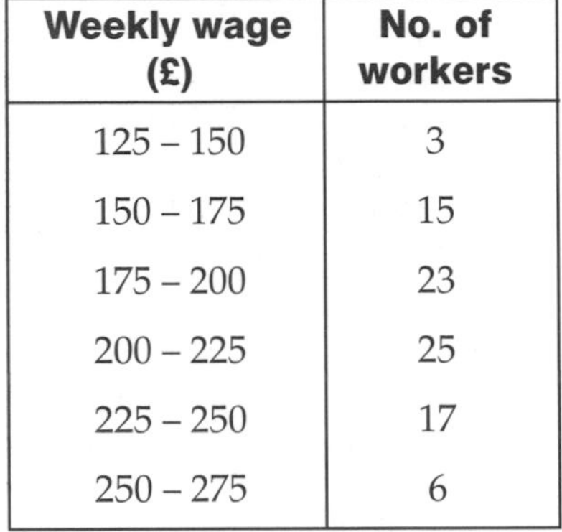

Weekly wage (£)	No. of workers
125 – 150	3
150 – 175	15
175 – 200	23
200 – 225	25
225 – 250	17
250 – 275	6

The wages could in principle take any value within a given range: the data are therefore continuous and can be displayed as a histogram.

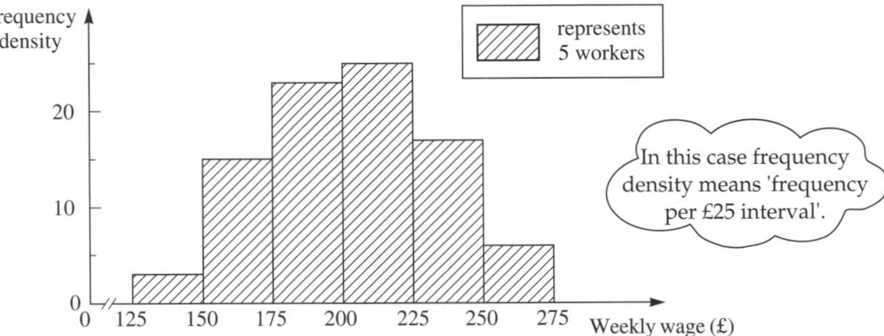

Although a histogram looks rather like a bar chart, there are two important differences.

- There are no gaps between the blocks of a histogram (unless there is an empty category). This is because the scale is continuous.
- The area (not the height) of each block represents the frequency for that category. The vertical axis is therefore labelled 'frequency density' rather than just 'frequency'.

From the histogram you can see that

- The class £200–£225 has the most workers: it is the *modal class*.
- Most of the workers earn between £150 and £250 per week.

Line diagrams

The table below shows the number of children in each household on Century Estate. Since these data are discrete (only whole numbers of children are possible), they can be represented on a line diagram.

No. of children	0	1	2	3	4	5
No. of households	27	29	25	12	5	2

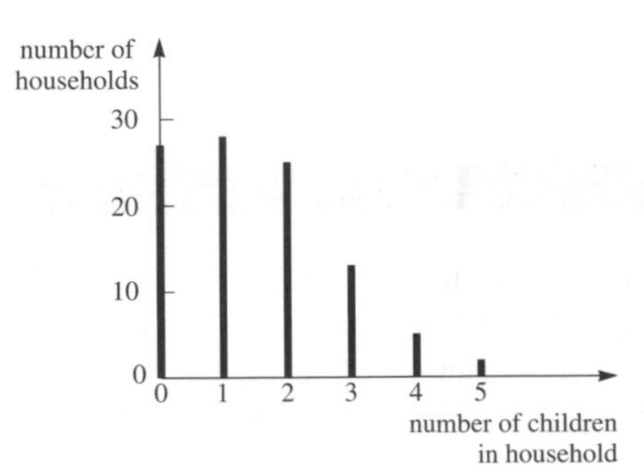

The line diagram is shown on the left. (You have already seen it in the activity on page 97.) You can see that

- about one quarter of the households have no children;
- of those households with children, most have one or two of them.

Exercise

1. The crop from each of 60 tomato plants was recorded and the results are summarised below.

Amount c of crop (gram)	Number of plants
$0 \le c < 500$	9
$500 \le c < 1000$	3
$1000 \le c < 1500$	5
$1500 \le c < 2000$	9
$2000 \le c < 2500$	12
$2500 \le c < 3000$	10
$3000 \le c < 3500$	6
$3500 \le c < 4000$	4
$4000 \le c < 4500$	2

(i) Draw a histogram to represent the data.
(ii) State the modal class.

2. The times taken by trainees to complete a task are recorded to the nearest minute as shown below.

Time taken (min)	No. of trainees
6–8	4
9–11	18
12–14	25
15–17	12
18–20	5
21–23	2

(i) In which class would 11 minutes 20 seconds go?
(ii) In which class would 11 minutes 35 seconds go?
(iii) What would be the boundary value between the 9–11 and 12–14 classes?
(iv) Interpreting the classes as $5\frac{1}{2}–8\frac{1}{2}, 8\frac{1}{2}–11\frac{1}{2}$, etc. draw a histogram to represent the data.
(v) State the modal class.

3. Julie collected 40 leaves from a plant and recorded their lengths to the nearest millimetre. Her results are shown below.

Length of leaves (mm)	Number of leaves
25–29	2
30–34	6
35–39	13
40–44	12
45–49	4
50–54	2
55–59	1

(i) Write down the actual class boundaries of the classes labelled (a) 25–29, and (b) 30–34.
(ii) Using the actual class boundaries, draw a fully-labelled histogram to display the data.
(iii) Draw a frequency polygon by joining the top midpoint of the blocks of the histogram.
(iv) Julie concludes that most of her leaves are between x and y millimetres in length. State what you consider to be the most appropriate values of x and y.

4. Jan did a survey to monitor the number of visitors each patient in a hospital ward received during a visiting session. The results were as shown.

0	1	1	4	2	1	3	0	1	2	0	4
5	2	1	0	4	2	0	7	1	3	1	3

(i) Draw a line diagram to represent these data.
(ii) How many patients have fewer than two visitors?
(iii) In your opinion how many seats should be put near each bed?

Activity

(a) Collect data showing the cost of a 3-bedroom house in your area. Draw a histogram to represent these data.

(b) Collect data showing the cost of second hand cars which are not more than four years old. Draw a histogram to represent these data.

(c) Collect data showing the number of people travelling in a car. (You might do this by observing all the cars passing your school or college entrance during a half hour period, for example.) Draw a line diagram to represent these data.

Cumulative frequency curves

The table gives the heights of 160 sixteen-year-old boys.

Height (cm)	140–150	150–160	160–170	170–180	180–190	190–200
No. of boys	3	25	34	51	40	7

From these data form a cumulative frequency table in which you enter the number of boys with *less than* the given height as follows:

At 140 enter 0; no boys have a height less than 140 cm;

At 150 enter 3; 3 boys have a height less than 150 cm;

At 160 enter 28; 3 + 25 = 28 boys have a height less than 160 cm;

At 170 enter 62; 3 + 25 + 34 = 62 boys have a height less than 170 cm;

and so on for 180, 190 and 200.

Height (cm)	140	150	160	170	180	190	200
Cum. frequency	0	3	28	62	113	153	160

The *cumulative frequency curve* (or *ogive*), based on this table, is shown below. The points plotted are (140,0), (150,3), (160,28) and so on.

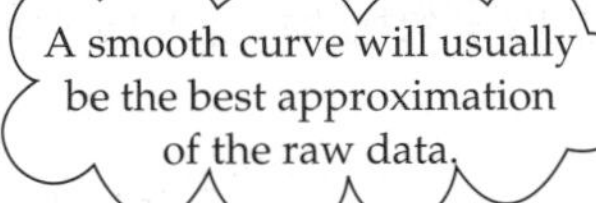

A smooth curve will usually be the best approximation of the raw data.

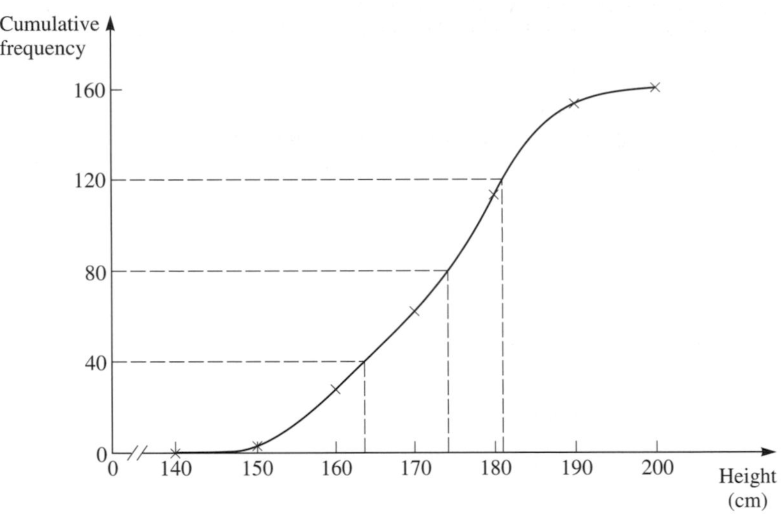

The curve conveys a great deal of information.

- 80 boys (50% of the group) have heights less than 174 cm. 174 cm is called the *median*, or the *50th percentile*.
- 40 boys (25% of the group) have heights less than 164 cm. 164 cm is called the *lower quartile*, or the *25th percentile*.
- 120 boys (75% of the group) have heights less than 182 cm. 182 cm is called the *upper quartile*, or the *75th percentile*.
- The difference between the upper and lower quartiles, 182 cm–164 cm = 18 cm is called the *interquartile range* (see page 120).

Exercise

1. Indira asked 120 car owners how much they paid each year in car insurance premiums.

Insurance premium (£)	No. of people
100–199	7
200–299	17
300–399	24
400–499	33
500–599	22
600–699	10
700–799	5
800–899	2

(i) Construct a cumulative frequency table showing the number of premiums less than £100, less than £200, less than £300, and so on. Use it to draw the cumulative frequency curve

Estimate from your curve

(ii) the premium corresponding to the 25th percentile;

(iii) the number of people paying premiums less than £250;

(iv) the premium corresponding to the 80th percentile;

(v) the number of people paying premiums greater than £650.

2. The ages of the mothers for all live births in the UK in 1993 and 2000 are summarised in the table below. (Sources: *1995, 2002 Annual Abstract of Statistics*)

Mother's age (years)	No. of Mothers (thousands)	
	1993	2000
under 20	51	52
20–24	171	120
25–29	267	192
30–34	194	203
35–39	66	95
40–44	11	16
45 and over	1	1

(i) Compare the two data sets.

The first and last classes are open-ended, but it is reasonable in this case to approximate the under 20 class as 15–19 and the 45 and over class as 45–49. Draw a cumulative frequency curve for 1993 using the categories < 15, < 20, … < 50. Use it to estimate

(ii) the number of births to women under 27;

(iii) the number of births to women 38 and over;

(iv) the age corresponding to the 75th percentile;

(v) the percentage of births to women 34 and over.

Activities

1. The chart on the right is used in the health service: it shows boys' weights from their 10th to their 16th birthdays. (Note that the term 'weight' is used here in its everyday sense: strictly it should say 'mass'.)

The chart contains a lot of information. For example, at the age of 12 years, the weight corresponding to the 30th percentile is 37kg. This means that 30% of boys weigh less than 37kg on their 12th birthday, and so 70% weigh 37kg or more.

From the chart, deduce the value of a, b, c, d, e, f, and g in the following statements.

(i) At 16, a% weigh less than 69 kg.

(ii) At 14, 10% weigh less than b kg.

(iii) At 15, 3% weigh c kg or more.

(iv) The median weight of boys aged $14\frac{1}{2}$ is d kg.

(v) At 11, e% weigh less than 48 kg.

(vi) At 13, 3% weigh less than f kg.

(vii) At g years, 30% weigh 46 kg or more.

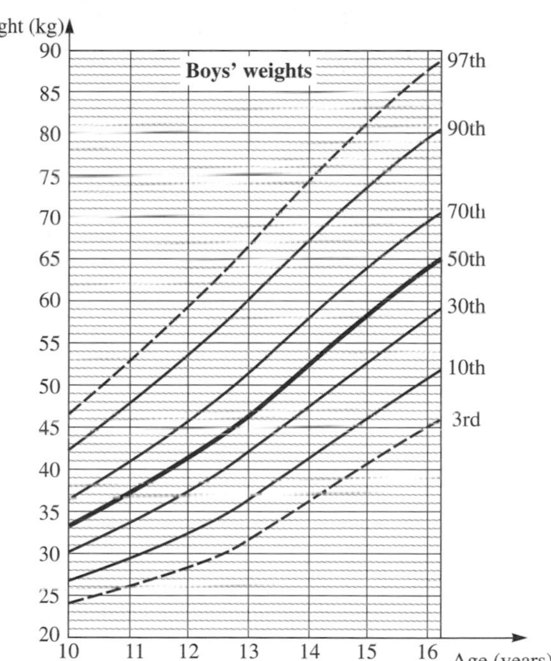

2. This box and whisker plot illustrates the data on the opposite page. The box covers the interquartile range, 164 to 182, and the whiskers cover the rest of the data.

Draw box and whisker plots for the data in questions **1** and **2**.

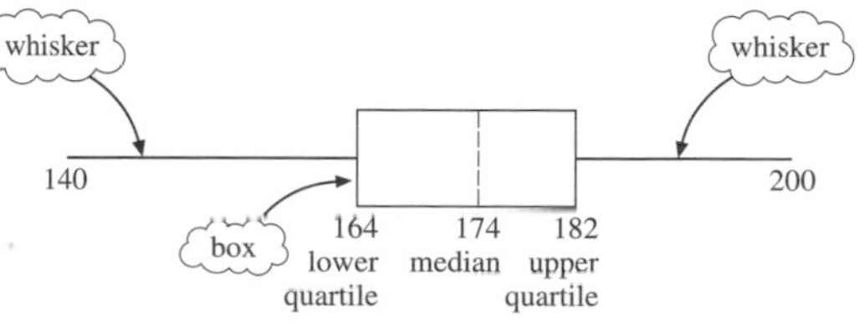

Mean, median and mode

You have collected sets of numerical data, used diagrams to display the results and now need a way of describing the data using numbers. You will have met statements such as

'The average lifespan of a female is 79 years.'
'The average number of children in a family is 1.9.'
'The average weight of a boy at birth is 3.25kg.'

The term 'average' suggests a single value which is in some way typical of the whole set of values. The most commonly used 'averages' are the *arithmetic mean* (usually called simply the *mean*), the *median* and the *mode*.

Chloe, Ann, Claire, Dee and Diana are arguing about the average number of television sets in a household. They have 1, 3, 9, 1 and 2 respectively.

Mean

Ann finds the *mean* by adding the values together and dividing by the number of values:

$$\text{mean} = \frac{1+3+9+1+2}{5} = \frac{16}{5} = 3.2$$

Ann argues that 3.2 is the best average value, since each data item contributes to the calculation of the mean.

Diana points out that with only a small number of items, the contribution of the largest item, 9, leads to a result which is higher than 4 of the 5 values. She says that 3.2 is an inappropriate average value.

Median

Diana finds the median: she arranges the values in order of size and selects the middle one.

1 1 2 3 9

The median is 2. If there is an even number of data items, the median is the mean of the middle two.

Diana argues that the median is entirely dependent on one or two central values and not influenced by extreme results such as Claire's 9.

Ann points out that if the 9 were replaced by a 4 then the median value would be unchanged, but that the new value of the mean would reflect this change.

Mode

Dee chooses the *mode*: the value which occurs most often, in this case, 1. Dee states that the mode is a recorded value and so it will always be meaningful. She says to Ann 'How can you interpret a mean of 3.2 television sets?' Ann retorts that the mode is not a 'central' value and asks Dee, 'How can you argue that the mode is a typical value when it is equal to the lowest recorded value?'

Overview

You can see that the mean, the median and the mode all have strengths and weaknesses. The choice of which to use will be governed by the situation. In practice the mean (often referred to as the average) is the most widely used.

Exercise

1. In a science experiment, six students each measured the acceleration due to gravity and their results (in ms^{-2}) were 9.9, 9.7, 9.7, 8.3, 10.1 and 10.0. Calculate the mean and explain whether you think it is a good measure of average in this context.

2. You are devising the wording for a job advertisement. It runs as follows.

   Salespersonrequired:
   Earn £.....
   per week!

   You have spoken to Brian, who does a similar job, and his earnings for each of the last 8 weeks were:

£100	£200	£100	£150
£300	£250	£2000	£100

   (i) For Brian's figures, calculate
   (a) the mean (b) the median (c) the mode
   (ii) Which of these figures, if any, would you insert in the advertisement? Explain your answer.

3. A leisure centre is open from 0900 to 2200 hours. On March 8th, the number of badminton courts booked in each of the 13 one hour slots is:

   3 6 5 4 6 5 3 3 2 4 6 6 5.

   (i) Suggest why there are no numbers higher than 6.
   (ii) Calculate (a) the mean, and (b) the mode.
   (iii) If you were planning a new leisure centre in a similar area, how many badminton courts would you build on the basis of these data? Explain your answer.

4. The number of items produced on two days by each of 4 production lines is recorded as follows.

	Line A	Line B	Line C	Line D
Tues	190	198	213	202
Weds	200	186	207	140

   (i) Calculate the mean number of items per production line for (a) Tuesday, (b) Wednesday.
   (ii) The mean figure for Wednesday is significantly lower than that for Tuesday. Suggest a possible explanation for the lower figure.

5. Dan and Kay each work at area offices but drive to head office for meetings. Their mileage claims for January to June are shown below.

	Jan	Feb	Mar	Apr	May	Jun
Dan	120	100	120	120	140	60
Kay	210	175	210	245	210	210

   (i) Calculate the mean monthly mileage for
   (a) Dan, (b) Kay.
   (ii) Suggest two possible explanations for Kay's mean monthly mileage being greater than Dan's.
   (iii) Suggest a possible explanation for Dan's low mileage figure for June.

6. Emily chooses, at random, ten one metre square areas in a rectangular field. She records the number of buttercups in each area as follows.

   2 3 0 7 5 7 1 7 10 3

   (i) Calculate
   (a) the mean (b) the median (c) the mode.
   (ii) Emily wishes to estimate the total number of buttercups in the field. Which of the measures in (i) do you think is most appropriate? Explain your answer.
   (iii) Emily measures the field to be 100m by 80m. Estimate the number of buttercups in the whole field.

7. The Monday to Friday sales of a daily paper are 1 251 000, 1 274 000, 1 264 000, 1 258 000, and 1 253 000. Calculate the mean number of sales per day.

8. The times spent in a doctor's surgery by six patients are shown below. (Times are in minutes and seconds.)

4 min 40 s	5 min 10 s	8 min 20 s
3 min 50 s	2 min 40 s	5 min 20 s

   Calculate the mean time that a patient spent in the surgery.

Investigation

(a) Find a set of five numbers whose mean is 5, whose median is 6, and whose mode is 2.

(b) Find a set of seven positive whole numbers whose median is 2 and whose mode is 3.

(c) How many sets of 5 positive whole numbers can you find for which the median is 6 and the mode is 8? In how many of these sets is the mean equal to 5.6?

Calculating mean: frequency table

Example

The number of cars parked outside 15 randomly chosen residences is summarised in the frequency table below. Calculate the mean number of cars per residence.

Number of cars	0	1	2	3
Number of residences	3	7	4	1

Solution

Mean number of cars per residence = $\dfrac{\text{total number of cars}}{\text{total number of residences}}$

From the table you can see that 3 residences have 0 cars, 7 residences have 1 car, and so on, so

total number of cars = $(3 \times 0) + (7 \times 1) + (4 \times 2) + (1 \times 3) = 18$.

Total number of residences = $3 + 7 + 4 + 1 = 15$,

$\Rightarrow$ mean number of cars per residence = $\dfrac{18}{15} = 1.2$.

Calculating mean: grouped frequency table

Example

The pocket money received by a group of 25 children is recorded below. Estimate the mean amount per child.

Amount	£1.00–£2.99	£3.00–£4.99	£5.00–£6.99	£7.00–£8.99
No. of children	8	12	3	2

Solution

Mean amount per child = $\dfrac{\text{total amount received}}{\text{total number of children}}$.

Notice that the question says 'estimate'. This is because you do not know the exact amount of pocket money for each child: you only know the interval within which it falls. This means you can only find an approximate value for the total amount received. To do this you treat it as if each child in the first class receives £1.99$\frac{1}{2}$ (the midpoint of that class), and each child in the second class receives £3.99$\frac{1}{2}$, and so on. In fact, since these values are themselves approximations, it is reasonable in this case to take the midpoints as £2, £4, £6 and £8.

Total amount received = (8 × £2) + (12 × £4) + (3 × £6) + (2 × £8)
= £98.

Total number of children = $8 + 12 + 3 + 2 = 25$.

Recall that this is an approximation.

$\Rightarrow$ Mean pocket money per child= $\dfrac{£98}{25}$=£3.92 i.e. about £3.90.

Exercise

1. The numbers of A level passes achieved by the students at a sixth form centre last year are summarised in the table.

No. of passes	0	1	2	3	4	5
No. of students	15	24	27	15	3	1

Calculate the mean number of A level passes per student.

2. A survey of 200 houses found the number of people living at each house to be as follows.

No. of people	No. of houses
1	54
2	70
3	35
4	32
5	5
6	2
7	2

(i) Calculate the mean number of people per house.
(ii) A town has 8000 houses. Use the table to predict how many houses in the town are occupied by 2 people. Discuss the limitations of your prediction.

3. A charity runs two day-centres called The Beeches and The Limes. Attendances over the last four weeks (Monday–Friday) are shown below.

People per day	8	9	10	11	12	13
The Beeches	1	2	5	6	5	1
The Limes	2	3	4	3	5	3

(i) Calculate the mean daily attendance at (a) The Beeches, and (b) The Limes.
(ii) Due to lack of resources, one centre will have to be closed. List briefly some of the things which should be considered when deciding which centre to close.

4. Maria records the resting pulse rates (in beats per minute) of the students attending her aerobics class.

Pulse rate at rest (b.p.m)	No. of students
50–54	2
55–59	5
60–64	6
65–69	5
70–74	4
75–79	2
80–84	1

Calculate the mean resting pulse rate of Maria's students.

5. The salaries of all the employees of a small business are listed below.

£9 000	£10 500	£11 500
£17 000	£22 500	£8 000
£9 000	£7 000	£11 500
£7 000	£13 000	£14 500
£9 000	£12 500	£8 500
£10 500	£14 500	£7 500

(i) Summarise these data in a frequency table using classes £6000–£7999, £8000–£9999, etc.
(ii) Choose an appropriate value to represent the midpoint of each class.
(iii) Using the frequency table from (i) and the midpoints from (ii), estimate the mean salary of the employees.
(iv) Explain why the figure you have found in (iii) is only an estimate.
(v) Calculate the exact mean salary using the original data from the table, and compare it with your estimate. Was it a good estimate?

Activity

Find out how to use the statistical keys on your calculator to find the mean from a frequency table.

Use this procedure to check your answers to questions **1**, **2** and **3**.

Measures of spread

Machine A	Machine B
39	32
40	45
40	40
41	43
40	40

The mean, median and mode each provide a single typical number to represent a data set but they do not tell the whole story. Look at the figures in the table. They give the numbers of matches in 5 boxes taken as samples from the output of two machines, A and B. The matchboxes are marked 'average contents 40 matches'.

It happens that the mean, median and mode of both data sets is 40 (check this for yourself) but you can see at a glance that the *spread* of the data from Machine B is much greater than that from Machine A. The manufacturer would certainly not be satisfied with the output from Machine B: a customer getting only 32 matches would have every right to complain and would not be consoled by someone else getting a box with well over 40. There are several ways of measuring spread.

Range

This is defined by range = highest value – lowest value

For Machine A: range = 41–39 = 2

For Machine B: range = 45–32 = 13.

Machine A has a much smaller range than Machine B.

The range depends only on the largest and smallest values, so it is influenced by extreme results. This can sometimes give a misleading picture, so other measures of spread may be preferred.

Interquartile range

You have already met the interquartile range on page 114. It measures the spread of the middle 50% of the values and so is not distorted by extreme values. You are better not using interquartile range with very small data sets, like those above for the matches.

Standard deviation

Another measure of spread is *standard deviation*. You will usually use your calculator to find this. Find out how to work out standard deviation on your calculator, and make sure you agree that the standard deviations in the matchbox example are 0.632 for Machine A and 4.427 for Machine B.

If you have to work out standard deviations by hand you use the formula

$$\text{SD} = \sqrt{\frac{\Sigma d^2}{n}}$$

Σx means the sum of all the x's.

where d is the deviation of each data value from the mean, and n is the size of the data set. The table on the left shows a helpful way of setting out the calculation for Machine B.

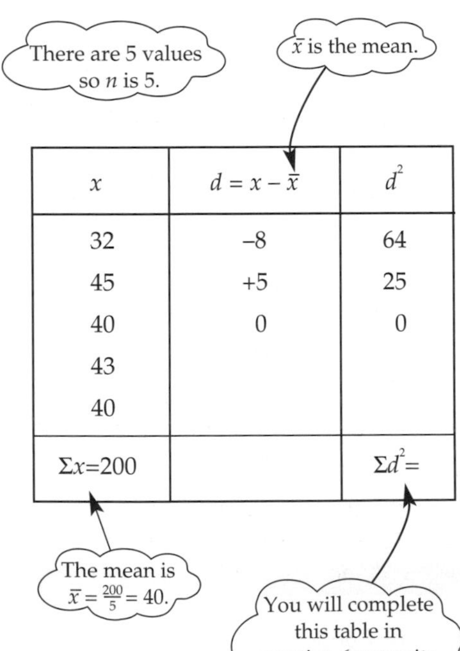

x	$d = x - \bar{x}$	d^2
32	–8	64
45	+5	25
40	0	0
43		
40		
Σx=200		Σd^2=

There are 5 values so n is 5.
$\bar{x}$ is the mean.
The mean is $\bar{x} = \frac{200}{5} = 40$.
You will complete this table in question 6 opposite.

Exercise

1. The maximum daily temperatures in °F recorded at two resorts over a six day period in July 1995 were as shown (Source: *The Daily Telegraph*).

Hastings	79	86	75	72	73	70
Casablanca	75	73	75	77	75	73

Calculate (i) the mean maximum temperature, and (ii) the range, for each resort.
Comment on your results.

2. Three machines are packing bulbs into packets. A quality inspector checks a sample of 8 packets from each machine and records the number of bulbs per packet as shown below.

Machine 1	254	251	257	248
	253	252	251	254
Machine 2	250	220	217	283
	262	275	234	253
Machine 3	262	250	273	255
	268	252	266	259

(i) Calculate (a) the mean, and (b) the range, for each sample.
(ii) The packets are marked 'average contents 250'. Which machine(s) would you stop and reset?

3. Basil carried out a survey to find out how much people paid for a 7-day package holiday. Eighty people responded and he recorded the results as shown below. Basil actually used intervals of £100.00–£149.99, £150.00–£199.99, and so on, but decided the format below was neater for presentation.

Cost of holiday(£)	No. of people
100–150	5
150–200	9
200–250	19
250–300	23
300–350	14
350–400	5
400–450	3
450–500	2

(i) Draw a cumulative frequency curve.
(ii) From your graph estimate (a) the median (b) the lower quartile (c) the upper quartile.
(iii) Calculate the interquartile range.

4. Megan recorded the cholesterol levels in 140 women as shown below.

Cholesterol (mg dl^{-1})	No. of women
120–140	4
140–160	7
160–180	13
180–200	24
200–220	40
220–240	25
240–260	15
260–280	7
280–300	5

(i) Draw a cumulative frequency curve.
(ii) From your graph estimate (a) the median (b) the lower quartile (c) the upper quartile.
(iii) Calculate the interquartile range.
(iv) Megan had deleted 3 results below 120 mg dl^{-1} and 3 results above 300 mg dl^{-1} as being 'too extreme'. How would the inclusion of these additional 6 results affect the interquartile range?

5. The table below gives the gross weekly earnings (£) of full-time adult male employees in Great Britain in the last quarter of 1997, 1998, 1999 and 2000. (Source: *2002 Annual Abstract of Statistics*)

	Lower quartile	Median	Upper quartile
1997	256.4	349.7	480.0
1998	265.3	362.8	499.0
1999	274.5	374.3	517.3
2000	283.9	386.6	532.8

From the table find
(i) the interquartile range for each year,
(ii) the year when the interquartile range is greatest,
(iii) the mean of the lower and upper quartiles for (a) 1997 (b) 1998 (c) 1999 (d) 2000.

The figure you found in each part of (iii) is greater than the median for that year. Give an interpretation of this.

6. Copy and complete the table on the opposite page for the calculation of Σd^2 for Machine B. Check that the standard deviation is indeed 4.427.

Scatter diagrams

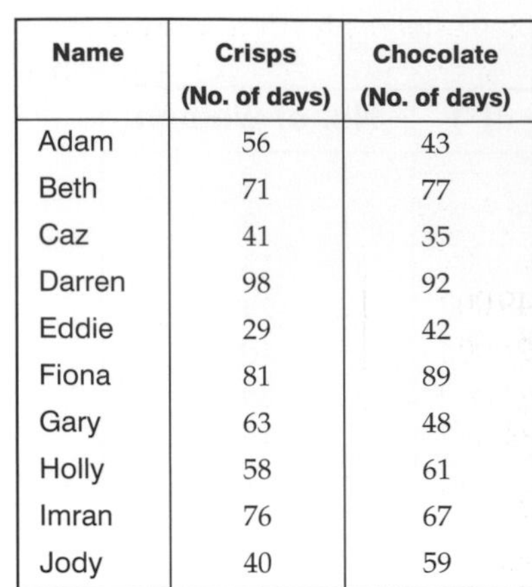

Name	Crisps (No. of days)	Chocolate (No. of days)
Adam	56	43
Beth	71	77
Caz	41	35
Darren	98	92
Eddie	29	42
Fiona	81	89
Gary	63	48
Holly	58	61
Imran	76	67
Jody	40	59

Jessica is a dietitian who wants to find out if there is any association among 16-18 year old students between eating crisps and eating chocolate. She asks 10 students to keep records over a 100-day period of the number of days on which they eat crisps and the number of days on which they eat chocolate. The results are shown on the left.

In this case each item of data consists of two variables – the number of days on which crisps were eaten (x) and the number of days on which chocolate was eaten (y). Jessica plots the data as shown below.

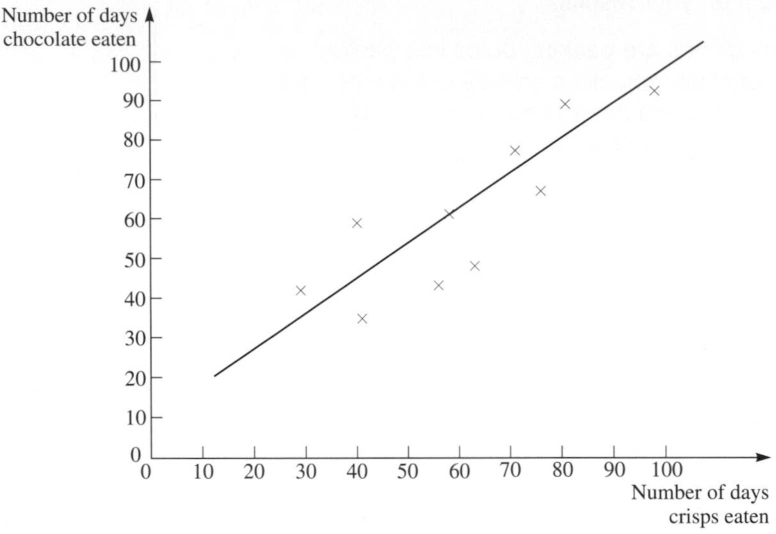

At this stage you can only draw the line of best fit by eye. If you study more statistics you will learn how to calculate its equation.

This diagram is called a scatter diagram. It shows that on the whole those who ate crisps often also ate chocolate often. There is an association between the two. The points are close to a straight line indicating linear correlation, so it is reasonable to draw a line of best fit.

Linear correlation

Positive correlation

The diagram above shows that as crisp-eating increased, chocolate-eating tended to increase. The line of best fit has positive gradient and this illustrates positive linear correlation.

Negative correlation

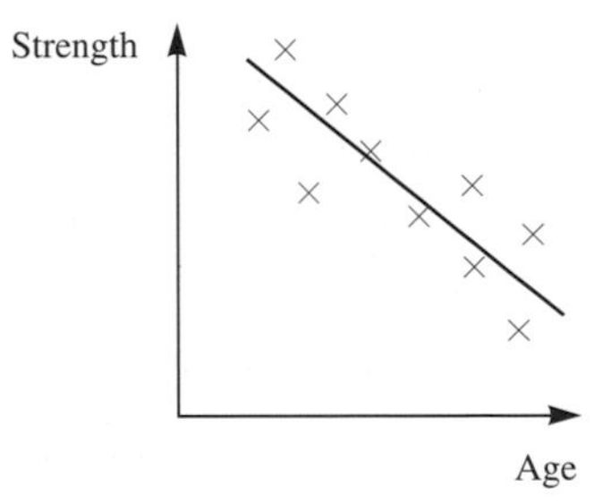

The diagram on the left shows the age of men over 50 and their muscle strength. As men get older their muscle strength tends to decrease. The line of best fit has negative gradient and this illustrates negative linear correlation.

No correlation

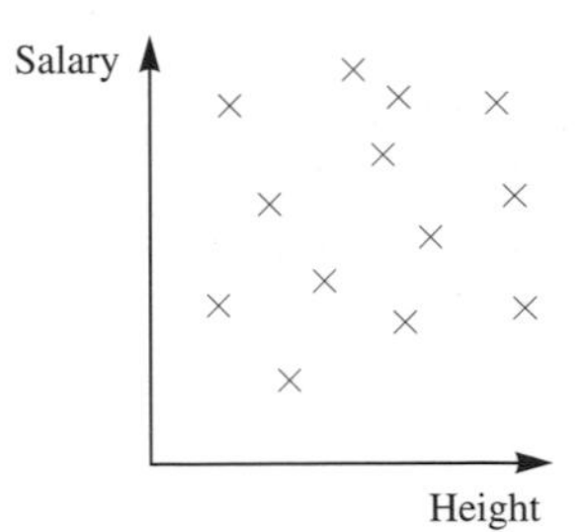

The diagram on the left shows the height and salary of employees. There are points everywhere so it appears that there is no association between height and salary. There is no obvious place for a line of best fit. In these circumstances there is no correlation. (Would you have expected any correlation?)

Exercise

1. The maximum temperature and the number of hours of sunshine at 13 British resorts on 23 May 1996 are shown below. (Source: *The Times*)

Resort	Temperature (°C)	Sunshine (hrs)
Bognor Regis	14	2.1
Clacton	16	7.5
Colwyn Bay	15	4.4
Cromer	17	4.0
Eastbourne	14	1.2
Exmouth	13	1.5
Hastings	15	3.0
Herne Bay	17	6.4
Kinloss	16	6.0
Margate	18	5.0
Skegness	18	7.5
Ventnor	12	0.2
Weymouth	13	0.0

(i) Plot these data on a scatter diagram.
(ii) What type of correlation, if any, do you think exists?
(iii) Draw a line of best fit.
(iv) Describe the association between the maximum temperature and the hours of sunshine at a resort.

2. Kieran takes a random sample of employees at his company and records their length of service with the company and their salary.

Service (years)	Salary (£000)
2	13.4
5	8.6
4	13.8
12	9.4
2	22.0
8	18.5
12	20.6
7	11.4
3	8.8
10	13.4
1	16.0
1	9.2

(i) Plot these points on a scatter diagram.
(ii) What type of correlation, if any, do you think exists?
(iii) What association does this imply between an employee's length of service and salary?
(iv) State with reasons, whether you think the above data are realistic.

3. Suja hopes to buy a particular model of car, and wants to know if the price of an older car is associated with the recorded mileage. She looks in her local newspaper and records details of 12 such cars as shown.

Price (£)	Mileage
4150	35000
2995	49000
2795	71000
2850	67000
5995	25000
4999	48000
5999	45000
1995	52500
2495	53000
3250	58500
3999	38000
4750	35000

(i) Plot a scatter diagram for these data.
(ii) What type of correlation, if any, exists?
(iii) What association does this imply between car price and mileage?
(iv) Draw a line of best fit.
(v) Use the line of best fit to predict the likely mileage of a car costing £3500.

4. Gemma decides to investigate whether spelling ability is associated with age. She selects a sample of 10 adults and asks them to spell 20 words. She records their age and test-scores as shown.

Age	20	42	57	48	23	51	30	34	38	55
Score	10	14	19	20	15	12	7	19	11	8

(i) Plot these data on a scatter diagram.
(ii) What type of correlation, if any, exists?

Gemma then asks 10 children to spell the same 20 words and obtains the following results.

Age	11	15	9	13	15	10	14	17	12	16
Score	6	18	5	7	9	3	10	19	8	12

(iii) Plot these data on a scatter diagram.
(iv) What type of correlation, if any, exists?
(v) Do you think the two sets of results above are realistic? Give your reasons.

Time series

From the diagrams on the left do you think Business A is doing well? Do you think Business B is in decline? The diagrams seems to suggest so, but you need to think carefully.

If Business A traditionally relies on improved sales figures in the autumn and a big turnover in the pre-Christmas period then the diagram is merely reflecting what is expected. But if Business A usually has fairly steady quarterly sales figures, it is possible that its performance really is improving.

If Business B normally has a peak in sales figures in the summer, then the lower figure for the last quarter should come as no surprise. If Business B expects steady quarterly figures, you might ask whether the summer sales this time were exceptional, or whether the first and final quarters' figures were affected by strikes or other one-off events.

In each case you need more information. You need the sales figures over a longer period of time. The sales figures form a *time series*. By analysing the time series you hope to be able to identify trends.

The sales figures for Business A over a 3-year period are shown in the table, and plotted on the graph beneath.

Year	1999				2000				2001			
Quarter	1	2	3	4	1	2	3	4	1	2	3	4
Sales (£000)	50	55	65	90	60	63	75	96	60	67	83	98

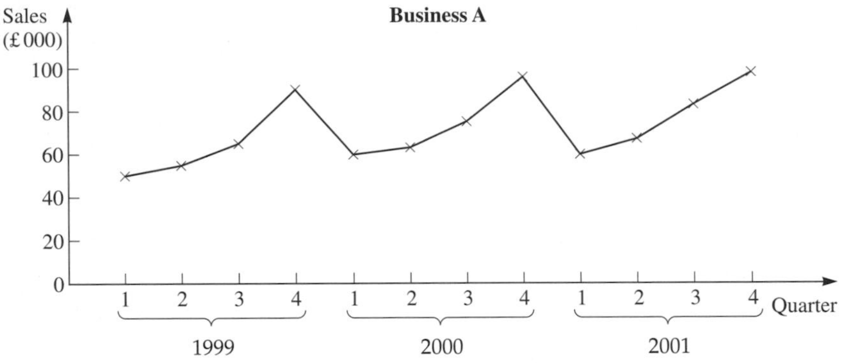

You can see that the figures follow a seasonal pattern, with sales peaking in the 4th quarter each year. A graph showing the sales pattern for only four quarters, as at the top of the page, can be misleading.

Many time series exhibit cyclical patterns. The length of a cycle can be very short (hourly or daily), or very long (several years). But in all cases, to see any underlying upward or downward trends you need to be able to smooth out the cyclical variations, and that is the subject of the next section.

Activity 1

Each part of this activity involves predicting trends over a given period of time. You can work alone if you wish, but it may be better to work in a small group and discuss the issues involved.

1.

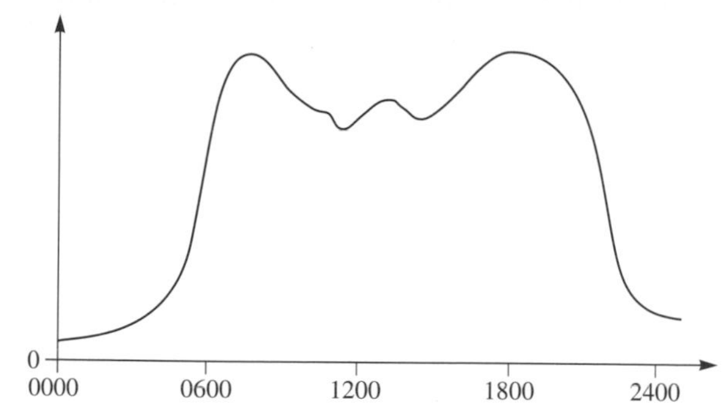

The diagram above represents one of the following.

(a) The domestic demand for electricity on a Saturday in January.

(b) The number of cars on the M25 on a Tuesday in October.

(c) The number of cars on the M25 on a Sunday in July.

(d) The number of people watching television on a Thursday in November.

(i) State, with reasons, which one you think it is most likely to be.

(ii) For each of the other three, sketch the shape of graph that you would expect.

2. Sketch a graph showing a likely weekly cycle for each of the following:

(i) foodstore takings;

(ii) bar takings in a public house;

(iii) daily newspaper sales at a railway station;

(iv) sales of Lotto tickets;

(v) sales of the local weekly paper which appears every Wednesday;

(vi) requests for doctor's appointments.

3. (i) Sketch a graph showing the likely annual cycle for each of the following:

(a) number of new motor cars registered;

(b) new house completions;

(c) number of deaths in the U.K.;

(d) number of births in the U.K.;

(e) sales of garden furniture;

(f) sales of school uniform;

(g) sales of calendars and diaries;

(h) aeroplane flights out of Gatwick;

(i) sales of mathematics textbooks;

(j) sales of photographic film.

(ii) Using resources like those shown below, try to compare your ideas for (a)-(d) with the actual figures.

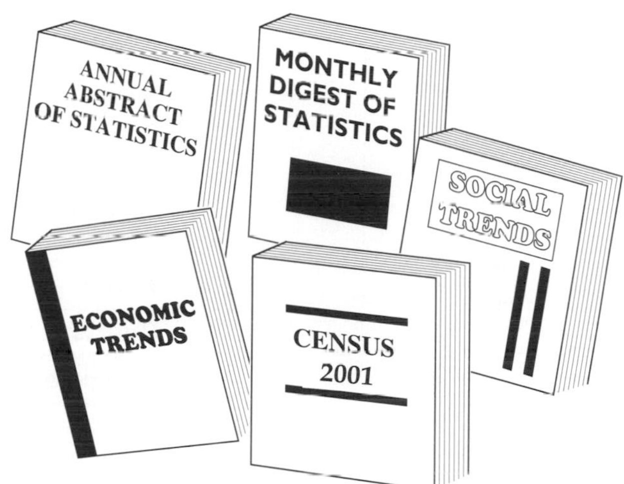

4. (i) State what you think would be the long term trend in the U.K. for each of the following:

(a) miles of railway track;

(b) number of beds in registered private nursing homes, hospitals and clinics;

(c) percentage of home ownership;

(d) amount of material recycled;

(e) number of elderly people in residential accommodation;

(f) number of blood donations;

(g) number of trade union members.

(ii) Using resources like those shown above, try to compare your ideas with the actual figures.

Activity 2

Using the 1999, 2000 and 2001 sales figures for Business A (opposite), calculate the mean quarterly sales for each year.

Use these figures to judge whether there is a long term trend in the sales figures.

Moving averages

In the last activity you should have found a long term upward trend underlying the seasonal variations of Business A's sales figures – but most businesses cannot wait three years for this kind of information. They need to know whether each quarter's performance represents an improvement or a decline, so that they can take appropriate action.

One common approach is to look not just at the figures for the most recent quarter, but at the average over the last four quarters (i.e. the last full cycle). This can be repeated each quarter to produce a set of *moving averages*. For Business A, the calculations (in thousands of pounds) are as follows.

The first moving average is the mean of the sales for quarters 1–4 of 1999. It is

$$\frac{50+55+65+90}{4} = 65.$$

The next moving average, for quarters 2, 3 and 4 of 1999 and quarter 1 of 2000, is

$$\frac{55+65+90+60}{4} = 67.5$$

Notice that the '50' for quarter 1 1999 is replaced by the '60' for quarter 1 2000.

This process is repeated until the data are exhausted. The complete set of moving averages is shown in the table.

Each moving average has been positioned in the middle of the period to which it relates.

Year	1999				2000				2001			
Quarter	1	2	3	4	1	2	3	4	1	2	3	4
Moving average		65	67.5	69.5	72	73.5	73.5	74.5	76.5	77		

The graph below shows the original quarterly figures together with the moving averages.

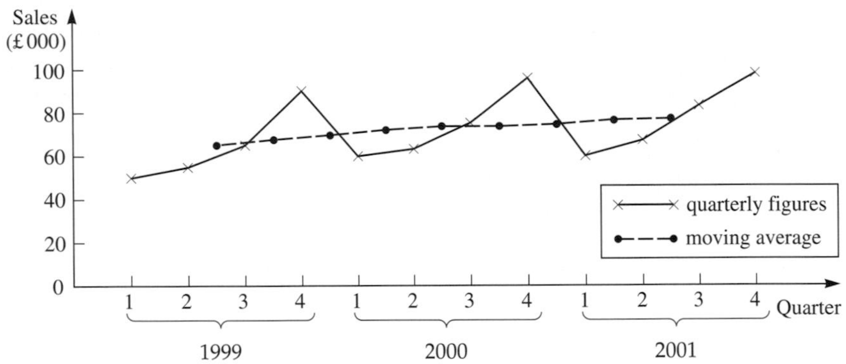

The moving averages show a gradual increase over the period. This suggests that the sales are indeed improving, but not as dramatically as the original graph on page 124 suggested.

Each moving average in this case involved 4 figures. It was a *4-point moving average*. If Business A were analysing its monthly figures, a 12-point moving average would be appropriate, since each average should correspond to a whole cycle. If you were given daily figures for something which followed a weekly cycle, you would probably need to use a 7-point moving average (or 6-point if, for example, there were no Sunday figures).

Exercise

1. The sales figures for Business B (see page 124) are given in the table below.

Year	Quarterly sales (£000)			
1999	50	85	80	45
2000	40	80	75	35
2001	45	75	65	30

   (i) Show that the average of the 4 quarterly figures for 1999 is 65.
   (ii) Show that the average of the last 3 quarters of 1999 together with the first quarter of 2000 is 62.5.
   (iii) Calculate the remaining 7 moving averages for these data.
   (iv) Draw a graph showing both the quarterly figures and the moving averages.
   (v) Comment on the trend.

2. A company's profits for a particular product for the period 1994–2002 are as shown.

Year	Profit (£m)
1994	9
1995	17
1996	19
1997	18
1998	18.5
1999	22.5
2000	15
2001	14
2002	19

   (i) Plot a suitable graph to represent these data.
   (ii) Calculate the 3-point moving averages for the period 1994–2002.
   (iii) Plot these points on your graph and join them.
   (iv) Comment on the trend.
   (v) Using the graph, what result would you predict for 2003? Explain how you calculated your predicted value.

3. The table below shows the number of visitors to Le Jardin and Le Château.

Year	'Le Jardin'	'Le Château'
1999	20 000	20 000
2000	19 000	22 000
2001	18 000	24 000

   (i) Predict the number of visitors to each in (a) 2002, (b) 2017.
   (ii) Suggest factors which may influence the number of visitors each year.
   (iii) How reliable do you think the predictions in (i) (a) and (b) are?

4. Chris and Jo have lived in their house for three years and the quarterly electricity bills are as follows.

	1st Qtr.	2nd Qtr.	3rd Qtr.	4th Qtr.
Year 1	£62	£35	£30	£55
Year 2	£68	£43	£37	£57
Year 3	£72	£56	£33	£65

   (i) Plot a suitable graph to represent these data.
   (ii) Suggest reasons which could explain
      (a) the 'up and down' nature of the figures,
      (b) the wide variation in the 2nd quarter figures from year to year.
   (iii) Calculate the 4-point moving averages.
   (iv) Plot them on your graph and join them together.
   (v) Comment on the trend.
   (vi) Discuss briefly whether an analysis of the number of units used would be a better method of predicting the amount of future bills.

Activity

(i) Collect data which form a time series, such as your electricity, gas or telephone bills. Ideally these will be quarterly and cover a period of three or more years.

(ii) Plot the original data, calculate the 4-point moving averages, plot these and try to predict your bills over the next year.

(iii) Identify factors that
   (a) have resulted in any unusually large or small bills in the past;
   (b) may influence the size of future bills.

Probability

Introduction

Lotto is hugely popular: many people dream that next week's draw might be the one that will change their lives.

Each week, 6 balls (ignoring the 'bonus ball') are drawn at random from a drum containing 49 balls. The balls in the drum are numbered 1–49, and to win the jackpot you have to predict correctly which 6 numbers will be drawn. It is a game of chance, and to understand it properly you need to know about probability.

The probability of an event is a measure of how likely it is to occur. Probability is measured on a scale from 0 to 1 where 0 represents 'impossible' and 1 represents 'certain'. For example,

- if you do not have a ticket then the probability of winning a lottery prize is zero;
- the probability that the first ball drawn will be a number less than 50 is 1.

Calculating probability

This week's Lotto main draw is in progress. Your entry is shown on the card on the left. The balls already drawn are shown below.

£1 BOARD B

1	2	3	4	5
6	7	8	9	10
11	12	13	14	15
16	17	18	19	20
21	22	23	24	25
26	27	28	29	30
31	32	33	34	35
36	37	38	39	40
41	42	43	44	45
46	47	48	49	

LUCKY DIP→ VOID

21 49 5 38 16

What is the probability that you will win the jackpot?

You want the final ball to be a 7. There are 44 balls left in the drum so the probability of 7 being drawn is 1 in 44. This can be written

$$\text{P(7 drawn)} = \tfrac{1}{44}.$$

You can also find the probability of not winning the jackpot. This is the probability that 7 is not drawn: 43 of the 44 remaining balls have numbers other than 7 so

$$\text{P(7 not drawn)} = \tfrac{43}{44}.$$

You have used the principle that

- if all outcomes are equally likely then the probability of a particular event is given by

$$\text{P(event)} = \frac{\text{number of outcomes for which the event occurs}}{\text{total number of possible outcomes}}$$

Notice that if you add the probability of not winning, $\frac{43}{44}$, to the probability of winning, $\frac{1}{44}$, you get the answer 1. This illustrates another important principle:

- the sum of the probabilities for all possible outcomes is 1.

This often provides a shortcut: it is sometimes easier to find the probability of an event *not* occurring than the probability of it occurring. To calculate the probability of the event you can then use the rule

$$\text{P(event occurs)} = 1 - \text{P(event does not occur)}.$$

Exercise

1. When an ordinary, fair die is thrown, what is the probability of obtaining:
   (i) 5?
   (ii) 1 or 2?
   (iii) a number other than 3?
   (iv) What is meant by a 'fair' die?

2. Alan has an ordinary pack of 52 playing cards. He takes a card at random. What is the probability that it is
   (i) a 9?
   (ii a club?
   (iii) a black 8?
   (iv) not a diamond?

3. A tennis club's membership consists of 27 males (including 5 juniors) and 23 females (including 6 juniors). The Club receives a ticket for the Wimbledon Championships and allocates it to a member chosen at random. What is the probability that the chosen member is
   (i) a male?
   (ii) a female junior?
   (iii) not a junior?

4. Danielle has a battery-driven clock on her mantelpiece. What is the probability that the battery will run out when the clock is showing a time
   (i) between six o'clock and nine o'clock?
   (ii) between two o'clock and quarter to three?

5.

For a fundraising event, 100 straws are prepared, each containing a different one of the numbers from 1 to 100. Corey buys the first straw. A prize is awarded if his number includes the 'lucky digit' 3.
   (i) What is the probability that he wins a prize?
   (ii) How is the probability affected if the 'lucky digit' is changed to (a) 2, (b) 1?

6. State whether you agree or disagree with each of the following statements. Give reasons for your answer.
   (i) 'The probability that a person, chosen at random, was born on a Sunday, is $\frac{1}{7}$'.
   (ii) 'Silver Knight is one of twenty horses entered for the Grand National. The probability of Silver Knight winning it is $\frac{1}{20}$'.
   (iii) 'There are four candidates at an election. The probability of each candidate winning the election is $\frac{1}{4}$'.
   (iv) 'The probability that the sun will shine in London on July 15th is $\frac{1}{2}$'.

Activity

(i) Using random number tables or the random key on a calculator, generate 25 random lottery entries. This is an example of a *simulation*.

(ii) The winning numbers are shown below.

8

For each of your entries, write down the number of 'matches' (i.e. correct numbers). For example, the entry 15 37 19 24 42 39 has two matches, 15 and 39.

(iii) Complete a summary table for your 25 entries.

No. of matches	0	1	2	3	4	5	6
Frequency							

(iv) Get together with several other students, and produce a summary table for the whole group.

(v) You are given one ticket for next week's lottery. From the data collected in your simulation, how would you predict the probability of having

   (a) 3 matches? (b) 6 matches?

(vi) Do you think that the probabilities you predicted in (v) (a) and (b) are realistic? Discuss this in your group.

Estimating probability

There are many situations in which an accurate figure for probability cannot be found, and the best you can do is make an estimate.

Suppose, for example, you want to know the probability of getting 3 'matches' in an entry for the Lotto. This can be calculated accurately using the principles of probability, but it requires mathematics beyond the scope of this book. Instead we shall use the data from a simulation involving 200 entries, shown below. (If you prefer, you can work through this example using the results of your own simulation in the last activity.)

No. of matches	0	1	2	3	4	5	6
Frequency	87	81	29	3	0	0	0
Relative frequency	0.435	0.405	0.145	0.015	0	0	0

The relative frequencies are given by $\dfrac{\text{class frequency}}{\text{total number of entries}}$

$$\text{e.g. } \frac{87}{200} = 0.435.$$

On the basis of these figures, you can estimate that

$$\text{P(3 matches)} = \frac{3}{200} = 0.015.$$

In other words, the relative frequency of an event in a simulation gives an estimate of its probability. The more data you generate, the more accurate your probability estimates are likely to be.

Sometimes it is possible to work out relative frequencies from past data rather than running a simulation: this approach is used for example by motor insurance companies to work out which groups of people represent a higher risk.

There are some situations, though, which cannot be tackled using relative frequencies, because there is no past data and no simulation is possible. For example, how would you estimate the probability that Lotto will still be running in five years' time?

In this situation you have to resort to making a subjective judgement based on any information available.

You could argue that Lotto is highly likely to continue because it raises money for charity, it is very popular, and no government would risk the uproar of stopping it. You might feel that a probability of 0.99 is appropriate. Essentially this is your best guess.

You might on the other hand argue that the government will decide to replace it because it will lose its novelty: people will realise that large wins are very rare and don't bring happiness, and that the money they spend on it would be better spent on other things. You may in this case estimate the probability that the lottery will still be running in five years' time to be about 0.2. Again, this is really a best guess.

Exercise

1. Mary owns a gift shop and each week she receives a consignment of 50 ornaments. Over the last 20 weeks she has recorded the number of breakages per consignment as follows.

Number of breakages	0	1	2	3	4 or more
Number of consignments	10	4	2	1	3

On the basis of these data estimate the probability that the number of breakages in the next consignment will be
(i) zero (ii) 1 (iii) 2 or more.

2. In the last 365 days a dispensing machine has been out of action for a total of 292 hours. This was due to short-term faults which occurred at random. Taking these figures to be typical, estimate the probability that when Paul approaches the machine to make a purchase, he will get served.

3. Kaniz records her journey time to work as follows.

Time (minutes)	No. of journeys
40 to 50	16
50 to 60	14
60 to 70	13
70 to 80	7

(i) How many days do her records cover?
(ii) On the basis of the information collected, estimate the probability that her next journey to work will take
 (a) more than fifty minutes;
 (b) less than an hour.
(iii) Explain why these might not be good predictions.

4. You work for an insurance company with 100 000 customers. Last year 13 500 customers made claims. You have identified some typical customers for marketing purposes and Mr Patel is one of these.
(i) Estimate the probability that Mr. Patel will make a claim next year.

Ms Webb also insures with the company and has made claims in each of the last three years.
(ii) Would you charge Ms Webb and Mr Patel the same premium next year? Justify your answer.

5. Sara and Nick go to the sports centre twice a week to play badminton, but never book ahead. If no court is available they go swimming instead. In the last 10 weeks they have played badminton 4 times on Tuesdays and 7 times on Saturdays. On Friday night Nick tells Sara that there is a 55% chance that they will play badminton tomorrow. Explain how Nick comes to this conclusion. Is he justified?

6. Rob, Tracy and Vicki are investigating randomness. They throw a die 60 times and produce the following results.

	1	2	3	4	5	6
Rob	10	10	10	10	10	10
Tracy	11	8	10	9	11	11
Vicki	5	10	9	12	8	16

Comment on each person's results.

Activities

1. Write some questions of your own (don't copy earlier questions exactly!) based on drawing a card from a pack of 52 playing cards. The answers should be the probabilities
(i) $\frac{1}{4}$ (ii) $\frac{1}{13}$ (iii) $\frac{1}{2}$ (iv) $\frac{12}{13}$ (v) $\frac{1}{52}$

2. Write questions based on birthdays whose answers are the probabilities
(i) $\frac{1}{365}$ (ii) $\frac{31}{365}$ (iii) $\frac{335}{365}$ (iv) $\frac{1}{7}$

3. You are to write three questions whose answers are the probabilities (a) $\frac{1}{7}$ (b) $\frac{3}{7}$ (c) $\frac{5}{7}$.

The questions are to be based on drawing coloured balls from a bag. There are red, blue, yellow and green balls available.
(i) What combination of coloured balls would you put in the bag?
(ii) What would be your three questions?

Two or more events

The basic ideas of probability developed in the seventeenth century from analysis of popular games of chance such as dice and cards. In this section, the ordinary pack of 52 playing cards will be used to illustrate the probability of two or more events occurring.

You need to distinguish between two types of event:

- 'either-or' events, such as drawing *either* a club *or* a diamond from a pack of cards
- 'first-then' events, such as drawing *first* a spade *then* another spade from a pack of cards.

Either-or events

Example

A card is drawn from an ordinary pack of 52 playing cards. What is the probability that it is

(i) either a club or a diamond?

(ii) either a heart or a 2?

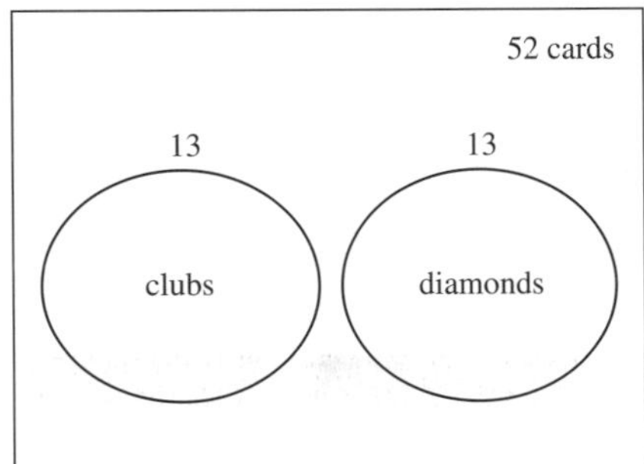

Solution

(i) The events 'club' and 'diamond' cannot happen simultaneously. They are described as *mutually exclusive*. The probabilities of either-or events with no overlap are added:

$$\begin{aligned} \text{P(club or diamond)} &= \text{P(club)} + \text{P(diamond)} \\ &= \tfrac{13}{52} + \tfrac{13}{52} \\ &= \tfrac{26}{52} \\ &= \tfrac{1}{2}. \end{aligned}$$

NOTE

Alternatively you could start by saying that there are 13 clubs and 13 diamonds, so that 26 cards out of 52 will produce the event 'club or diamond'.

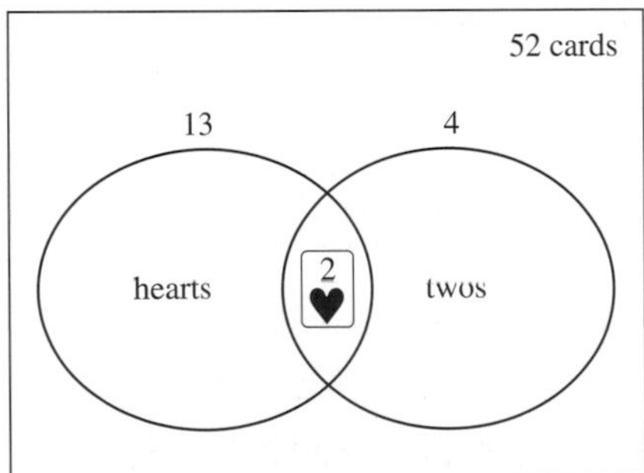

(ii) The events 'heart' and '2' do have an overlap. They have the 2 of hearts in common. The probabilities of either-or events with an overlap are added, then the probability of the overlap is subtracted to avoid double-counting:

$$\begin{aligned} \text{P(heart or 2)} &= \text{P(heart)} + \text{P(2)} - \text{P(both heart and 2)} \\ &= \tfrac{13}{52} + \tfrac{4}{52} - \tfrac{1}{52} \\ &= \tfrac{16}{52} \\ &= \tfrac{4}{13} \end{aligned}$$

NOTE

Alternatively you could start by saying that there are 13 hearts (including the 2 of hearts) and 3 other twos, so that 16 (=13+3) cards out of 52 satisfy the event 'heart or 2'.

Exercise

1. A card is drawn at random from a pack of 52 playing cards. Calculate the probability that it is
   (i) a king or a queen,
   (ii) a red card or an 8,
   (iii) a heart or a court card (king, queen or jack).

2. The Maths and English marks of a small group of students are shown in the table below. (The pass mark is 40.)

Student	Maths	English
Mitesh	61	38
Ellen	83	66
Paul	32	41
Conrad	67	59
Emma	21	33
Kim	50	28

   An inspector visits and chooses one of the group at random to interview to get a student's view of the course. Calculate the probability that the student chosen has achieved a pass in
   (i) Maths;
   (ii) English;
   (iii) both Maths and English;
   (iv) Maths or English (or both).

3. In the 1860s Mendel investigated seed colour and texture in peas. He found that under certain conditions, 4 types of seed - yellow and round, yellow and wrinkled, green and round, green and wrinkled - occurred in the ratio $9:3:3:1$ respectively. Find the probability that a seed chosen at random under these conditions is
   (i) green and round;
   (ii) yellow;
   (iii) wrinkled;
   (iv) round but not green;
   (v) yellow or round (or both).

4. Kieran interviews all 100 employees of a company. He finds that 70 read 'Company News', 55 read 'Union Update' and everyone reads at least one of these. Calculate the probability that an employee chosen at random reads both.

5. Jane is a geneticist who has been studying a pair of genes, which are found on a particular chromosome. She has found that the combination AB occurs on 50% of these chromosomes, Ab on 25%, aB on 15%, ab on 10%. Calculate the probability that a randomly chosen chromosome
   (i) possesses A;
   (ii) possesses b;
   (iii) possesses either A or B (or both).

Activity

This activity is best carried out in pairs.

(i) Collect data on the sum of the scores when two dice are rolled. Roll the dice 108 times, or simulate 108 rolls using a computer package or the random number key on your calculator.

(ii) Use relative frequency as an estimate of probability to complete the table below.

Sum	Probability estimates	Sum	Probability estimates
2	$\frac{}{108}$	8	$\frac{}{108}$
3	$\frac{}{108}$	9	$\frac{}{108}$
4	$\frac{}{108}$	10	$\frac{}{108}$
5	$\frac{}{108}$	11	$\frac{}{108}$
6	$\frac{}{108}$	12	$\frac{}{108}$
7	108		

(iii) Complete the table below to show the 36 possible outcomes when 2 dice are rolled and the scores added.

	1	2	3	4	5	6
1	2	3	4			
2	3					
3						
4					9	
5						
6		8				

Use the table to calculate the probabilities of all the outcomes (from 2 to 12). Compare your answers with those you found experimentally.

First-then events

A card is drawn at random from an ordinary pack of 52 cards. It is replaced, and then a second card is drawn. What is the probability that both cards are spades?

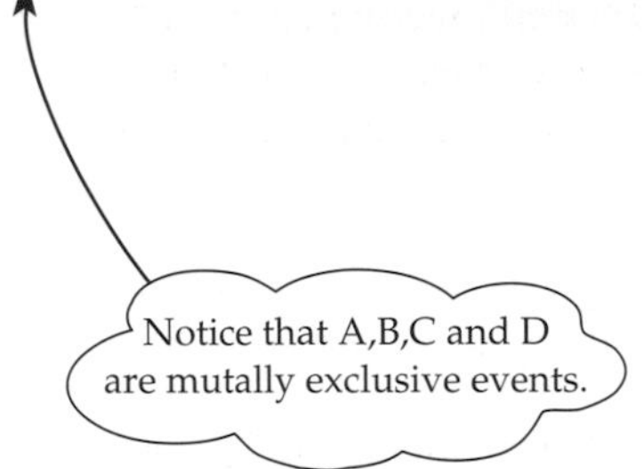

	1st draw	2nd draw
A	spade	spade
B	spade	not spade
C	not spade	spade
D	not spade	not spade

Notice that A,B,C and D are mutally exclusive events.

This is an example of a first-then event. To calculate the probability, it is best to classify the possible outcome of each draw as 'spade' or 'not spade'. When the two cards are drawn, the possible outcomes are as shown in the table. Only outcome A satisfies the requirement that both cards are spades.

Since the first card is replaced before the second card is drawn, the outcome of the second draw is *independent* of the first card drawn. Each time, the probabilities are

$$\text{P(spade)} = \tfrac{13}{52} = \tfrac{1}{4} \text{ and P(not spade)} = \tfrac{39}{52} = \tfrac{3}{4}.$$

The probabilities of first-then events are multiplied, so the probability of outcome A is

$$\text{P(spade then spade)} = \text{P(spade)} \times \text{P(spade)}$$
$$= \tfrac{1}{4} \times \tfrac{1}{4} = \tfrac{1}{16}.$$

The probability that both cards are spades is therefore $\frac{1}{16}$.

If there are two or more ways in which a first-then event can occur, there is an extra stage to the calculation.

Example

In the situation above, what is the probability that just one card is a spade (i.e. one card is a spade and one is not)?

Solution

From the table, outcomes B and C both produce one 'spade' and one 'not spade' so there are two ways in which this event can occur. Outcome B is a first-then event, so its probability is found as follows.

$$\text{P(spade then not spade)} = \text{P(spade)} \times \text{P(not spade)}$$
$$= \tfrac{1}{4} \times \tfrac{3}{4} = \tfrac{3}{16}.$$

Similarly for outcome C,

$$\text{P(not spade then spade)} = \text{P(not spade)} \times \text{P(spade)}$$
$$= \tfrac{3}{4} \times \tfrac{1}{4} = \tfrac{3}{16}.$$

Now since the required event occurs if *either* outcome B *or* outcome C occurs, the probabilities of these two are added:

$$\text{P(just one card is a spade)} = \text{P(outcome B)} + \text{P(outcome C)}$$
$$= \tfrac{3}{16} + \tfrac{3}{16} = \tfrac{6}{16} = \tfrac{3}{8}$$

NOTE

The probability of neither card being a spade (outcome D) is

$$\text{P(not spade then not spade)} = \text{P(not spade)} \times \text{P(not spade)}$$
$$= \tfrac{3}{4} \times \tfrac{3}{4} = \tfrac{9}{16}.$$

Outcomes A, B, C and D are the only possible outcomes (it is certain that one of them will occur), and they are either-or events with no overlap. You can check that their probabilities add up to 1.

Exercise

1. In a large batch of components it is estimated that 5% are faulty. Two components are drawn at random to make a piece of equipment. What is the probability that
   (i) both components are good?
   (ii) one component is good and one is faulty?

2. Seven in ten people have natural immunity to a particular disease, and do not need a vaccination against it. Immunity is distributed randomly in the population. Helen and Kalpesh go for tests to see if they need a vaccination. What is the probability that
   (i) both have natural immunity?
   (ii) just one of them has natural immunity?
   (iii) both need vaccinations?

3. Most people do not have the haemophilia gene. Those who do have it may be carriers or may exhibit symptoms. Harriet is a carrier and Jason does not have the gene. They plan to have children and are advised that each time they have a baby there will be four possibilities, each with a probability of $\frac{1}{4}$:
   A a boy who does not have the gene;
   B a boy who exhibits symptoms;
   C a girl who does not have the gene;
   D a girl who is a carrier.

   (i) What is the probability of their first child not having the gene?
   (ii) If they have two children, what is the probability that
      (a) neither child has the gene;
      (b) only one child has the gene?

4. A fair die is rolled twice. Calculate the probability of obtaining
   (i) a 6 followed by a 4
   (ii) a 5 followed by a 1 or a 2
   (iii) the same score on each throw.

5. Carmen and Des each manage a branch of an estate agency. In the last ten weeks their sales were as follows.

Carmen	2	3	0	1	2	4	3	2	1	3
Des	0	2	1	3	1	1	2	2	0	3

   Assuming these figures are typical of all weeks, what is the probability that next week
   (i) Carmen will make more than 2 sales?
   (ii) Des will make no more than 1 sale?
   (iii) both Carmen and Des will make 3 or more sales?

6. In a large island population the number of people in blood groups A, B, O and others is in the ratio 10:2:12:1. What is the probability that the next two people arriving at the island's blood transfusion centre are
   (i) both from group A?
   (ii) one from group A and one from group O?
   (iii) both from groups other than B?

7. Ian and Esme have just taken out a mortgage. According to life expectancy charts Ian has a 90% chance of still being alive at the end of the loan period while Esme's chances of survival are 95%. What is the probability that both die before the end of the loan period?

Activity

A fund-raising committee has asked you to devise a game, based on probability, which they can run at a summer fête to raise money. You must decide the nature of the game, entry fee, winning outcomes, prizes, likely number of participants, expected profit, financial risks, expenses, and so on. Write a report for the committee giving your recommendations along with the relevant calculations.

Tree diagrams

Tree diagrams can be used to solve first-then problems like those in the last exercise, but they are most useful for *dependent* events, when the outcome of the second event is influenced by that of the first.

Example

A card is drawn from an ordinary pack of 52 cards and is not replaced. A second card is then drawn. What is the probability that (i) both cards are spades? (ii) just one card is a spade?

Solution

The outcome of the second draw is influenced by the outcome of the first draw: if the first card was a spade there will only be 12 spades among the remaining 51 cards.

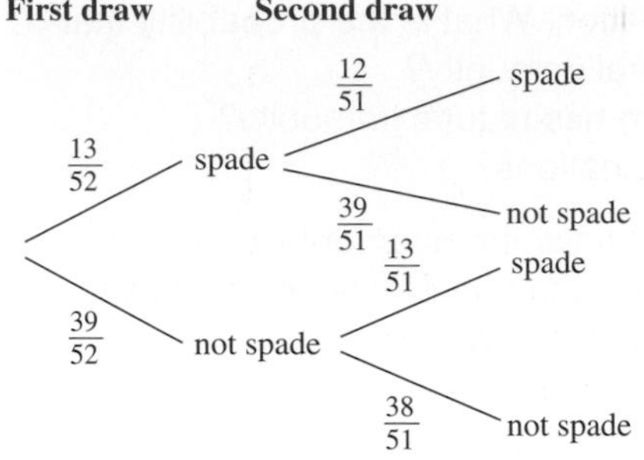

(i) Two spades are obtained if you follow the top 'branch' of the tree diagram. The two events are first-then events so you multiply their probabilities.

$$\text{P(two spades)} = \tfrac{13}{52} \times \tfrac{12}{51} = \tfrac{1}{17}$$

(ii) Just one spade is obtained if you follow either of the middle two branches. The probability for each branch is found by multiplying the probabilities on it, as above. The two branches represent either-or events with no overlap, so their probabilities are then added:

$$\text{P(just one spade)} = \tfrac{13}{52} \times \tfrac{39}{51} + \tfrac{39}{52} \times \tfrac{13}{51} = \tfrac{13}{34}$$

Example

Neil knows that if a night is clear the probability that the next is clear is 0.8, while if a night is cloudy the probability that the next is cloudy is 0.4. Last night was clear, so Neil books the next two nights at the observatory. What is the probability that neither night is clear?

Solution

The probabilities given in the question, marked*, were put in first, and the others were deduced using the fact that the total probability adds up to 1.

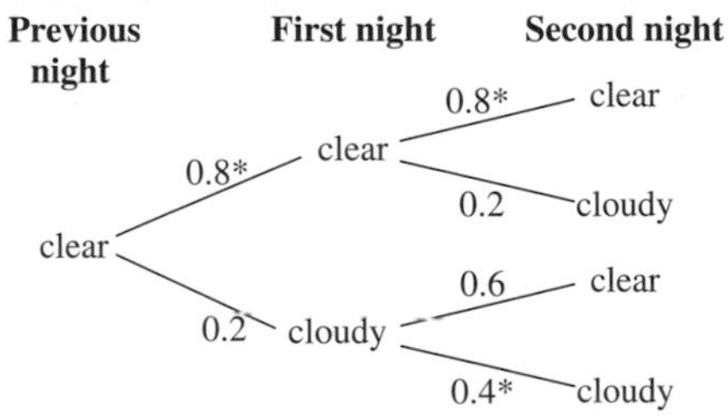

The probability that Neil doesn't get a clear night is given by the bottom branch of the tree diagram:

$$\text{P(no clear night)} = \text{P(cloudy, cloudy)} = 0.2 \times 0.4 = 0.08$$

Exercise

1. A cold spell is forecast for two nights in early May. There is a 40% chance of frost on the first night and a 30% chance of frost on the second night. A single night of frost will be enough to damage the blossom. What is the probability that the blossom is damaged during this cold spell?

2. Mr. Singh's shop is protected by two burglar alarms which operate independently. In the event of a break-in, the probability that the first one goes off is 90%, and the probability that the second goes off is 80%. What is the probability that during a break-in
   (i) both alarms go off?
   (ii) just one alarm goes off?
   (iii) neither alarm goes off?

3. A store manager offers customers free repairs for 5 years if they buy both a television set and a video recorder. He estimates that there is a 10% chance of the television developing a fault and a 20% chance of the video recorder developing a fault within this period. Calculate the probability that a customer will need repairs to (i) both items, (ii) just one item.

4. 

   Anthea has bought five tickets for this raffle. A total of 100 tickets have been sold. What is the probability that Anthea wins (i) both prizes? (ii) just one prize?

5. At a school fête, the 'Wheel of Fortune' is divided into 12 equal sectors as shown, so that when the pointer is spun there is an equal probability of it stopping on each sector.

   For 10p you can buy a ticket printed with one of the numbers 1–12. The pointer is then spun once, and if it stops on your number you win £1. What is the probability that
   (i) if you play once you make a profit?
   (ii) if you play twice you make a profit?

6. A bus driver encounters two sets of traffic lights. The probability that the first set is green is $\frac{2}{5}$. If the first is green then the probability that the second is green is $\frac{4}{5}$, while if the first is red the probability that the second is green is $\frac{3}{5}$. There is no amber light. Calculate the probability that the bus driver has to stop at
   (i) both sets of lights; (ii) neither set of lights.

7. Avril and Barbara are to play a tennis match in which the first player to win two sets wins the match. If the probability of Avril winning any particular set is $\frac{3}{5}$, find the probability that
   (i) Avril wins the match 2–0;
   (ii) the match reaches 1–1;
   (iii) Barbara wins 2–1.

Activity

Assume that the probability of a new baby being a boy (B) is $\frac{1}{2}$, and being a girl (G) is also $\frac{1}{2}$. A family of 2 children can be BB, BG, GB or GG. Each of these has probability

$$\tfrac{1}{2} \times \tfrac{1}{2} = \tfrac{1}{4}.$$

It follows that P(2 boys) = P(BB) $= \frac{1}{4}$

P(1 boy, 1 girl) $= \text{P(BG)} + \text{P(GB)}$

$= \frac{1}{4} + \frac{1}{4} = \frac{2}{4} \left(\text{or } \frac{1}{2}\right)$

[Note: part (iii) is easier if you do not cancel these.]

and P(2 girls) $= \text{P(GG)} = \frac{1}{4}$

(i) List the outcomes and find the probabilities for a family of (a) 3 children, and (b) 4 children.

(ii) Copy the table and write your results from part (i) into the appropriate cells.

(iii) Find the pattern of probabilities and hence write in the probabilities for a family with 5 children.

No. of children	Number of girls					
	0	1	2	3	4	5
1	$\frac{1}{2}$	$\frac{1}{2}$				
2	$\frac{1}{4}$	$\frac{2}{4}$	$\frac{1}{4}$			
3						
4						
5						

Chapter

Trigonometry

Trigonometry is the study of triangles, and this includes the relationships between the lengths of their sides, their angles and their areas. Understanding triangles allows you to analyse many real situations.

Measuring lengths and angles

There are many ways of measuring lengths and angles.

- **Tape measures** and **rulers** are for direct measurement of lengths.
- **Calipers** grip the object to be measured.
- **Feeler gauges** are inserted into small gaps.

- **Microscopes** are used for measuring very small lengths. A typical microscope scale is in micrometres (previously called microns). One **micrometre** is one millionth of a metre.
- **Sonic tapes** and **radar** both use the time taken for a signal to return after bouncing off an object to measure its distance. Sonic tapes are used by estate agents for measuring rooms. Radar is used for longer distances (kilometres or more).
- Devices like the mileage indicator on a car which measure distance by counting how many times a wheel turns are called **odometers**.

- A **plumb line** is used for checking whether a line is vertical, a **spirit level** for horizontal. When used with a suitable scale these can also give angles to the vertical and horizontal.
- **Protractors** are used for measuring angles on paper.
- **Theodolites** (on land) and **sextants** (at sea) are used for measuring vertical angles. Sextants can also be used horizontally.
- Horizontal directions or bearings are measured by **compass**. A compass bearing is measured clockwise from North. A magnetic compasss measures from magnetic North, a gyroscope from true North.

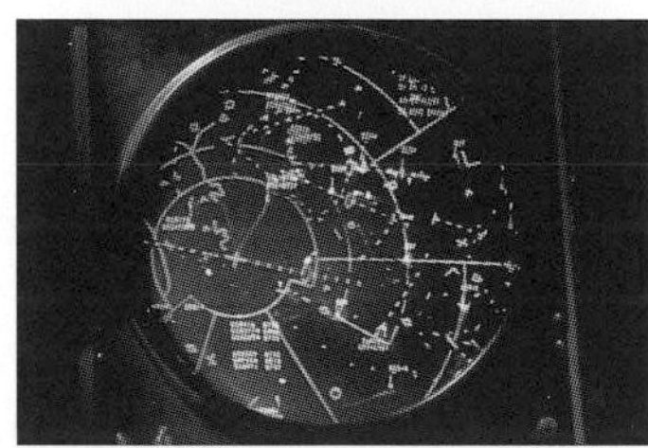

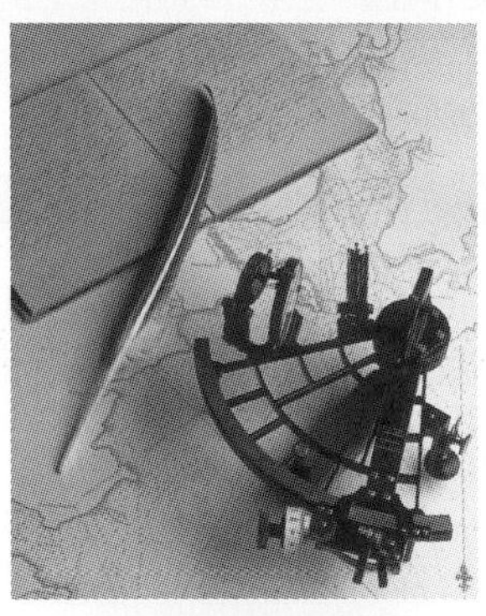

For discussion

For each of the items or situations illustrated below,

(i) decide which measuring instrument or instruments would be most appropriate;
(ii) decide whether the length would best be given in micrometres, millimetres, centimetres, metres or kilometres;
(iii) give an estimate of the length.

1. The dimensions indicated on this chest of drawers

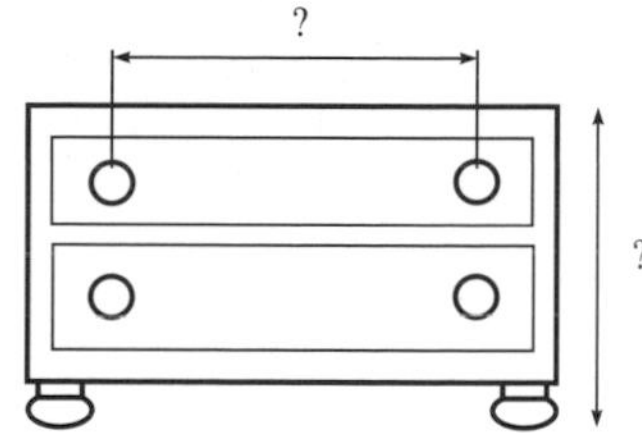

2. The length and thickness of a cheque card

3. The height of a house

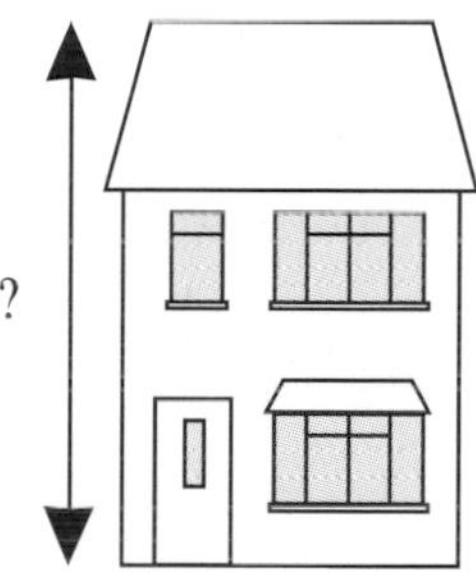

4. The diameter of a human hair

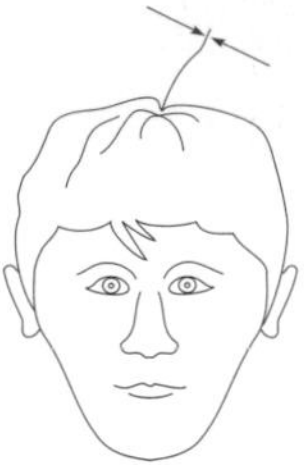

5. The length of a bacterium

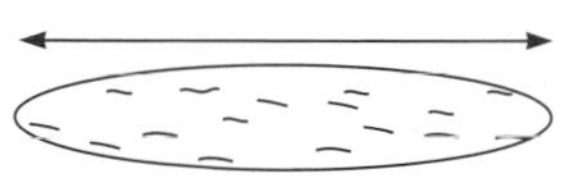

6. The gap on a spark plug

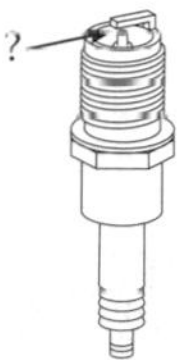

7. The span of your hand

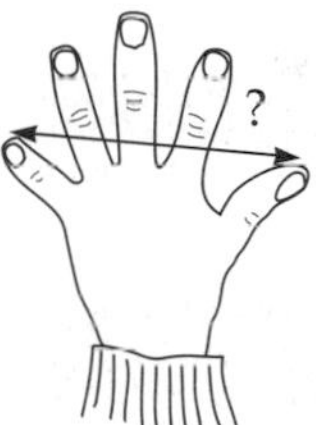

8. The distance from your home to Lands End

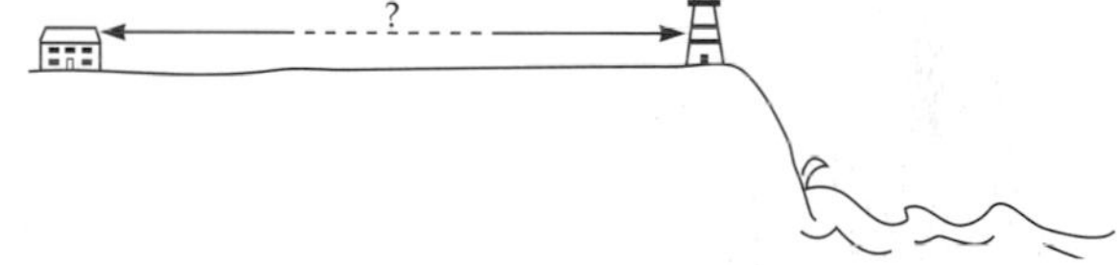

Investigation

Accuracy of a sonic tape

Sonic tapes are available from D.I.Y. shops. There are substantial variations in price, according to quality.

Investigate the accuracy of a cheap sonic tape by comparing its measurements with those obtained using a good quality steel tape measure.

Take measurements of several rooms of different sizes, with different amounts of furnishing.

Estate agents are now bound by the Property Misrepresentations Act when describing the houses they are selling. The level of accuracy required is not defined but a working rule is that a room's measurements should be accurate to $\frac{1}{10}$ of a metre. Does your sonic tape meet this requirement?

Triangles

A triangle is a 3-sided figure. In *any* triangle the three angles add up to 180°.

Some triangles have special names.

Equilateral triangle

- All three sides are equal;
- All three angles are 60°.

Isosceles triangle

- Two sides are equal;
- Two angles are equal.

Right-angled triangle

- One angle is 90°.
- The side opposite the right angle is called the hypotenuse.

Similar triangles

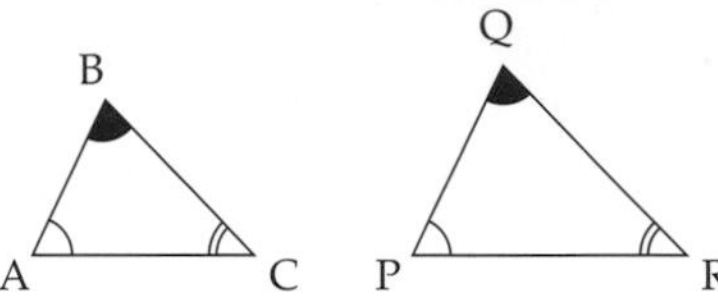

In the diagram above, ∠A = ∠P, ∠B = ∠Q and ∠C = ∠R. This means that triangles ABC and PQR are *similar* and that the ratio of corresponding sides is constant.

So $\frac{AB}{PQ} = \frac{BC}{QR} = \frac{AC}{PR}$.

Example

Triangles ABC and PQR are similar.

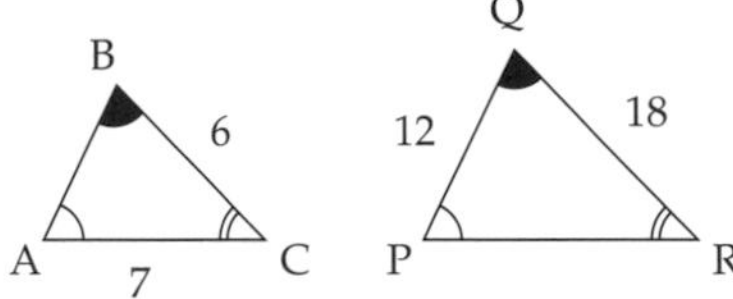

Find (i) AB and (ii) PR.

Solution

The ratio of one pair of corresponding sides, BC:QR, is known so you work with this. A good method is to write the ratio of sides starting with the side you want to find. (This will minimise the amount of manipulation required.)

(i) $\frac{AB}{PQ} = \frac{BC}{QR}$

So $\frac{AB}{12} = \frac{6}{18}$ giving $AB = 12 \times \frac{6}{18} = 4$.

(ii) $\frac{PR}{AC} = \frac{QR}{BC}$

So $\frac{PR}{7} = \frac{18}{6}$ giving $PR = 7 \times \frac{18}{6} = 21$.

Exercise

1. Find the unknown angles in the triangles below.

(i)

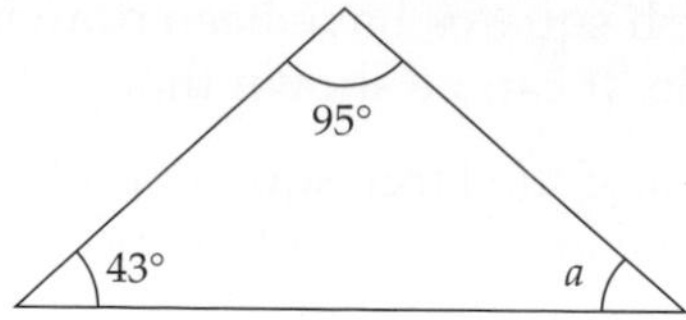

(ii)

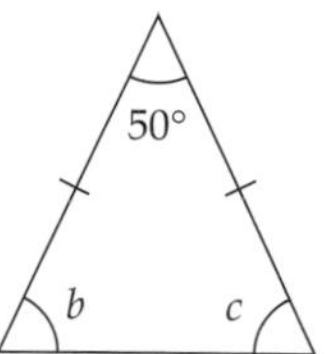

(iii)

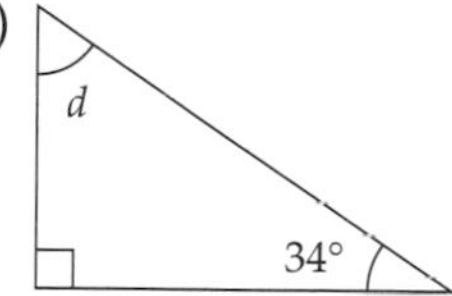

(iv)

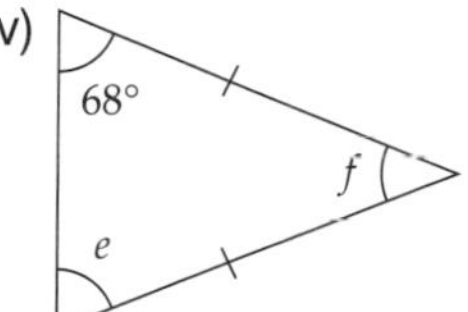

(v)

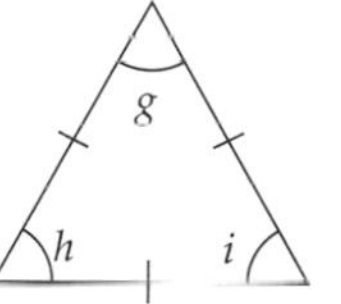

(vi)

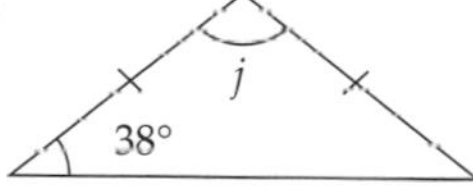

2. For each of these pairs of similar triangles, find the lengths marked with letters. All lengths are in centimetres.

(i)

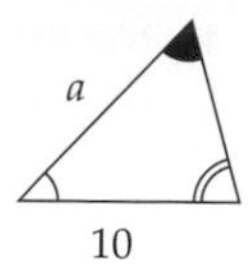

(ii)

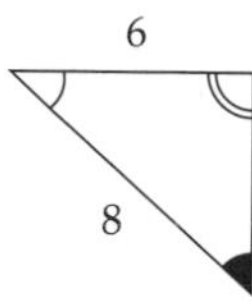

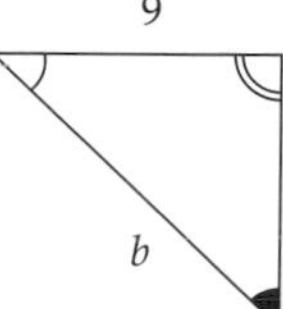

(iii)

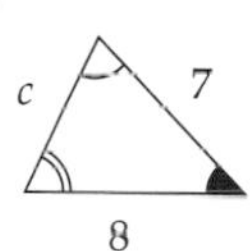

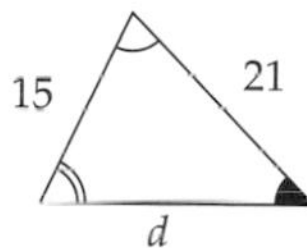

(iv)

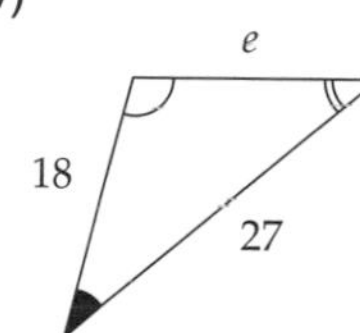

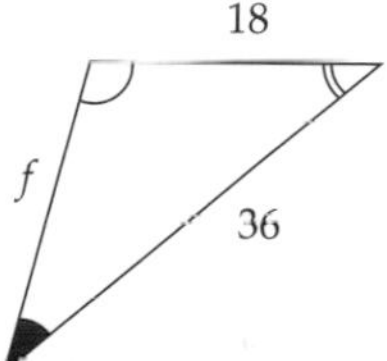

(v)

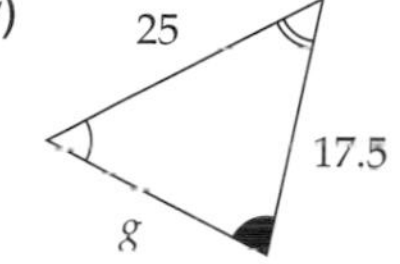

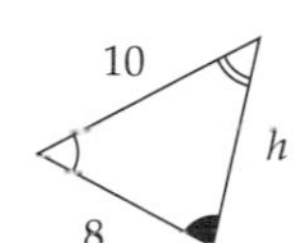

(vi)

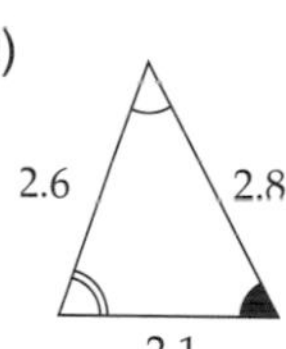

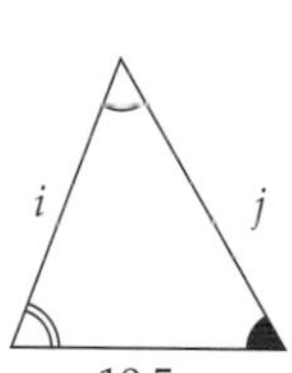

Pythagoras' Theorem

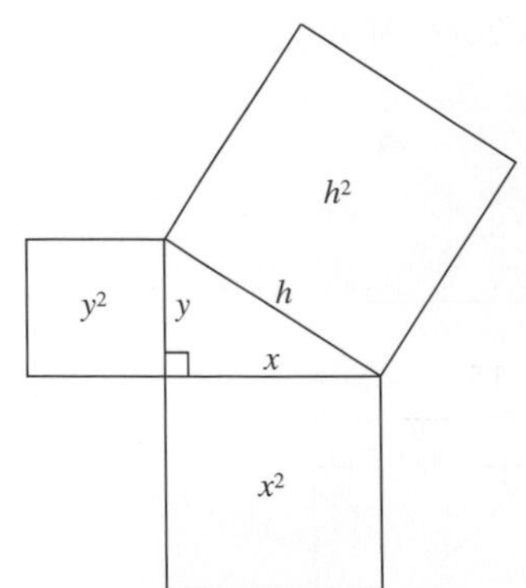

Look at the diagram on the left, in which squares have been drawn on the three sides of a right-angled triangle. It can be shown that

- for any right-angled triangle the area of the large square (on the hypotenuse) is equal to the sum of the areas of the other two.

This may be written using algebra as

$$h^2 = x^2 + y^2$$

This is Pythagoras' Theorem.

When you know the lengths of two sides of a right-angled triangle, you can use Pythagoras' Theorem to find the third. This is done in the next two examples.

Example

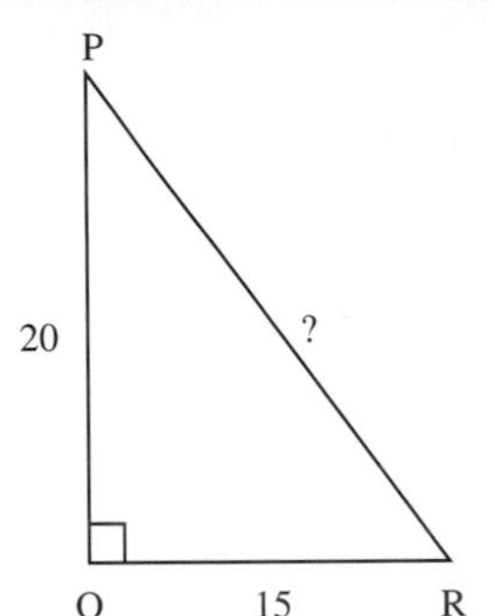

In Δ PQR, ∠ Q = 90°, PQ = 20 cm and QR = 15 cm.

Find the length of PR.

Solution

Since the right angle is at Q, PR is the hypotenuse.

$$PR^2 = PQ^2 + QR^2$$
$$= 20^2 + 15^2$$
$$= 625$$
$$PR = \sqrt{625} = 25$$

Notice that PQ means the length of the line PQ.

PR is 25 cm long.

Example

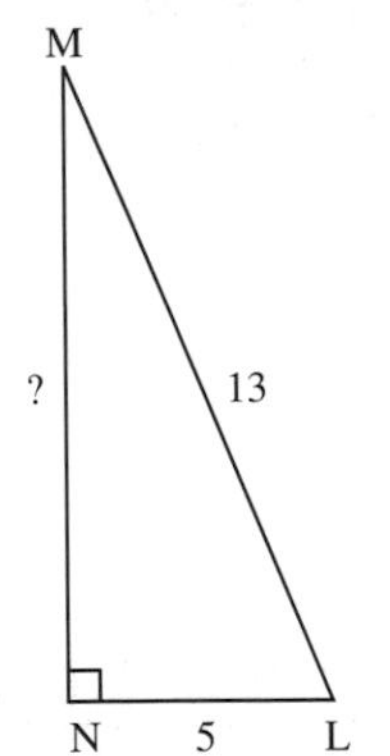

In Δ LMN, ∠ N = 90°, LM = 13 m and LN = 5 m.

Find the length of MN.

Solution

Since the right angle is at N, LM is the hypotenuse.

$$LM^2 = MN^2 + NL^2$$
$$13^2 = MN^2 + 5^2$$
$$\text{and so}\quad MN^2 = 13^2 - 5^2 = 169 - 25 = 144$$
$$MN = \sqrt{144} = 12$$

MN is 12 m long.

NOTE

The converse of Pythagoras is also true.

If in triangle ABC,
$AB^2 + BC^2 = AC^2$
then ∠ABC = 90°.

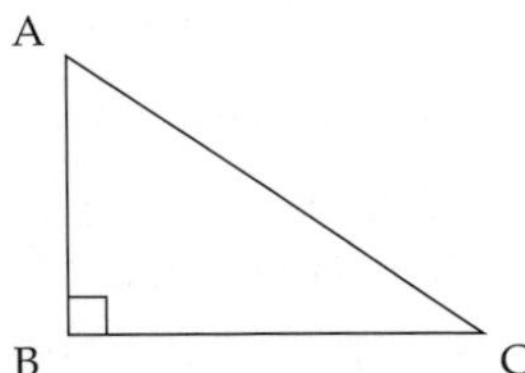

Exercise

In each of the following triangles find the length of the side indicated. Lengths are in centimetres.

1.

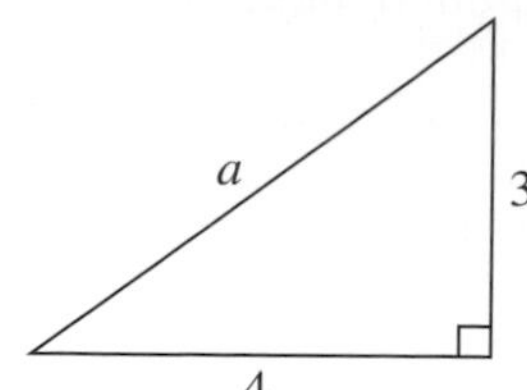

2.

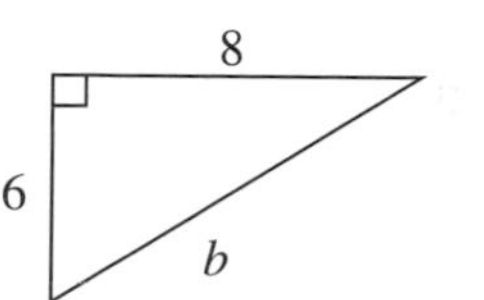

3.

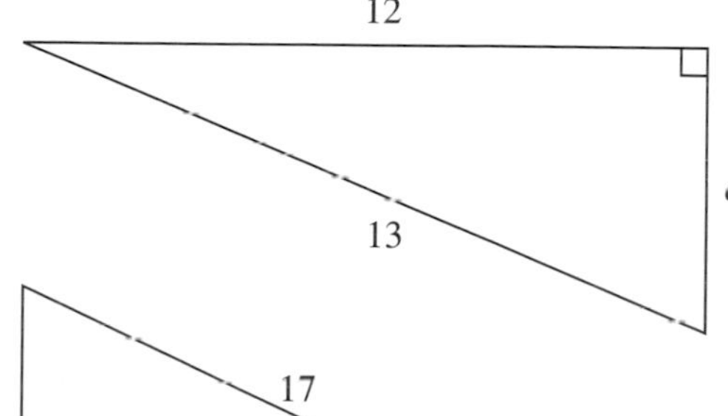

4.

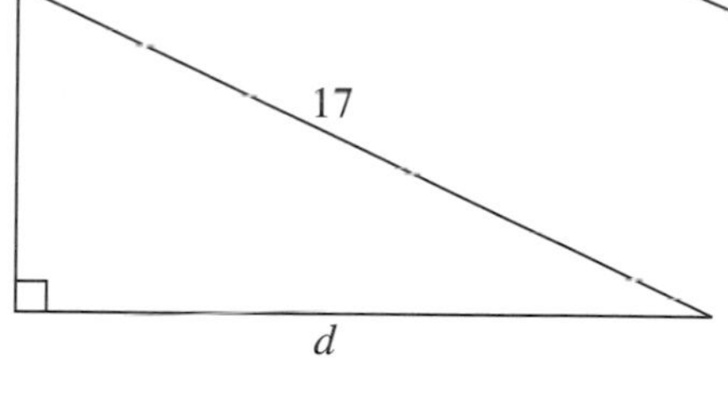

5.

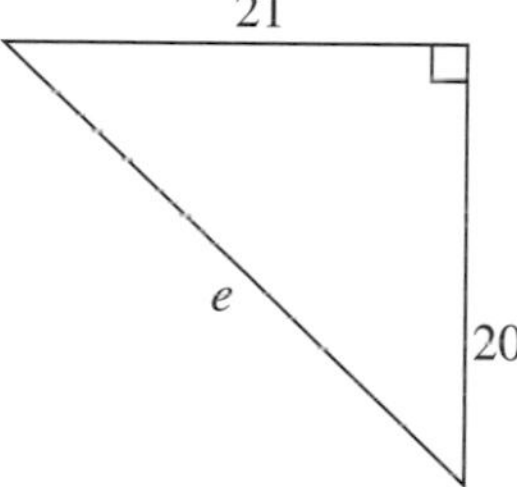

6.

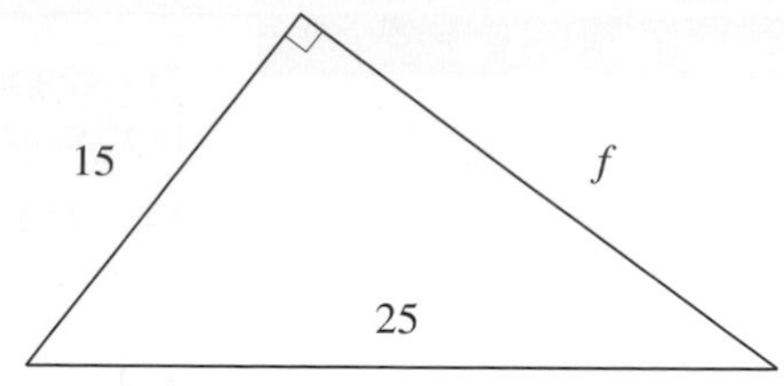

7.

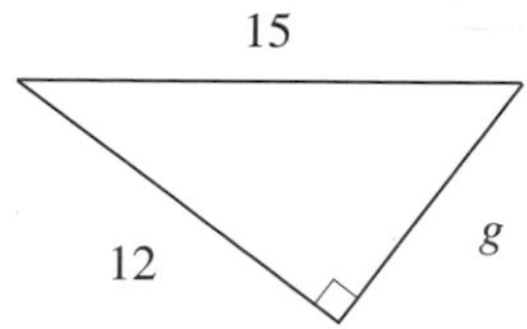

8.

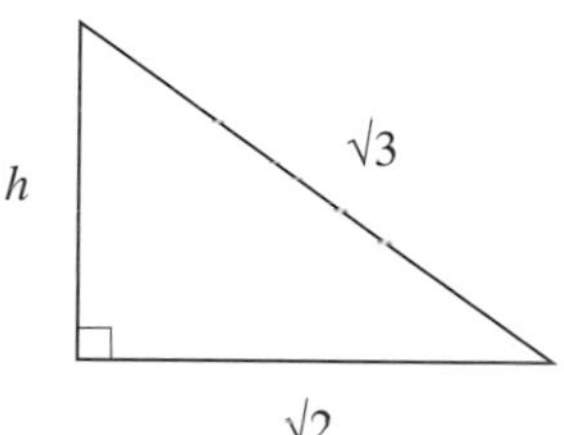

9.

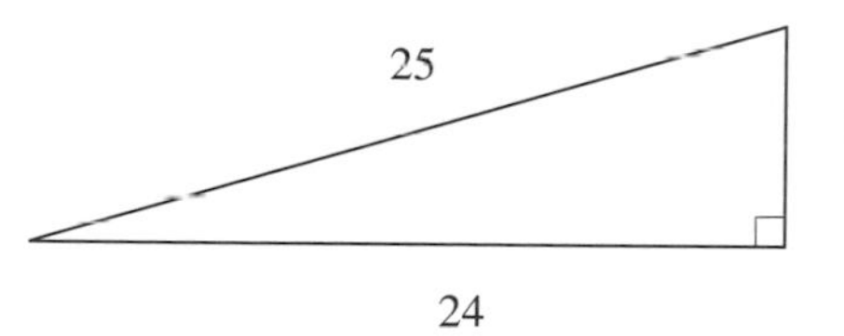

10. Which of the following triangles PQR are right-angled?
(i) PQ = 12, QR = 16 and PR = 20
(ii) PQ = 5, QR = 6 and PR = 8
(iii) PQ = 2.5, QR = 6 and PR = 6.5
(iv) PQ = 50, QR = 14 and PR = 48

Activity

Proving Pythagoras' Theorem (The Chinese method)

The diagram shows a square ABCD drawn inside a larger square PQRS. The lengths of the sides are shown.

(i) Show that the area of PQRS is $(x + y)^2$

The area of PQRS may also be found as
Area ABCD + Area of the 4 triangles (AQB etc.).

(ii) Show that the area of PQRS may also be written as

$$h^2 + 2xy$$

(iii) Hence show that $h^2 = x^2 + y^2$

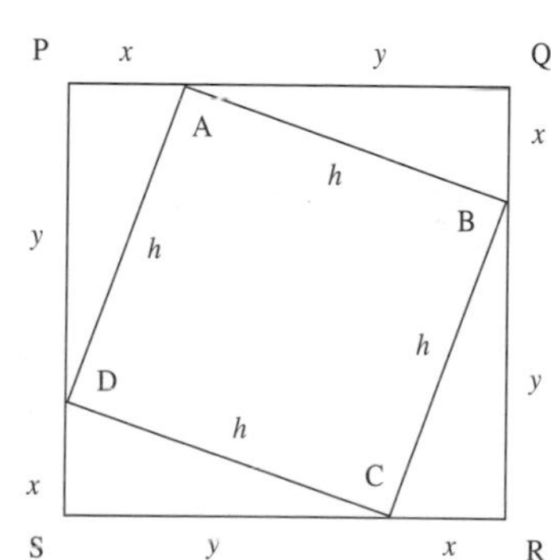

Pythagoras' Theorem: practical problems

Example

The diagram shows a house 15 m long and 8 m wide. The side walls are 6 m high and the ridge of the roof is 9 m above the ground. Find

(i) the slant length of the roof;
(ii) the area of the roof;
(iii) the cost of slating the roof at £20 per square metre.

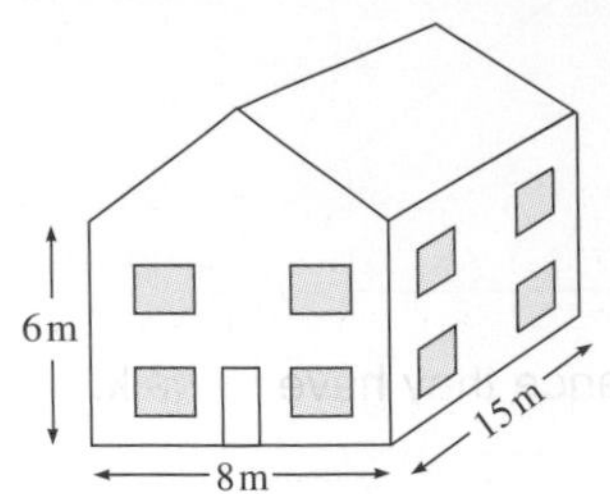

Solution

The diagram shows the gable end of the house, with a vertical line TNM drawn from the ridge of the roof.

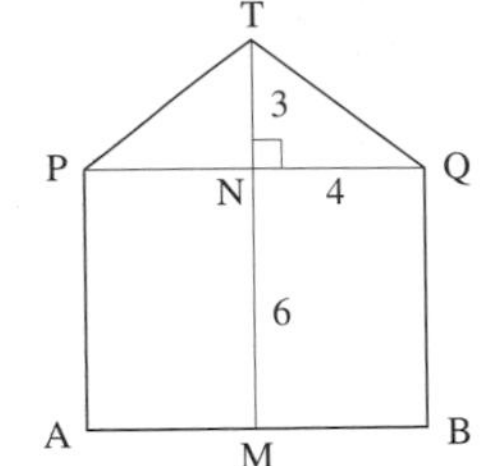

(i) Since TN is vertical and PQ is horizontal, $\angle$ TNQ is a right angle. Since Δ TPQ is isosceles with TP = TQ, the symmetry of the situation tells us that PN = NQ = 4. We also know that TN = 9 – 6 = 3.

By Pythagoras' Theorem in Δ TNQ,
$$TQ^2 = NQ^2 + TN^2$$
$$= 4^2 + 3^2 = 25$$
and so $TQ = \sqrt{25} = 5$

The slant length, of the roof is 5 metres.

Notice how the isosceles triangle is split into two right angled triangles.

(ii) The roof consists of 2 rectangles, each 15 m × 5 m

Area = 2 × 15 × 5 square metres = 150 square metres.

(iii) The cost is 150 × £20 = £3000.

Example

The diagram shows the frame of a roof-beam used in a large warehouse. The line BC is horizontal, and dimensions are in metres. How high is A above BC?

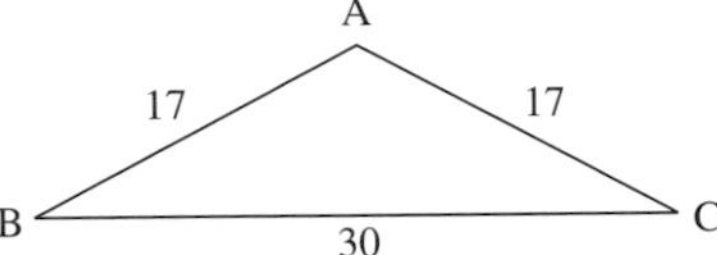

Solution

The diagram on the left shows the vertical line AM from A onto BC. Since BC is horizontal, $\angle$ AMC = 90°. We want to find the length of AM.

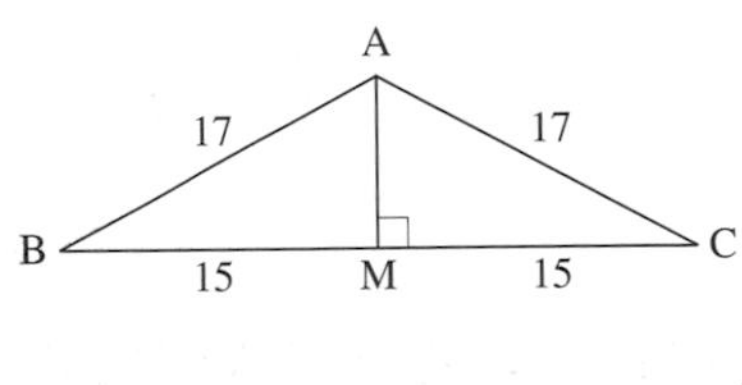

Since Δ ABC is isosceles, the situation is symmetrical, so

$$MC = \tfrac{1}{2} \times 30 = 15.$$

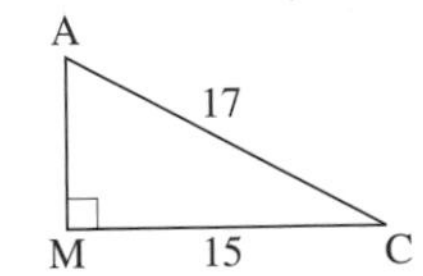

By Pythagoras' Theorem in Δ AMC,
$$AC^2 = AM^2 + MC^2$$
$$17^2 = AM^2 + 15^2$$
$$AM^2 = 289 - 225$$
and so $AM = \sqrt{64} = 8.$

AM is 8 m, and this is the height of A above BC.

Exercise

1. A ship is anchored in 40 fathoms of water. The length of the anchor chain is 104 fathoms and it is taut. How many yards from the ship is the anchor? $(1\,\text{f} = 2\,\text{yd})$

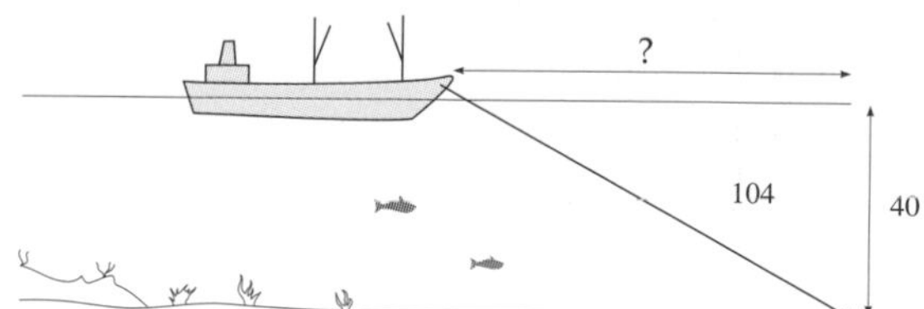

2. A radio mast is secured by a wire 25 m long which goes from the top to a point 7 m from the base of the mast. How high is the mast?

3. The diagram shows part of a roof-truss. The horizontal part, AB, is 16 m long and VA and VB are both 10 m long. How far above AB is V?

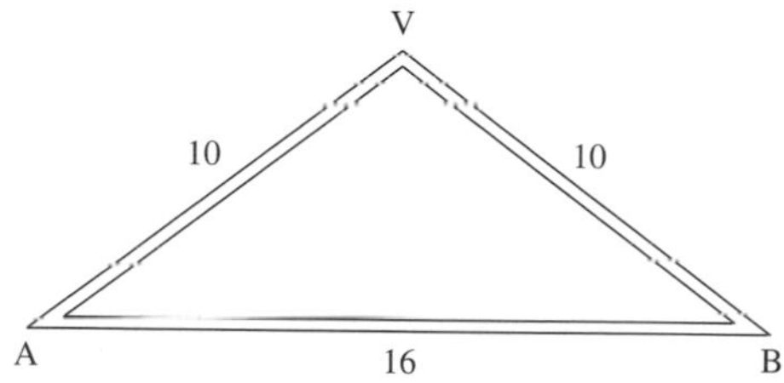

4. The first diagram shows an elastic string stretched horizontally between two points A and B, 12 cm apart. In the second diagram the middle point of the string is pulled down 2 cm. By how much has the length of the string been extended?

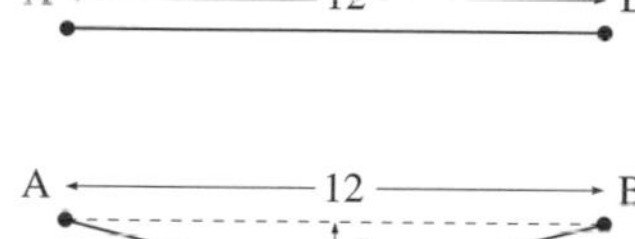

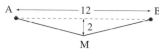

5. A box is 12 cm long, 9 cm wide and 8 cm high. Find the length of the longest nail which will fit into it
   (i) lying flat on the bottom;
   (ii) with one end at a bottom corner, the other at the opposite top corner.

6. On a rough diagram, plot the points A and B and then work out the distance AB.
   (i) A(4,2), B(1,6)
   (ii) A(–5, –1), B(7,4)

7. A party of hikers is at point A on the map and needs to go to point B. However they choose first to look at an old church at C. Use Pythagoras' Theorem to calculate the extra distance they have to walk.

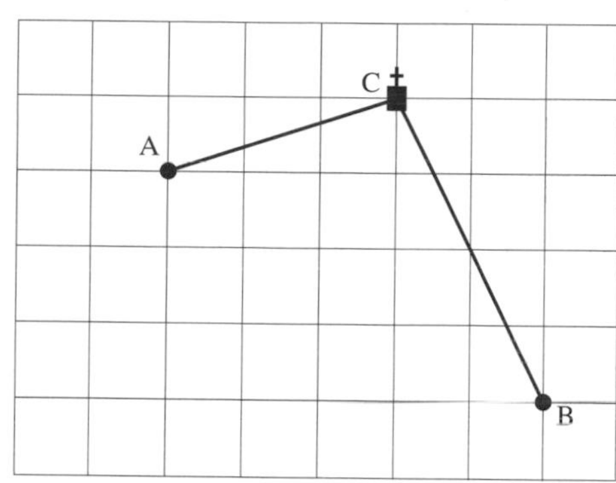

grid squares 1 km × 1 km

8. The diagram shows a dog-leg hole on a golf course. The hole begins at the tee, T, and ends at the flag, F. Most players aim to play their first shot due East along the line TP and then approach the flag with another shot. The distance TP is 240 m and PF is 70 m. The point P is due north of F. Ella plays her ball 200 m along the line TP, ending at A.
   (i) How far is Ella's ball from the flag?
   (ii) How far is a direct shot from T to F? Why would players be reluctant to try this shot even if they were strong hitters?

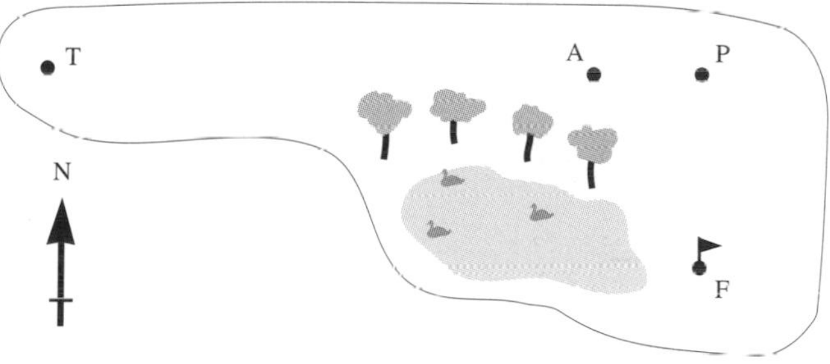

Investigation

The diagram shows a box in which you might buy sandwiches in a supermarket or on a train. What does this tell you about the size and shape of sandwich bread? How does it differ from ordinary bread?

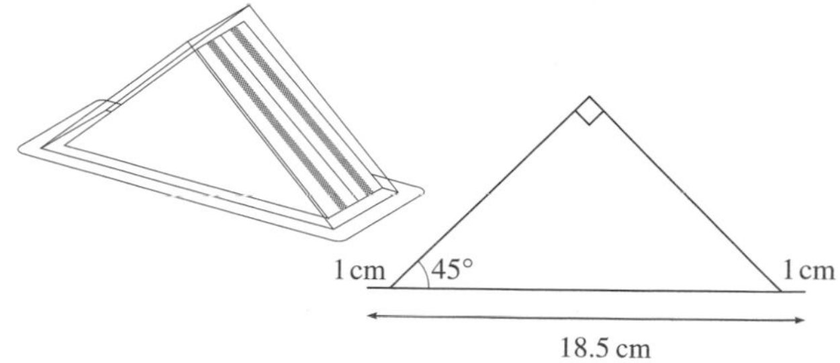

Sine, cosine and tangent

The right-angled triangles below are similar.

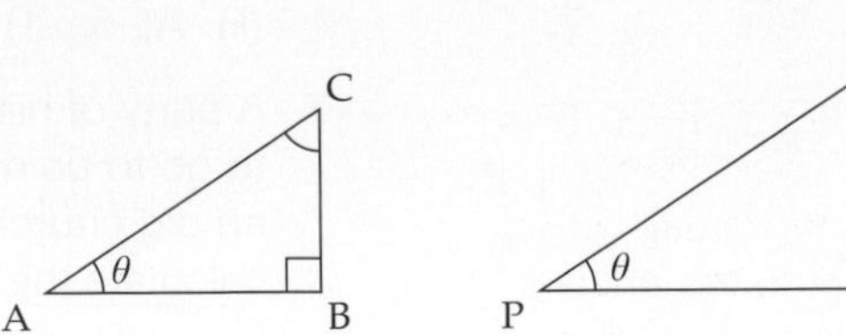

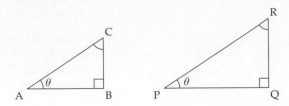

Using the result about the ratio of corresponding sides (see page 140) gives

$$\frac{BC}{QR} = \frac{AC}{PR}, \quad \frac{AB}{PQ} = \frac{AC}{PR} \text{ and } \frac{BC}{QR} = \frac{AB}{PQ}.$$

Each of these can be rearranged to give

$$\frac{BC}{AC} = \frac{QR}{PR}, \quad \frac{AB}{AC} = \frac{PQ}{PR} \text{ and } \frac{BC}{AB} = \frac{QR}{PQ}.$$

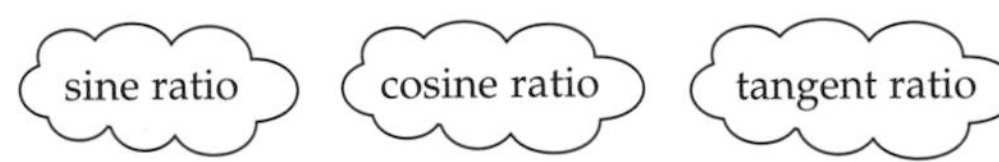

These facts make it possible to define three important ratios in right-angled triangles. First, though, you need a way of identifying the sides of a right-angled triangle. The triangles below are all similar (although their orientations are different), and one of the angles is marked θ. Their three sides are marked as follows

O : the side **O**pposite the angle θ
H : the **H**ypotenuse
A : the third side (**A**djacent)

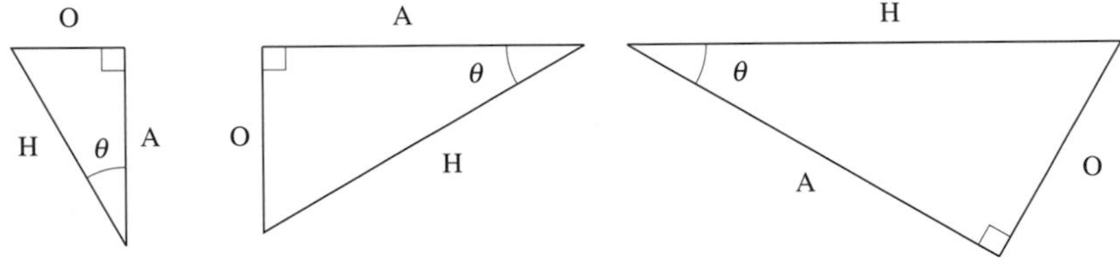

The three important ratios are $\frac{O}{H}$, $\frac{A}{H}$ and $\frac{O}{A}$.

Because the triangles are similar, each ratio has the same value in all of the triangles. The ratios are given special names *sine*, *cosine* and *tangent*, shortened to sin, cos and tan.

- $\frac{O}{H} = \sin\theta$
- $\frac{A}{H} = \cos\theta$
- $\frac{O}{A} = \tan\theta$

In the triangles above, $\theta = 30°$. If you measure the sides you will find that $\sin 30° = 0.5$, $\cos 30° = 0.87$ and $\tan 30° = 0.58$ (to 2 significant figures).

Activity

Using your calculator

Different calculators have different commands for finding sin, cos and tan. Here are two typical key sequences for finding $\tan 60°$.

(i) [tan] [60] [enter] (ii) [60] [tan] [rtn]

Make sure you know how to find $\tan 60°$ on your calculator. You should come out with the answer 1.732...(Your calculator must be in degree mode (deg), not radian (rad) or grade (grad).)

If you know that $\sin\theta = \frac{1}{2}$ and you want to find the angle θ, you need arcsin $\frac{1}{2}$, or 'the angle whose sin is $\frac{1}{2}$'. Again the key sequence for this varies with the model of calculator: it may be [inv] [sin] [0.5] or [arcsin] [0.5] or even [$\sin^{-1}$] [0.5]. Again, make sure you can find arcsin $\frac{1}{2}$ on your calculator: you should get $30°$.

Exercise

1. (*Do not use your calculator in this question.*) For each of the triangles below, measure the angles to the nearest degree and also the three sides.

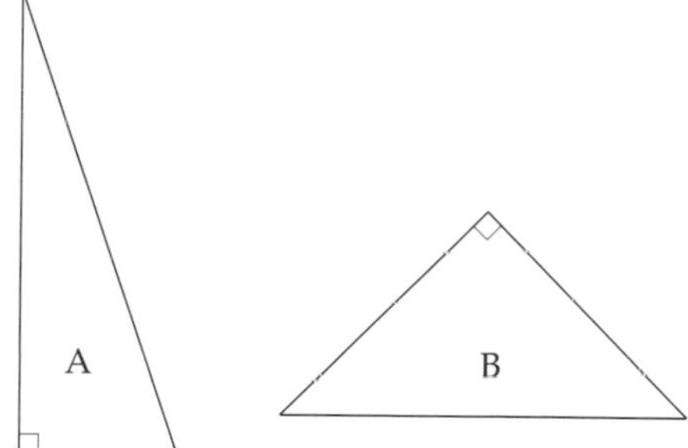

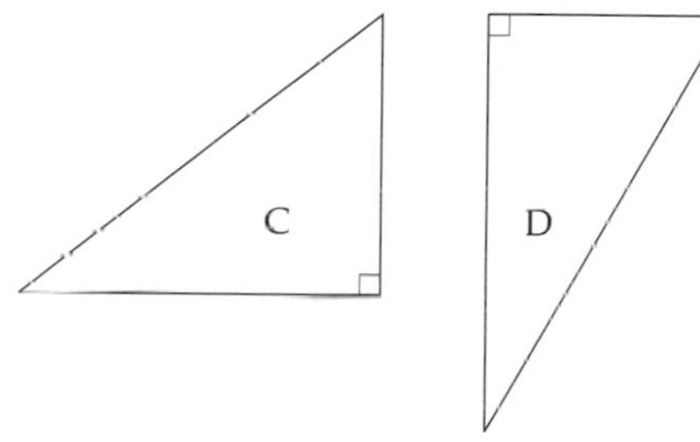

Using this information, write down approximate values for

(i) $\tan 52°$ (ii) $\cos 45°$ (iii) $\sin 30°$
(iv) $\tan 45°$ (v) $\sin 71°$

2. Use your calculator to find the values of
   (i) $\tan 52°$
   (ii) $\cos 45°$
   (iii) $\sin 30°$
   (iv) $\tan 45°$
   (v) $\sin 71°$

   Compare your answers with those in question **1**.

3. Use your calculator to find the angles $a - j$ (to the nearest tenth of a degree) such that
   (i) $\sin a = 0.5$
   (ii) $\cos b = 0.707$
   (iii) $\tan c = 0.6241$
   (iv) $\sin d = \frac{6}{17}$
   (v) $\cos e = \frac{3}{5}$
   (vi) $\sin f = \frac{1}{\sqrt{2}}$
   (vii) $\tan g = 2.5$
   (viii) $\cos h = \frac{\sqrt{3}}{2}$
   (ix) $\cos i = \frac{\sqrt{3}}{\sqrt{5}}$
   (x) $\tan j = 0$

Activities

1. (i) Draw the graph of $y = \sin x$ for values of x from $0°$ to $90°$. Use a scale of 1 cm for $10°$ for x and 10 cm for 1 unit for y.
   (ii) On the same sheet of graph paper draw the graph of $y = \cos x$ for values of x from $0°$ to $90°$.
   (iii) What is the largest value that $\sin x$ or $\cos x$ takes in this region?
   (iv) Are the two curves related to each other?
   (v) What happens when you try to draw the graph of $y = \tan x$ for values of x from 0 to $90°$?

2. Some people use the 'word' SOHCAHTOA to help them remember the definitions

$$\sin\theta = \frac{\text{O}}{\text{H}}, \quad \cos\theta = \frac{\text{A}}{\text{H}}, \quad \tan\theta = \frac{\text{O}}{\text{A}}.$$

Make up a sentence in which the first letter of each word gives you the same information. You might start '*Some old...*'

Finding angles in right-angled triangles

The ratios sine, cosine and tangent allow you to calculate the angles in a right-angled triangle if you know the lengths of two of its sides. The next two examples show you how this is done.

Example

In Δ LMN, LM = 11 cm, LN = 5 cm, ∠ N = 90°. Find ∠ M.

Solution

First draw the triangle (as shown on the left) and mark the sides O, A, and H such that O is opposite the angle you want to find.

From the diagram, the two sides given are O and H, so you use sin.

$$\sin M = \frac{\text{O}}{\text{H}} = \frac{5}{11}$$

$$\Rightarrow M = \arcsin\left(\tfrac{5}{11}\right)$$

arcsin, invsin, $\sin^{-1}$ all mean 'the angle whose sin is…'

Using a calculator, $\arcsin\left(\frac{5}{11}\right) = 27.0°$, so ∠ LMN is 27°.

Example

Samantha is rowing a boat, and looking at Rosie who is waving at her from the cliff-top. The distances involved are shown in the diagram.

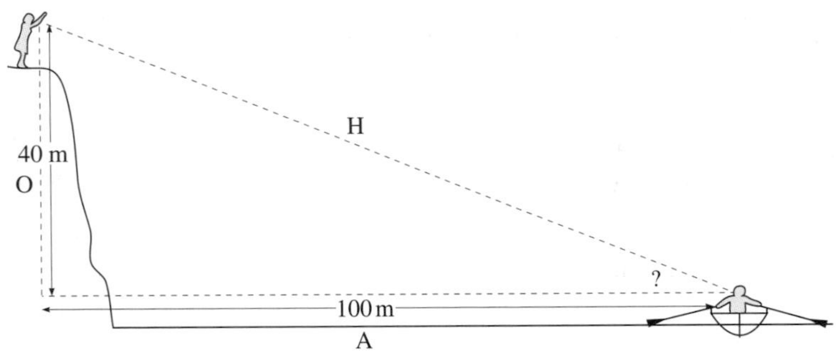

Find the angle between Samantha's line of vision and the horizontal.

Solution

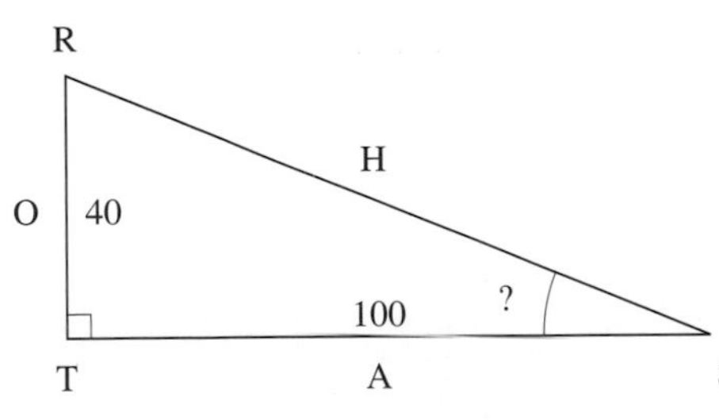

In the diagram on the left, R and S are the positions of Rosie's eye and Samantha's eye. The point T is directly below R and at the same horizontal level as S, so ∠ RTS = 90°.

The sides O, A and H are marked. Since the lengths of sides O and A are given you use tan:

$$\tan S = \frac{40}{100}$$

$$\Rightarrow S = \arctan 0.4 = 21.8°$$

$$S = 21.8°$$

NOTE

In situations like this the angle S is called the angle of elevation. *Rosie has to look down to see Samantha so the angle between Rosie's line of vision and the horizontal is called the* angle of depression. *Notice that they are equal.*

Exercise

Use trigonometry to find the angles $a - l$ in the triangles below. Lengths are in centimetres.

1.

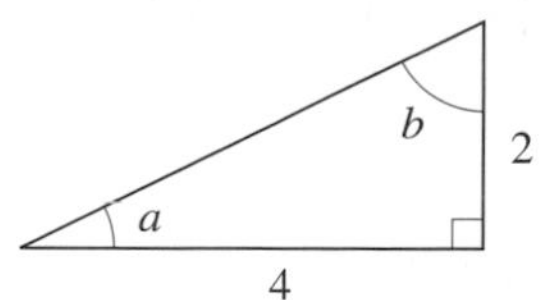

2.

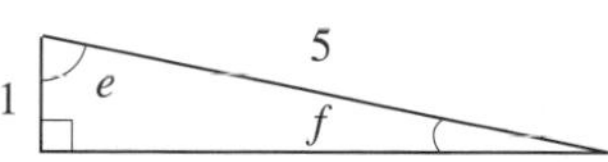

3.

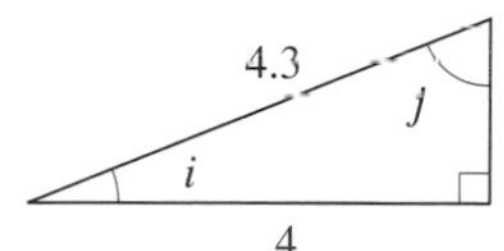

4.

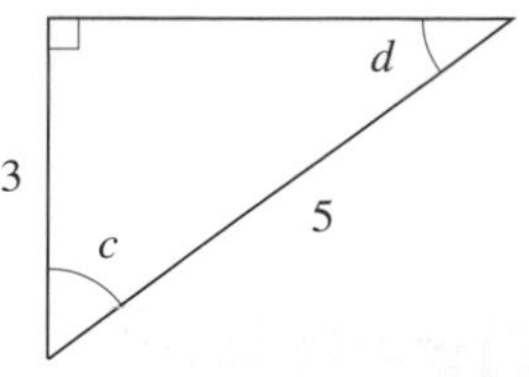

5.

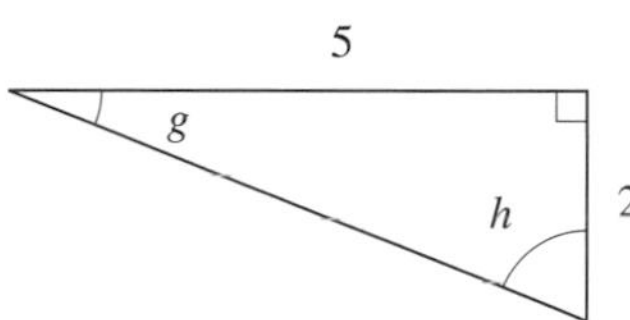

6.

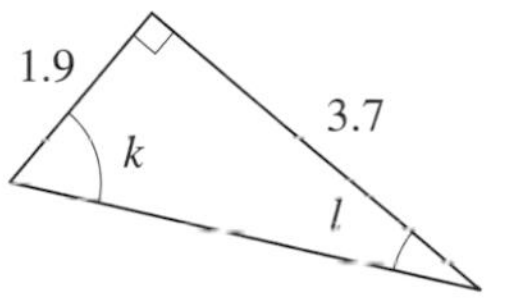

Activities

Gradients

The gradient of a line is a measure of its steepness. Look at the diagram below.

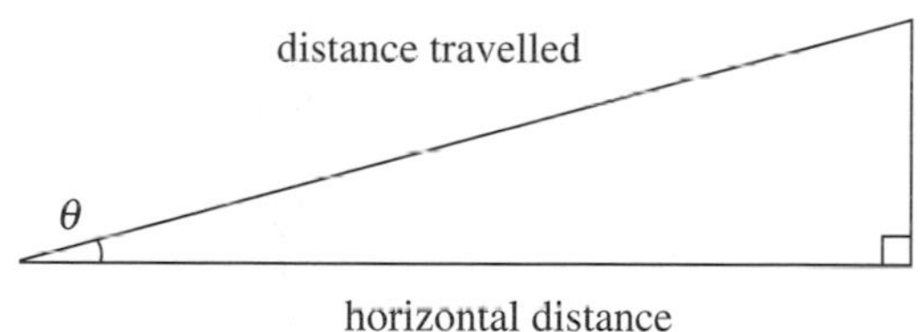

In mathematics,

$$\text{gradient} = \frac{\text{height gained}}{\text{horizontal distance}} = \tan\theta.$$

1. Find the angle, θ, between the horizontal and lines with gradients
   (i) $\frac{1}{2}$ (ii) 1 (iii) 2

On roads and railways,

$$\text{gradient} = \frac{\text{height gained}}{\text{travelled distance}} = \sin\theta.$$

2. Find the angle, θ, between the horizontal and roads or railways with gradients
   (i) $\frac{1}{250}$ (ii) 1 in 10 (iii) 5%

3. A road with gradient 1 in 8 is steep. What is the difference in the angle θ in this case using the two methods of calculation?

4. What is the meaning of (a) a zero gradient (b) a negative gradient?

5. The diagram below shows a map on which the vertical interval between contour lines is 50 m. A hiker walks in a straight line from P to Q. Estimate the gradient at the steepest part of his journey.

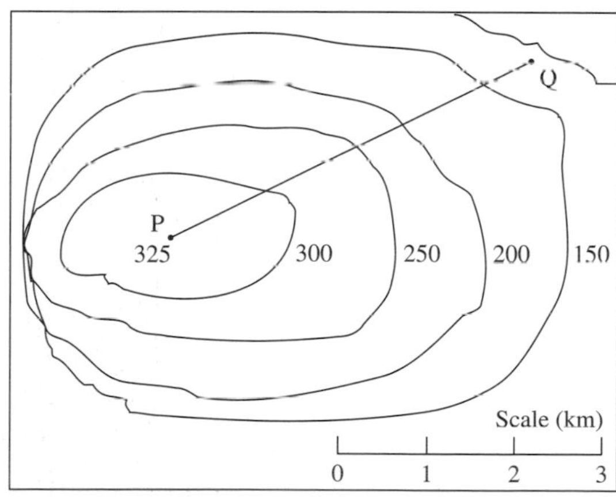

Finding sides in right-angled triangles

The ratios sin, cos and tan may also be used to find the lengths of sides in right-angled triangles, as in the following examples.

Example

Find the length of PQ in triangle PQR shown on the left.

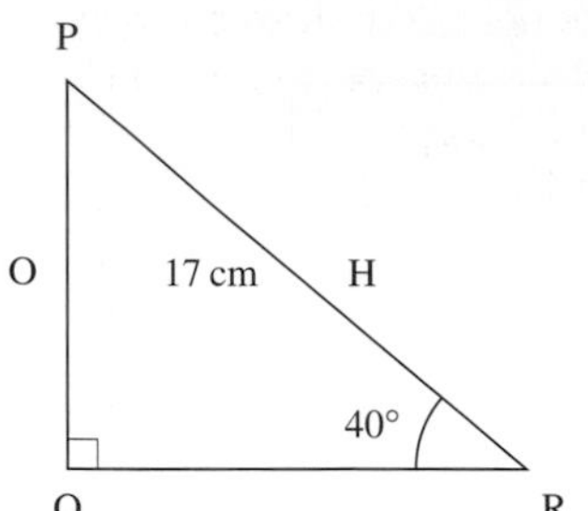

Solution

The side PQ is **O**pposite the given angle of 40°, and the side PR is the **H**ypotenuse:

$$\sin 40^\circ = \frac{O}{H} = \frac{PQ}{17}$$

Rearranging gives $PQ = 17 \sin 40^\circ$

The length of PQ is 10.9 cm.

Example

Find the length of LN in the triangle LMN shown on the left.

Solution

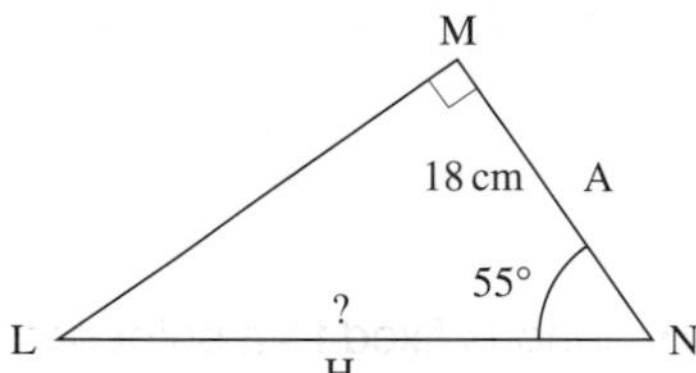

The side MN is **A**djacent to the given angle of 55° and LN is the **H**ypotenuse:

$$\cos 55^\circ = \frac{A}{H} = \frac{18}{LN}$$

Rearranging gives $LN = \dfrac{18}{\cos 55^\circ}$

The length of LN is 31.4 cm.

Example

A leisure company is planning an artificial ski slope. The initial design is represented by the triangle XYZ (left). The available land allows the horizontal distance, YZ, to have length 200 m; the slope is to make an angle of 15° (angle XYZ) with the horizontal. How high will the top of the slope be?

Solution

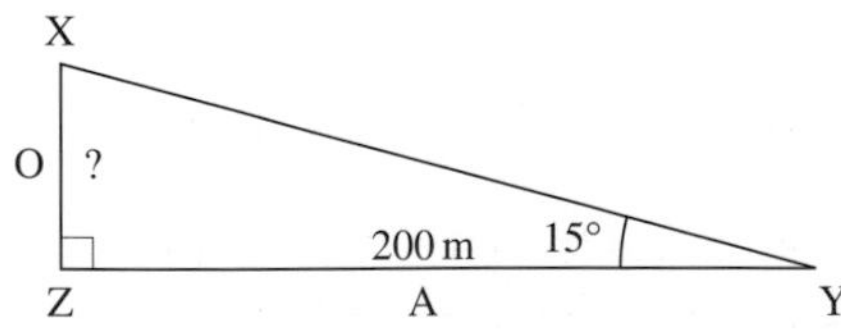

The height is XZ which is **O**pposite the given angle, and YZ (200 m) is **A**djacent:

$$\tan 15^\circ = \frac{O}{A} = \frac{XZ}{200}$$

Rearranging gives $XZ = 200 \tan 15^\circ$

$XZ = 53.6$ m.

The height is 53.6 m.

Exercise

In questions **1** to **10** find the side marked x.

1.

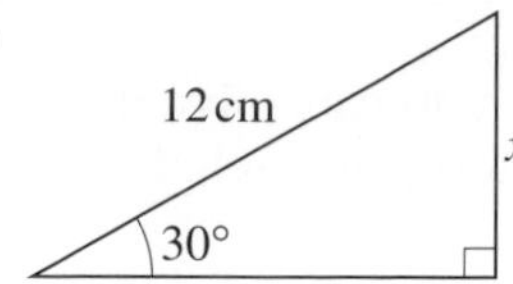

2.

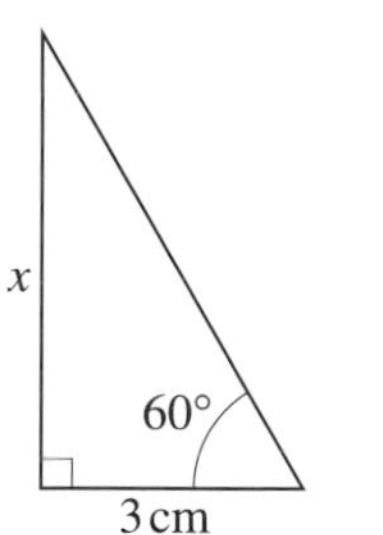

3.

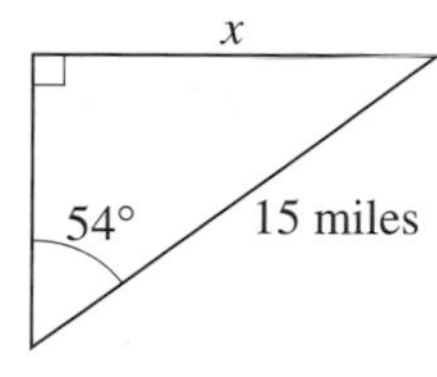

4.

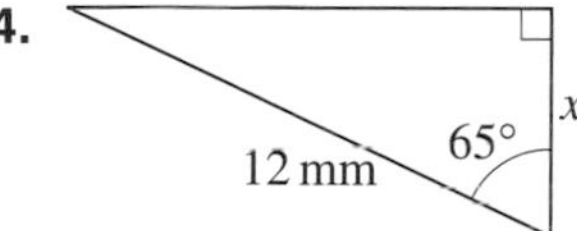

5.

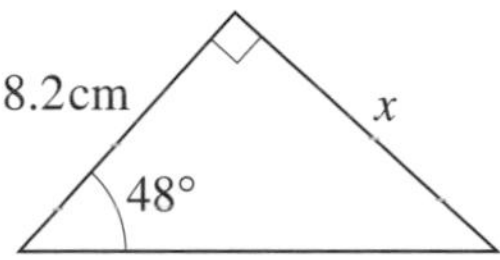

6.

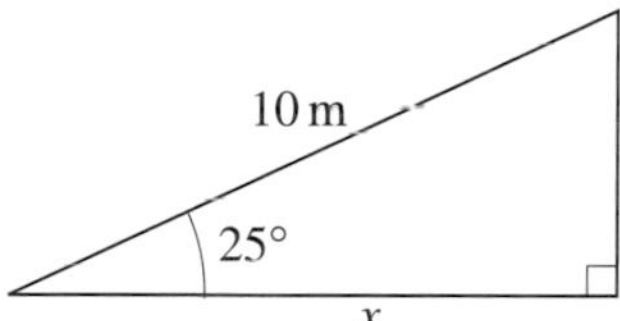

7.

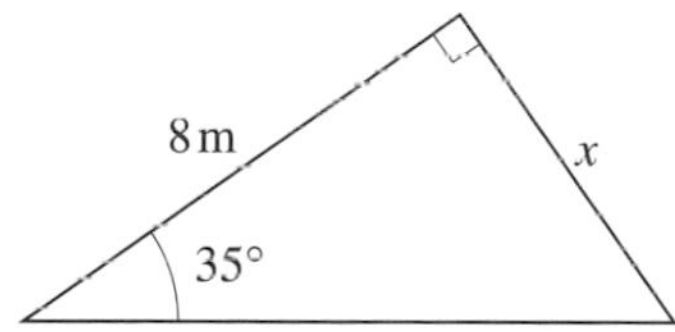

8.

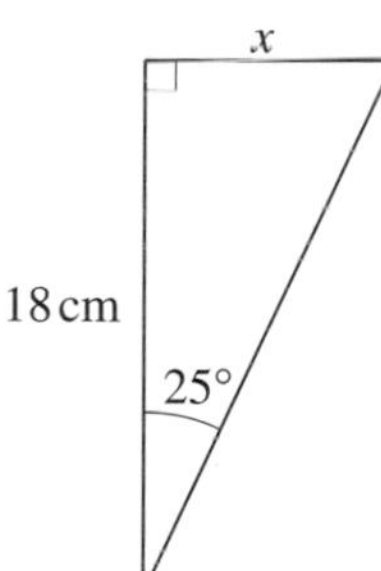

9.

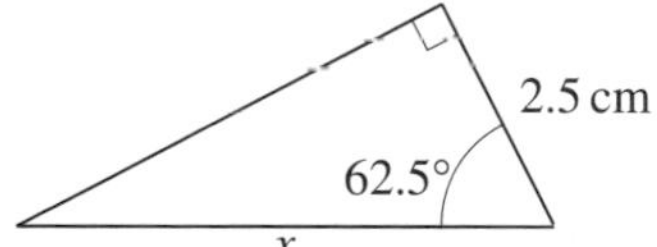

10.

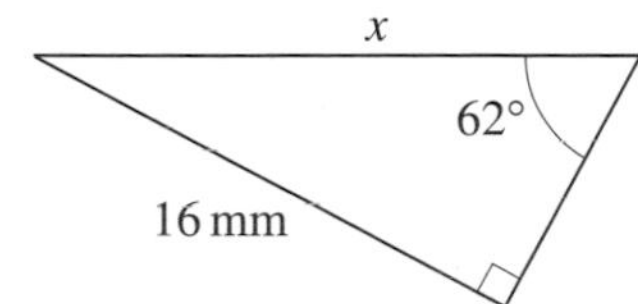

11. The diagram shows a chisel. What is the angle θ?

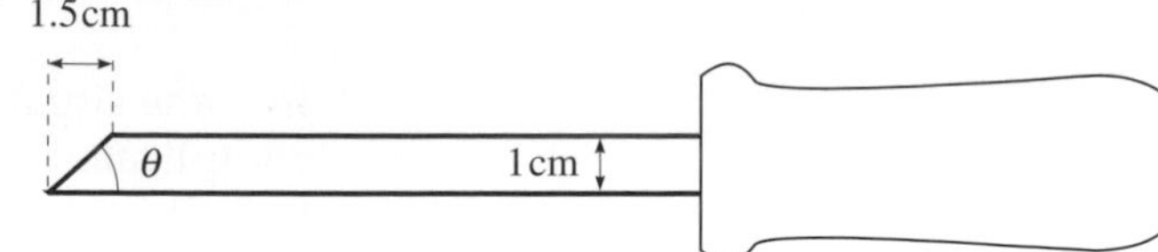

12. A surveyor observes the top of a building from a horizontal distance of 45 m. The angle of elevation is 12.4°. The observations are taken from a height of 1 m. The ground between the surveyor and the building is level. How high is the building?

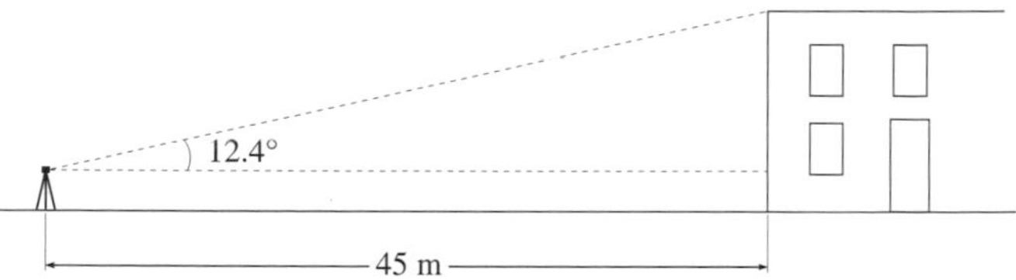

13. A boat leaves a harbour on a part of the coast which runs north-south, and sails on a bearing of 048°. How far away from the coast is the boat when it has travelled 5 km?

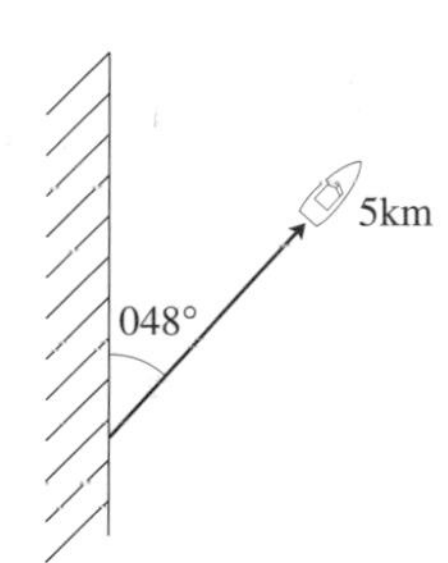

14. One end of a rope 40 m long is fixed to a point near the top of a tree. The rope is then pulled tight and the other end is secured to a point on the ground (which is level). The rope makes an angle of 11° with the horizontal. How high up the tree is it fastened?

15.

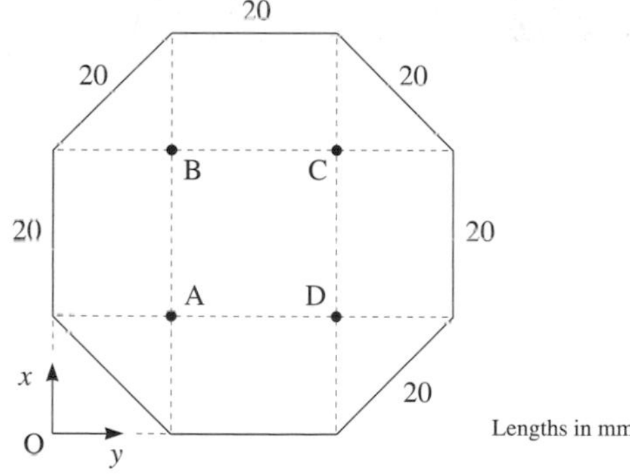

The object in the diagram is to be cut from a sheet of metal. It will then be used in the manufacture of a radio. The cutting is to be done by a robot which needs to have the co-ordinates of all 8 corners programmed into it, and also the positions of the four points A, B, C and D, where holes must be drilled. Find the necessary information, taking the point O as the origin.

Analysing more complicated situations

In some situations, like that in the example below, you can solve more complicated problems by dividing the diagram into right-angled triangles.

Example

A derrick on a ship consists of a mast MT and a jib MJ, both of length 8 m, hinged at M. When not in use the jib is kept in a vertical position alongside the mast as shown on the left. The diagram on the right shows the jib in use, about to lift a load at L, 5 m from M.

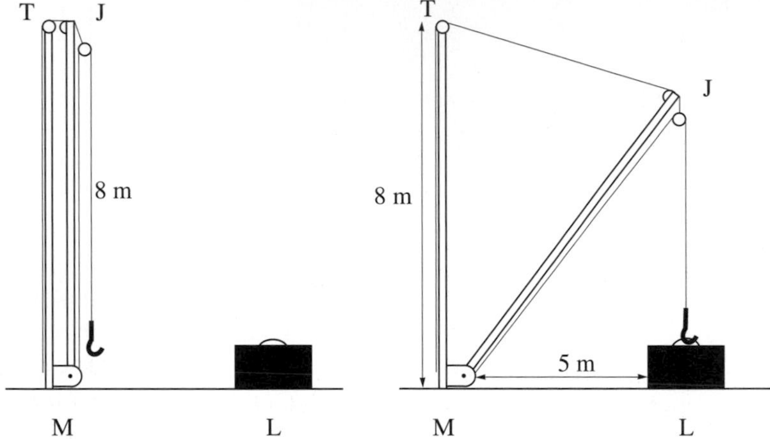

T, J, 8, 8, β, α, M, 5, L

In the second position, find:

(i) the angle TMJ (the angle the jib makes with the vertical);

(ii) the length TJ (the length of rope that had to be let out for J to be vertically above L).

Solution

This situation is represented by the diagram on the left.

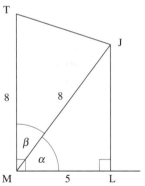

(i) In triangle MLJ

$$\cos\alpha = \tfrac{5}{8} \Rightarrow \alpha = 51.3^\circ$$

So $\angle \text{TMJ} = \beta = 90^\circ - \alpha$

$$= 38.7^\circ$$

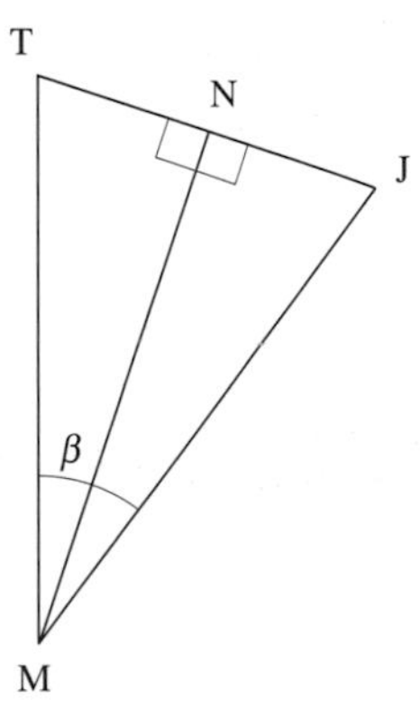

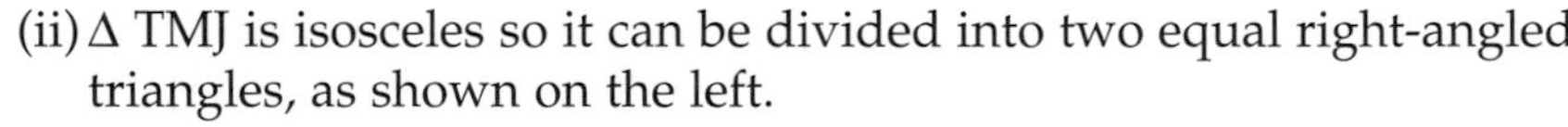
(ii) Δ TMJ is isosceles so it can be divided into two equal right-angled triangles, as shown on the left.

$$\text{Angle JMN} = \tfrac{1}{2}\beta = 19.4^\circ.$$

$$\sin 19.4^\circ = \frac{\text{NJ}}{8}$$

$$\Rightarrow \text{NJ} = 2.65$$

$$\text{TJ} = 2 \times 2.65 = 5.30 \text{ metres.}$$

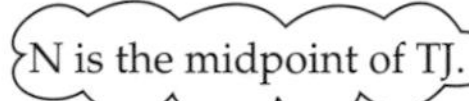

The jib makes an angle of about 39° with the vertical. A length of 5.3 m of rope has been let out.

Exercise

1. The diagram shows a stepladder. Each side is 1.5 m long. The sides are prevented from slipping outwards by a string 0.6 m long attached 1 m from the top of each side. Find
   (i) The angle each leg makes with the vertical.
   (ii) The height of the top above the ground.
   (iii) The distance between the feet of the ladder.

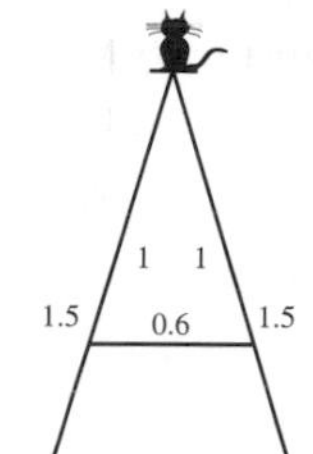

2. Two girls are hiking on moorland. At one point they separate, and Sue walks at $5\,\mathrm{km\,h^{-1}}$ due north, Vronnie at $5\,\mathrm{km\,h^{-1}}$ on 070°.
   (i) How far apart are they after 2 hours?

   Sue then walks due east, still at the same speed. Vronnie has a sleep and then walks due north also at the same speed, and the two arrive at the same point at the same time.
   (ii) How far does Sue walk before she meets Vronnie?
   (iii) For how long does Vronnie sleep?

3. In the diagram, PQRS is a symmetrical kite. $\angle QPS = 100°$, $\angle QRS = 40°$, and $QS = 50\,\mathrm{cm}$.

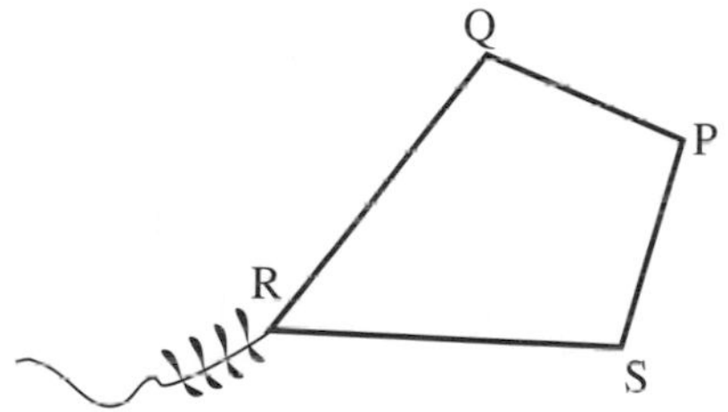

   (i) Copy the diagram and join PR and QS.
   (ii) Find $\angle PQS$ and $\angle RQS$.
   (iii) Find PR.
   (iv) Find the area of the kite.

4. The diagram shows a girder suspended horizontally by two ropes of lengths 5 m and 8 m from points A and B, which are both at the same height. (It is going to form part of a bridge across AB.) The 5 m rope makes an angle of 30° with the vertical.
   (i) How far below A and B is the girder?
   (ii) What angle does the 8 m rope make with the vertical?
   (iii) The girder is 1 m longer than AB. How long is the girder?

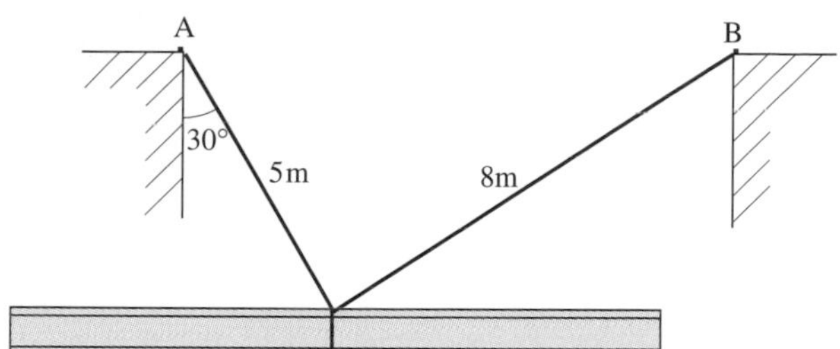

5. The diagram shows Halley and Robin rescuing Tabby the cat from on top of a building. Tabby is being hauled down in a basket. At the time illustrated the ropes RT and TS make angles 25° and 60° with the horizontal and have lengths 30 m and 15 m. Both Robin and Halley are holding their ends of the rope 1 m above the ground.

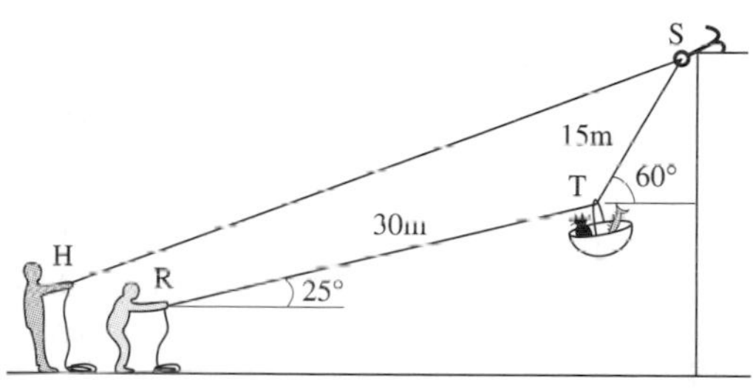

Tabby was lured into the basket using a fish.

   (i) How high is the building?
   (ii) How far is Robin from the building?
   (iii) Halley is 40 m from the building. What angle does the rope HS make with the horizontal?

Investigation

People who need to read maps or charts (e.g. hikers and sailors) often need to estimate bearings and angles. One way of doing this is to line objects up with features on your hand, when at arm's length.

Estimate how many degrees of the horizon are covered by your outstretched hand held at arm's length. What about the angles between fingers, knuckles, etc.?

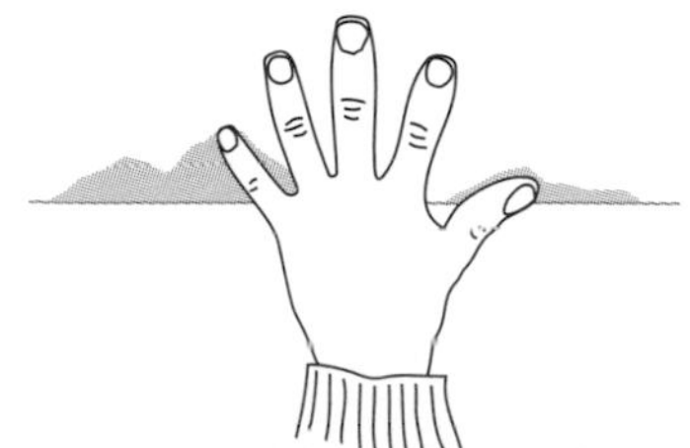

Vectors

Vectors are quantities with both size and direction. They are sometimes represented by lines in diagrams, as illustrated by the examples on this page.

Example

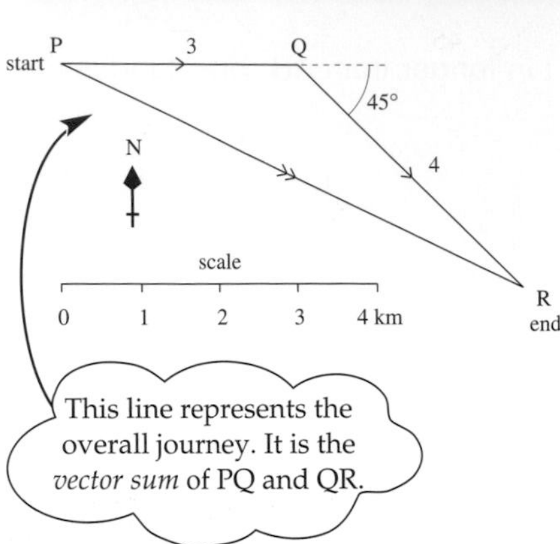

Sunil walks 3 km due east and then 4 km south east. How far is he then from his starting point?

Solution

Sunil's journey is represented by the lines PQ and QR in the diagram on the left. The line PQ = 3 cm, and it represents 3 km due east. The line QR = 4 cm, and it represents 4 km south east. The line PR, shown with a double arrow, represents Sunil's total journey.

Scale drawing suggests that PR is about 6.5 cm long and $\angle$ QPR about 26°. So although Sunil walked 7 km he ended up only 6.5 km from his starting place, on a bearing of 116°.

NOTE

A more accurate way to find the length of PR is to calculate it using trigonometry. You do this in question 1, opposite.

Example

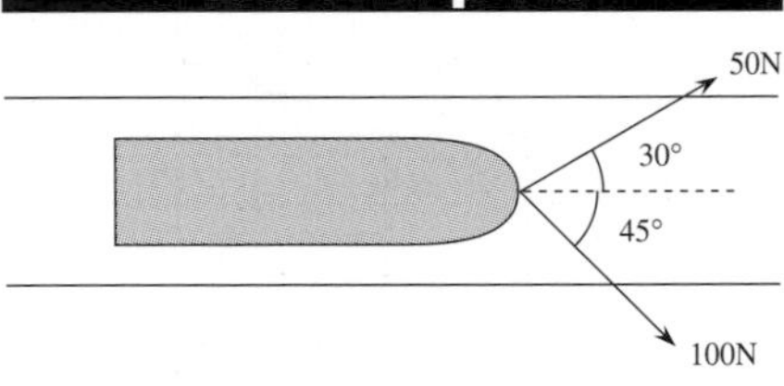

A barge is pulled by two people on opposite banks of a canal. As shown in the diagram (left), one pulls with a force of 50 N at 30° to the line of the canal, the other with 100 N at 45° to the line of the canal. Describe the total of the two forces.

Solution

Since the two forces are not in the same direction, we cannot just add their magnitudes. We have to treat them as vectors, and find their vector sum. Again this is done by drawing the two vectors end-to-end. The two forces are represented by the lines with single arrows in the diagram below.

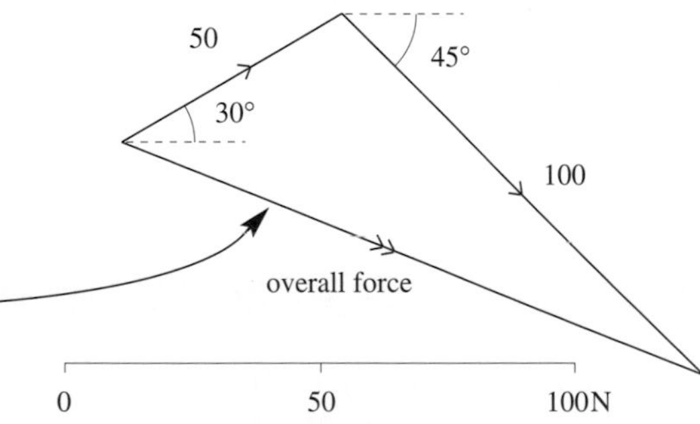

50
45°
30°
100
overall force
0
50
100N
This line represents the overall force on the barge. It is the *vector sum* of the 50N and 100N forces.

Scale drawing suggests the total force is about 123 N at an angle of 22° to the line of the canal.

The calculation for this is question **2** (opposite).

Exercise

1. *(This question refers to the situation in the first example on the opposite page.)*

   In the diagram, find

(i) QS	(ii) SR	(iii) PS
(iv) PR	(v) ∠ SPR	(vi) ∠ NPR

2. *(This question refers to the situation in the second example on the opposite page.)*

   In the diagram, find

(i) AE	(ii) BD	(iii) AF	(iv) BE
(v) DC	(vi) FC	(vii) AC	(viii) ∠ CAE

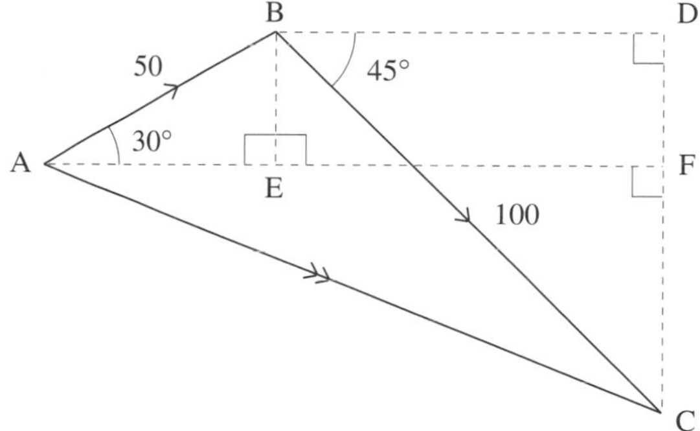

Investigation

Six cross-country walkers start at point X. Each is given a different route consisting of the journeys **P**, 5 kilometres due north, **Q**, 3 kilometres on 120° and **S**, 4 kilometres on 060°.

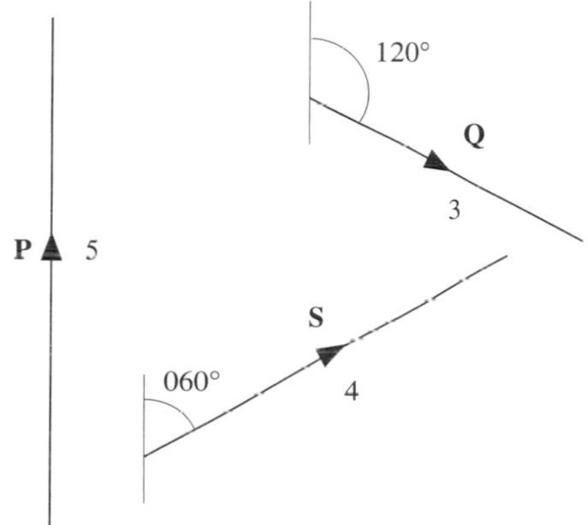

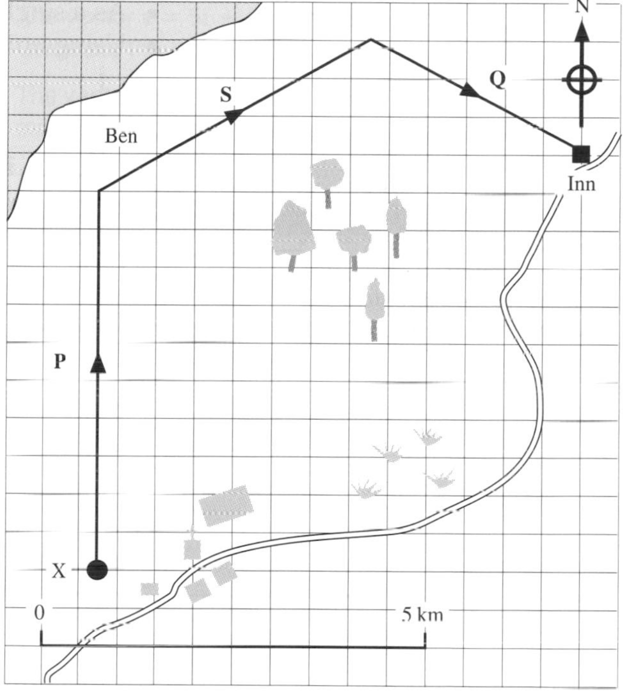

- Amy has first **P**, then **Q**, then **S**.
- Ben has first **P**, then **S**, then **Q**. (His route is shown on the map.)
- Carl has first **Q**, then **P**, then **S**.
- Di has first **Q**, then **S**, then **P**.
- Ellie has first **S**, then **Q**, then **P**.
- Faroza has first **S**, then **P**, then **Q**.

Where does each end up?

Operations on vectors

Addition

The diagram shows $\mathbf{a} = 2\mathbf{i} + 3\mathbf{j}$ with $\mathbf{b} = 4\mathbf{i} - \mathbf{j}$ following on.

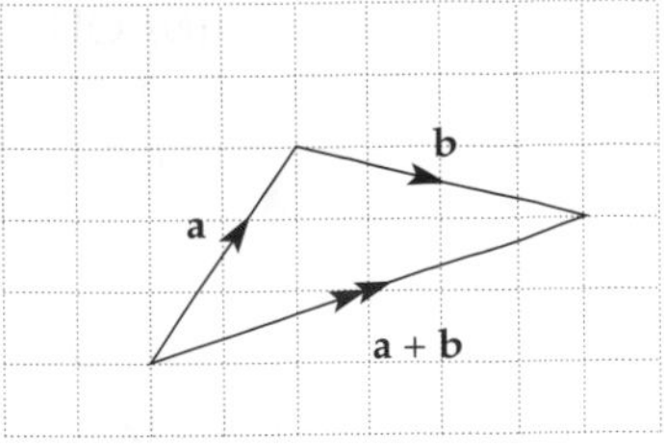

You can see that

$$\begin{aligned}\mathbf{a} + \mathbf{b} &= (2\mathbf{i} + 3\mathbf{j}) + (4\mathbf{i} - \mathbf{j}) \\ &= 6\mathbf{i} + 2\mathbf{j}\end{aligned}$$

$2\mathbf{i} + 4\mathbf{i} = 6\mathbf{i}$

$3\mathbf{j} - \mathbf{j} = 2\mathbf{j}$

Scalar multiplication

The vector $\mathbf{c} = 3\mathbf{i} + 2\mathbf{j}$ is shown below together with $2\mathbf{c}$, $\frac{1}{2}\mathbf{c}$ and $-\mathbf{c}$.

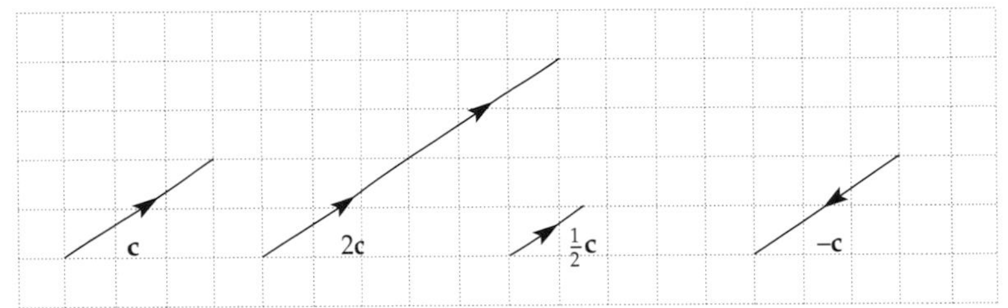

$2\mathbf{c}$, $\frac{1}{2}\mathbf{c}$ and $-\mathbf{c}$ are all *scalar multiples* of $\mathbf{c}$.

You can see that $\quad 2\mathbf{c} = 2(3\mathbf{i} + 2\mathbf{j}) = 6\mathbf{i} + 4\mathbf{j} \qquad \frac{1}{2}\mathbf{c} = \frac{1}{2}(3\mathbf{i} + 2\mathbf{j}) = 1\frac{1}{2}\mathbf{i} + \mathbf{j} \qquad -\mathbf{c} = -(3\mathbf{i} + 2\mathbf{j}) = -3\mathbf{i} - 2\mathbf{j}$

Subtraction

The diagram shows $\mathbf{a} = 2\mathbf{i} + 3\mathbf{j}$ with $-\mathbf{b} = -(4\mathbf{i} - \mathbf{j})$ following on.

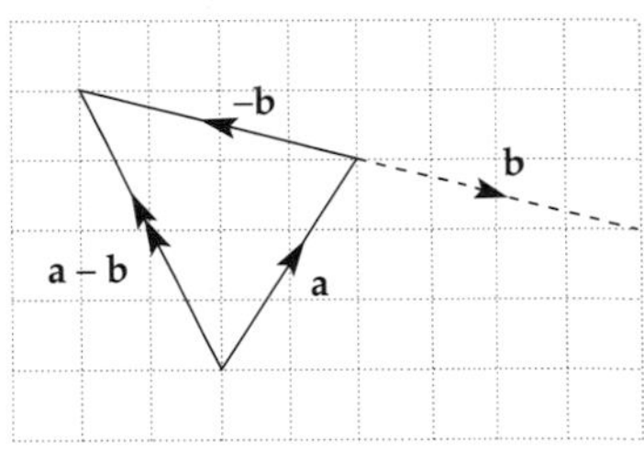

You can see that

$$\begin{aligned}\mathbf{a} - \mathbf{b} &= (2\mathbf{i} + 3\mathbf{j}) - (4\mathbf{i} - \mathbf{j}) \\ &= -2\mathbf{i} + 4\mathbf{j}\end{aligned}$$

$2\mathbf{i} - 4\mathbf{i} = -2\mathbf{i}$

$3\mathbf{j} - (-\mathbf{j}) = 4\mathbf{j}$

Here are two more examples.

- Given $\mathbf{c} = 5\mathbf{i} + 2\mathbf{j}$ and $\mathbf{d} = -4\mathbf{i} + 3\mathbf{j}$ then $2\mathbf{c} - 3\mathbf{d} = 2(5\mathbf{i} + 2\mathbf{j}) - 3(-4\mathbf{i} + 3\mathbf{j}) = 10\mathbf{i} + 4\mathbf{j} + 12\mathbf{i} - 9\mathbf{j} = 22\mathbf{i} - 5\mathbf{j}$
- Given $\mathbf{e} = \begin{pmatrix}3\\1\end{pmatrix}$ and $\mathbf{f} = \begin{pmatrix}5\\-7\end{pmatrix}$ then $2\mathbf{e} - \mathbf{f} = 2\begin{pmatrix}3\\1\end{pmatrix} - \begin{pmatrix}5\\-7\end{pmatrix} = \begin{pmatrix}6\\2\end{pmatrix} - \begin{pmatrix}5\\-7\end{pmatrix} = \begin{pmatrix}1\\9\end{pmatrix}$

Exercise

1. Given $\mathbf{a} = 3\mathbf{i} + 7\mathbf{j}$, $\mathbf{b} = 5\mathbf{i} - 2\mathbf{j}$ and $\mathbf{c} = 4\mathbf{i} + \mathbf{j}$ simplify
   (i) $\mathbf{a} + \mathbf{b}$ (ii) $\mathbf{a} - \mathbf{c}$
   (iii) $\mathbf{b} + \mathbf{c}$ (iv) $3\mathbf{a}$
   (v) $\mathbf{c} - \mathbf{a}$ (vi) $\mathbf{b} - \mathbf{c}$
   (vii) $2\mathbf{b}$ (viii) $\mathbf{a} + \mathbf{b} + \mathbf{c}$.

2. Using $\mathbf{b}$ and $\mathbf{c}$ from question 1 work out the magnitude and direction of
   (i) $\mathbf{b} + \mathbf{c}$ (ii) $\mathbf{b} - \mathbf{c}$.

3. Vectors $\mathbf{x}$, $\mathbf{y}$ and $\mathbf{z}$ are defined by
   $\mathbf{x} = \begin{pmatrix} 2 \\ 7 \end{pmatrix}$, $\mathbf{y} = \begin{pmatrix} -1 \\ 3 \end{pmatrix}$ and $\mathbf{z} = \begin{pmatrix} 5 \\ -2 \end{pmatrix}$.
   Simplify
   (i) $\mathbf{x} + \mathbf{y}$ (ii) $\mathbf{y} - \mathbf{z}$
   (iii) $4\mathbf{x}$ (iv) $\mathbf{z} - \mathbf{x}$
   (v) $\mathbf{x} + \mathbf{z}$ (vi) $-2\mathbf{y}$
   (vii) $\frac{1}{2}\mathbf{z}$ (viii) $\mathbf{x} - \mathbf{y} + \mathbf{z}$.

4. (i) The vectors $\mathbf{x} = 5\mathbf{i} + \mathbf{j}$ and $\mathbf{y} = a\mathbf{i} + b\mathbf{j}$ are equal. Write down the values of a and b.
   (ii) The vectors $\mathbf{t} = k\mathbf{i} - 5\mathbf{j}$ and $\mathbf{u} = 2\mathbf{i} + n\mathbf{j}$ are equal. Write down the values of k and n.

5. Find the values of a and b in the following:
   (i) $\begin{pmatrix} 4 \\ b \end{pmatrix} + \begin{pmatrix} -7 \\ 2 \end{pmatrix} = \begin{pmatrix} a \\ -1 \end{pmatrix}$
   (ii) $\begin{pmatrix} 2 \\ -3 \end{pmatrix} - \begin{pmatrix} a \\ -4 \end{pmatrix} = \begin{pmatrix} -3 \\ b \end{pmatrix}$

6. Vectors $\mathbf{p}$, $\mathbf{q}$ and $\mathbf{r}$ are defined by $\mathbf{p} = 4\mathbf{i} + 3\mathbf{j}$, $\mathbf{q} = 3\mathbf{i} - \mathbf{j}$ and $\mathbf{r} = -2\mathbf{i} + 5\mathbf{j}$.
   Simplify
   (i) $\mathbf{p} + 2\mathbf{q}$ (ii) $3\mathbf{q} - \mathbf{r}$
   (iii) $2\mathbf{p} + 3\mathbf{r}$ (iv) $5\mathbf{r} - \mathbf{p}$
   (v) $4\mathbf{q} - 2\mathbf{p}$ (vi) $\mathbf{p} + 4\mathbf{q} + 2\mathbf{r}$
   (vii) $\mathbf{q} + 7\mathbf{p} - 5\mathbf{r}$ (viii) $\frac{1}{2}(3\mathbf{p} + \mathbf{r})$.

7. Given $\mathbf{c} = \begin{pmatrix} 5 \\ 8 \end{pmatrix}$, $\mathbf{d} = \begin{pmatrix} -3 \\ -5 \end{pmatrix}$ and $\mathbf{e} = \begin{pmatrix} -1 \\ 4 \end{pmatrix}$ simplify
   (i) $2\mathbf{c} + \mathbf{d}$ (ii) $\mathbf{c} - 3\mathbf{e}$
   (iii) $4\mathbf{e} - 3\mathbf{d}$ (iv) $4\mathbf{d} + 5\mathbf{e}$
   (v) $\mathbf{c} + 3\mathbf{d} + 2\mathbf{e}$ (vi) $\mathbf{e} - (2\mathbf{c} - 4\mathbf{d})$
   (vii) $\frac{1}{4}(\mathbf{c} + 3\mathbf{e})$ (viii) $\frac{1}{2}(3\mathbf{c} - 2\mathbf{d} + \mathbf{e})$.

8. Using the vectors $\mathbf{c}$, $\mathbf{d}$ and $\mathbf{e}$ defined in question 7 calculate the magnitude and direction of
   (i) $\mathbf{c} + 4\mathbf{d}$ (ii) $3\mathbf{e} - 2\mathbf{d}$.

9. Given $\mathbf{a} = 4\mathbf{i} + \mathbf{j}$, $\mathbf{b} = -3\mathbf{i} + 8\mathbf{j}$ and $\mathbf{c} = -\mathbf{i} - 9\mathbf{j}$
   (i) work out $\mathbf{a} + \mathbf{b} + \mathbf{c}$,
   (ii) draw a diagram to show $\mathbf{a} + \mathbf{b} + \mathbf{c}$.
   (iii) What can you conclude?

10. Find the values of t and u in the following:
    (i) $2\begin{pmatrix} t \\ 4 \end{pmatrix} + 3\begin{pmatrix} 1 \\ -5 \end{pmatrix} = \begin{pmatrix} 7 \\ u \end{pmatrix}$
    (ii) $4\begin{pmatrix} u \\ 2 \end{pmatrix} - \begin{pmatrix} t \\ u \end{pmatrix} = 3\begin{pmatrix} -1 \\ 1 \end{pmatrix}$

Activities

1. (i) Draw the vectors $4\mathbf{i} + 3\mathbf{j}$ and $8\mathbf{i} + 6\mathbf{j}$.
   (ii) What can you conclude about their directions?
   (iii) Write down two vectors parallel to
   (a) $7\mathbf{i} - \mathbf{j}$ (b) $-3\mathbf{i} + 2\mathbf{j}$

2. (i) Draw these vectors:
   $\mathbf{a} = 4\mathbf{i} + 3\mathbf{j}$ $\mathbf{b} = -3\mathbf{i} + 4\mathbf{j}$ $\mathbf{c} = 3\mathbf{i} - 4\mathbf{j}$.
   (ii) Compare the directions of $\mathbf{a}$ and $\mathbf{b}$.
   (iii) Compare the directions of $\mathbf{a}$ and $\mathbf{c}$.
   (iv) Write down two vectors perpendicular to $5\mathbf{i} - 2\mathbf{j}$.

Investigation

The skipper heads his boat across the river at $10\,\text{km h}^{-1}$, directly towards the opposite bank.
The river is $400\,\text{m}$ wide and flows at $4\,\text{km h}^{-1}$.

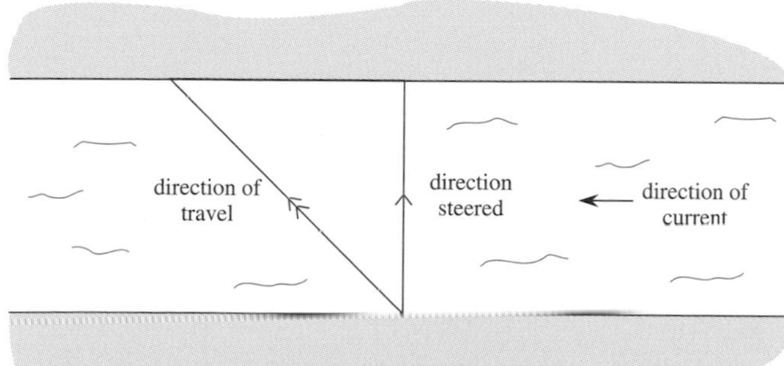

(i) Sketch the vector diagram showing the velocities in this situation.
(ii) Calculate the angle between the direction in which the boat is steered and that in which it travels.
(iii) How long does the boat take to cross the river?
(iv) How far downstream does the boat travel while crossing the river?

Drawing 3-dimensional objects

Trigonometry is important when you need to make or interpret drawings of 3-dimensional objects. There are several ways of representing solid objects on a flat page (or computer screen), and two of the most common are shown below.

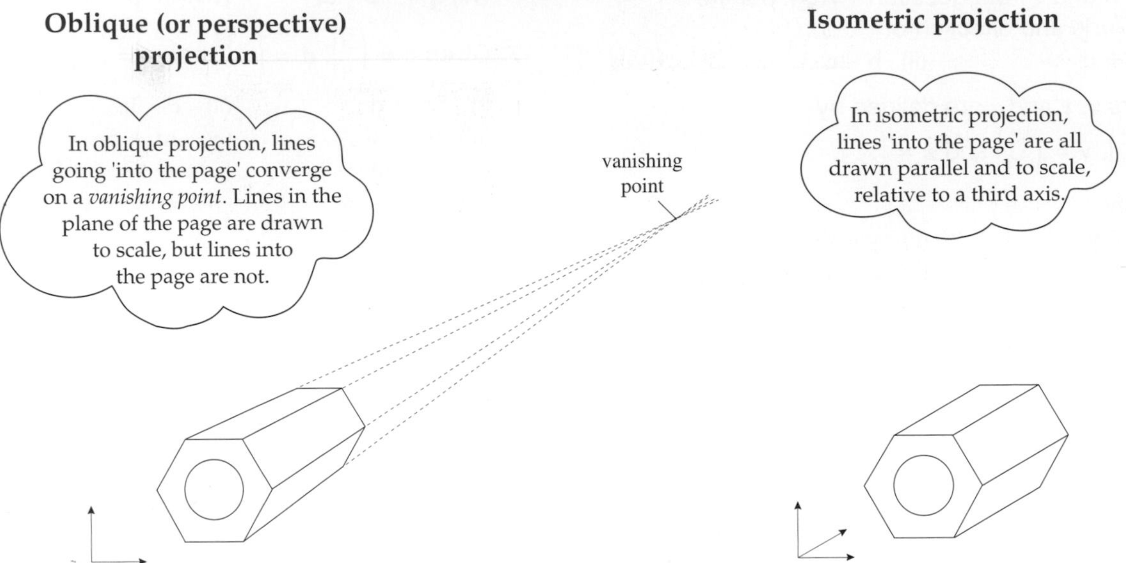

Oblique (or perspective) projection
In oblique projection, lines going 'into the page' converge on a *vanishing point*. Lines in the plane of the page are drawn to scale, but lines into the page are not.
vanishing point
Isometric projection
In isometric projection, lines 'into the page' are all drawn parallel and to scale, relative to a third axis.

Oblique projection produces a more realistic effect, but isometric projection is used in this chapter because it allows three dimensions of an object to be drawn to scale. Isometric projection is usually used in technical drawings, for the same reason.

Plans and elevations

Even in an isometric drawing, the object has to be distorted: some right angles do not appear to be 90°, and some lengths cannot be drawn to scale. *Plans* and *elevations*, on the other hand, show the exact shape of the object viewed from particular directions, drawn to scale. They are called *true shape diagrams.*

The plan of an object is its top view. The views from the front, the back and the sides are called elevations. The diagrams below show the front elevation, the right side elevation and the plan of the body above.

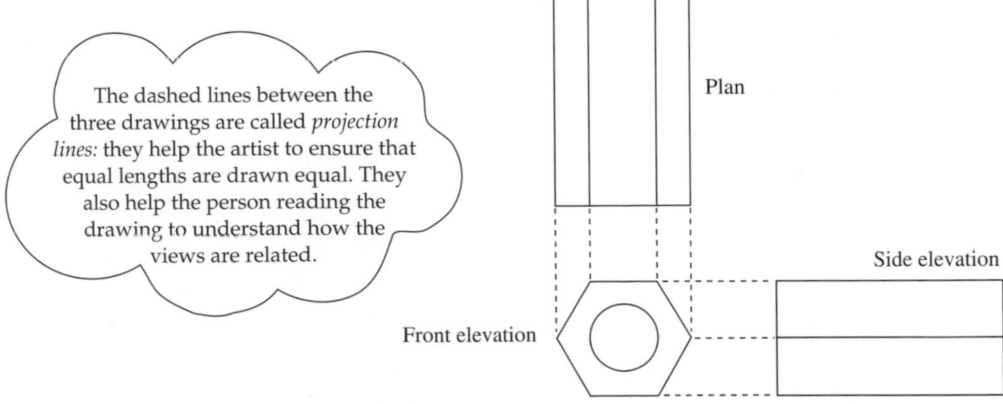

The dashed lines between the three drawings are called *projection lines:* they help the artist to ensure that equal lengths are drawn equal. They also help the person reading the drawing to understand how the views are related.
Plan
Side elevation
Front elevation

Exercise

1. (i) These diagrams show the cross sections of a number of objects of uniform cross section. Sketch a 3-dimensional view of a length of each object.

A girder A picture-frame moulding A track for sliding doors

(ii) List four more objects you might see in everyday life that have uniform cross section.

2. For each of these 3-dimensional objects, draw accurately (i) the plan, (ii) the front elevation, (iii) both side elevations.

(a)

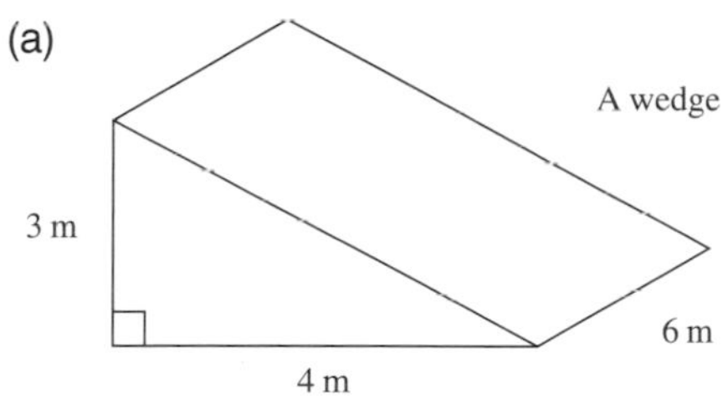

A wedge

(b)

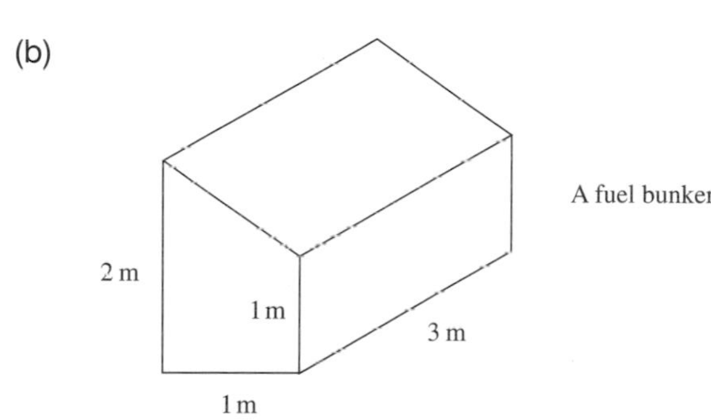

A fuel bunker

3. For each of these 3-dimensional objects sketch the plan, front elevation and side elevation (from the side indicated).

(a)

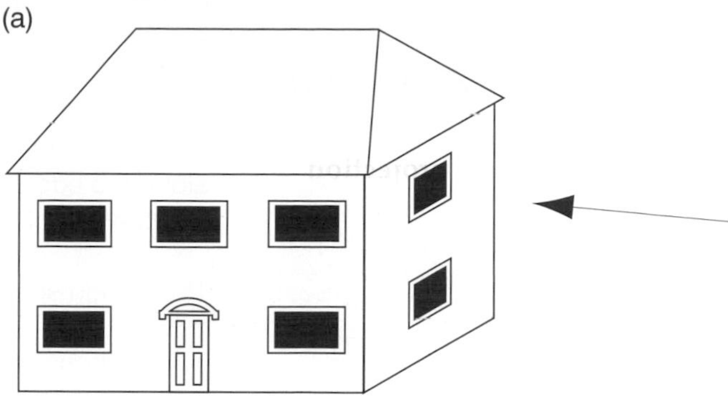

A house

(b)

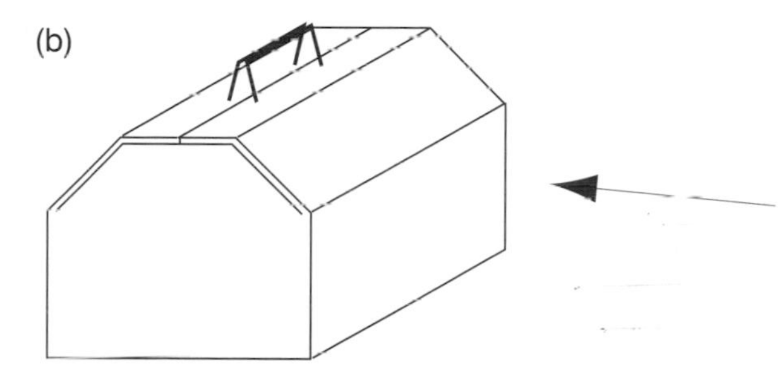

A tool box

(c)

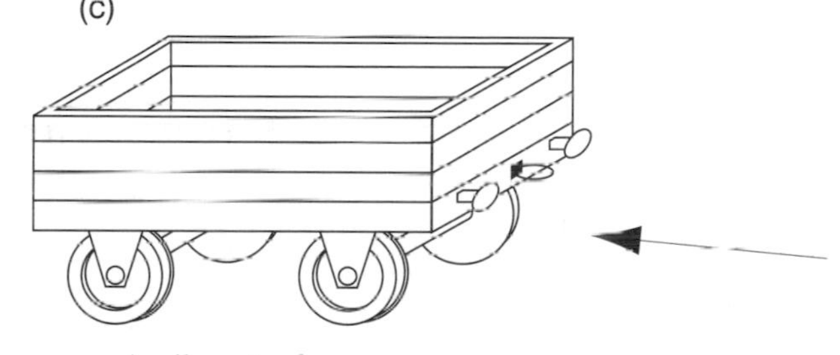

A railway truck

Activity

The diagram is a scale drawing of the net of a packing box. When cut up, folded and glued it becomes a closed box. Draw the plan, front elevation and one side elevation of the assembled packing box.

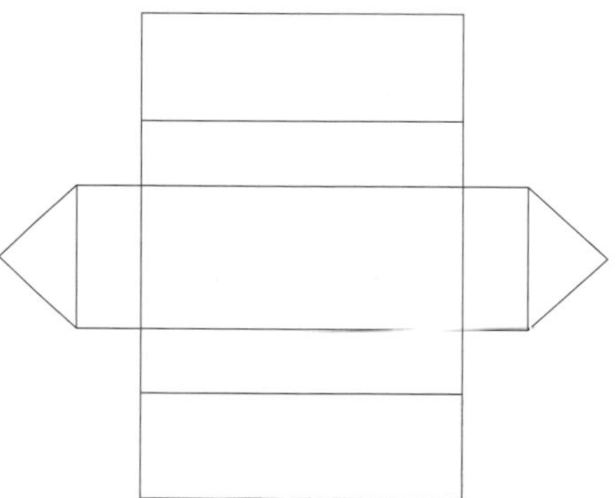

Working in 3 dimensions

When you need to calculate the sides, angles and areas of 3-dimensional objects, always start by drawing the relevant true shape diagrams, as in the next example.

Example

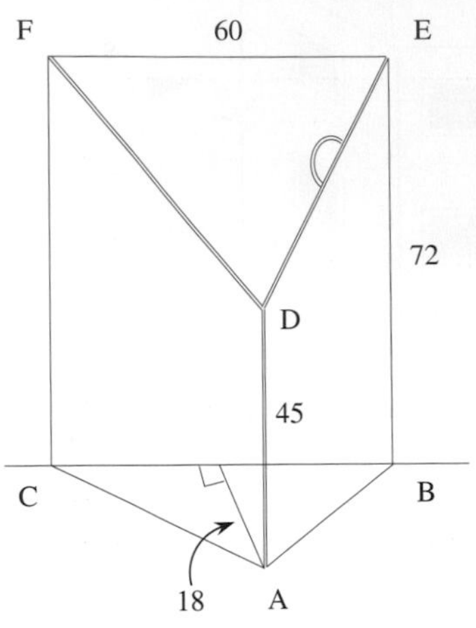

The diagram on the left shows a common design of collapsible fireguard. It is made up from three flat shapes, ABED, ACFD and DEF. There are hinges along the edges AD and DF; when the fireguard is in use the edge DE is held together by a bolt.

The design specification of one such fireguard is:

- It should cover an area 60 cm × 72 cm.
- The point A should be 18 cm from the front of the fire.
- The point D should be 45 cm above the floor.

Find the lengths of the other sides of the flat shapes, e.g. (i) AB (ii) DE, so that instructions can be given for their manufacture.

Solution

(i) First we draw the true shape diagram of triangle ABC.

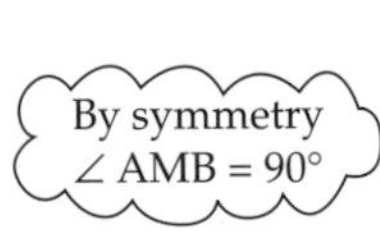

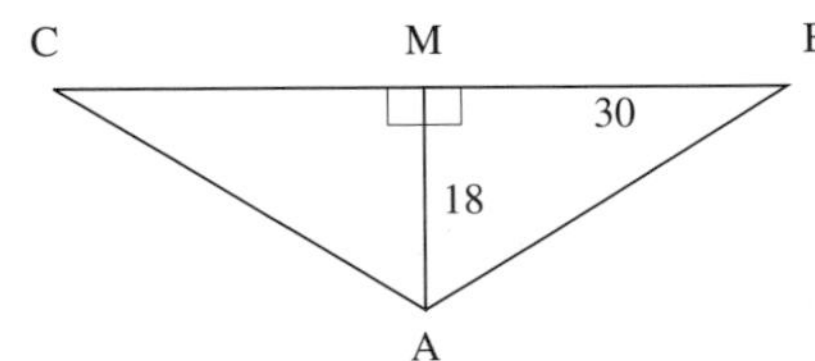

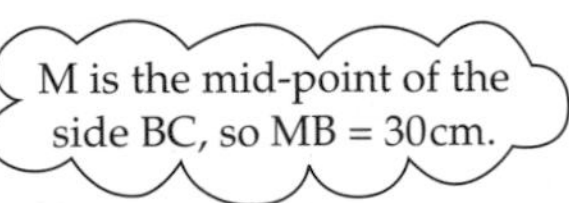

M is the mid-point of the
side BC, so MB = 30cm.

Using Pythagoras' Theorem in triangle ABM,

$$AB^2 = 30^2 + 18^2$$

$\Rightarrow$ the length of AB is 35.0 cm.

By symmetry, AC is the same.

(ii) To find DE draw the true shape diagram of ABED (left) and add the point N such that ABND is a rectangle.

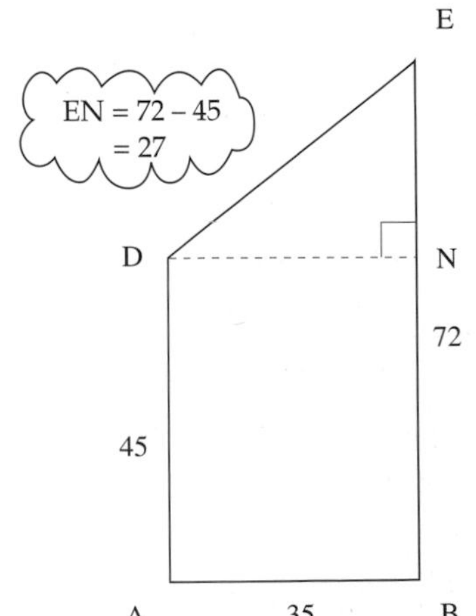

Using Pythagoras' Theorem in triangle DEN,

$$DE^2 = 35^2 + 27^2$$

$$\Rightarrow \quad DE = 44.2\text{ cm.}$$

By symmetry, FD is the same.

These results allow the fireguard to be made according to the design specification.

Exercise

1. The diagram shows a swimming pool. The dimensions marked on it are in metres.
   (i) (a) Draw a true shape diagram of ABCD.
   (b) Find the length AC and the angle CAB.
   (ii) (a) Draw true shape diagrams of AEGC and EFGH.
   (b) Find the lengths GE and EF, and the angle GEF.
   (iii) A swimmer enters the water at X and swims directly to F.
   (a) How far is it from X to F?
   (b) At what angle to the horizontal does she swim?

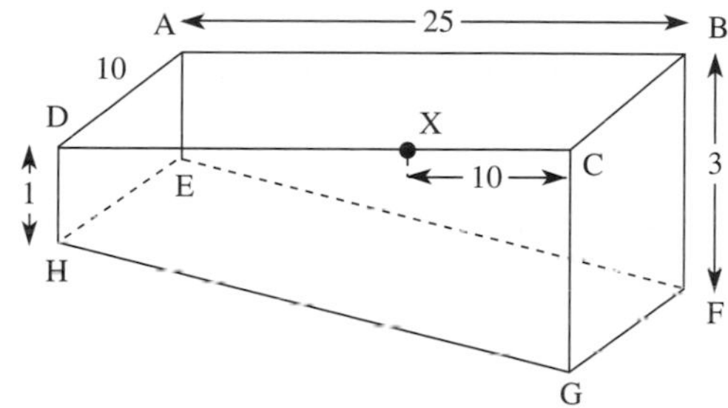

2. A mast, TP, 10 metres tall stands on level ground at P. It is supported by three staywires TA, TB and TC, all of the same length. Triangle ABC is equilateral with side 8 m.
   (i) Sketch a 3-dimensional view of the mast and staywires.
   (ii) Draw a true shape diagram of triangle ABC and calculate the length AP.
   (iii) Find $\angle$ ATP.
   (iv) Find $\angle$ BTC.

3. The diagram shows a room. The floor ABCD is rectangular, and the room is 3 m high. Streamers are stretched from the point P on the ceiling PQRS to E, F, G and H half way up the walls.
   (i) Find the angle each of these streamers makes with the horizontal. You may assume that each follows a straight line and does not sag.
   (ii) Find the angle between the streamers PF and PG.

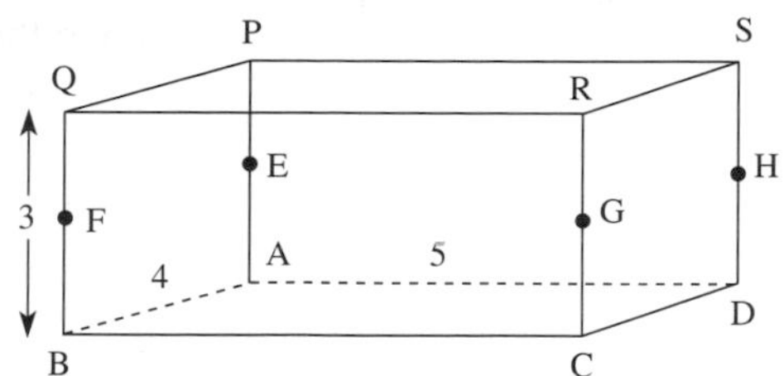

4. The diagram shows the roof of a house, and a symmetrical dormer window ABCD. The line AB is horizontal, E is the mid-point of CD, AC = 3.8 m and CD = 1.8 m. The angle of slope of the roof, $\angle$ EAB, is 40°. Calculate
   (i) AE (ii) AB (iii) BE (iv) $\angle$ ACB

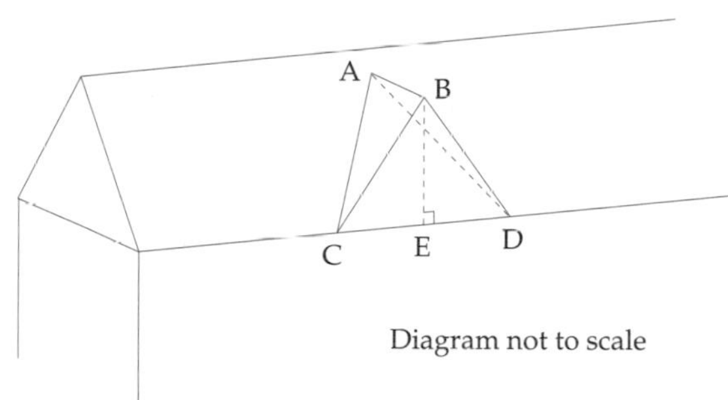

Diagram not to scale

Investigation

The diagram shows a proposal for the design of a fast-food carrier. It is required to carry a can of drink with radius 3.2 cm and height 12 cm, and a rectangular food package 12 cm × 10 cm × 8 cm. The sloping faces make an angle of 60° with the horizontal.

(i) Find appropriate lengths for the various sides.
(ii) Draw the net on a sheet of card, including suitable flaps. Cut it out and glue the container together. Does it work?

Angles of any size

The diagram shows a rotor arm OP of length 1 cm which is free to rotate with one end fixed at the origin O. What are the co-ordinates of P when it is in the position shown, at 60° to the x axis?

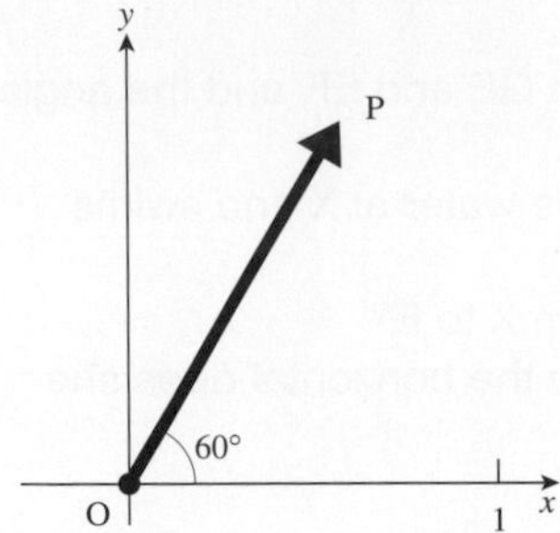

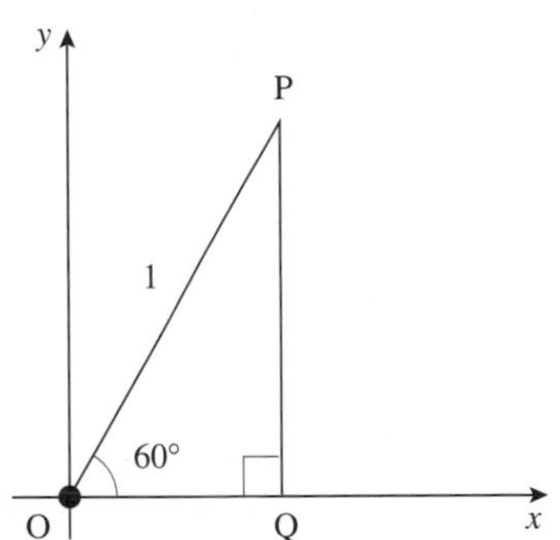

To answer this question draw in the line PQ from P to the x axis, as shown in the diagram on the left. Angle OQP = 90°. In triangle OPQ,

$$\cos 60^\circ = \frac{OQ}{OP} \Rightarrow OQ = 1\cos 60^\circ,$$

and

$$\sin 60^\circ = \frac{PQ}{OP} \Rightarrow PQ = 1\sin 60^\circ.$$

So the co-ordinates of the point P are (cos 60°, sin 60°).

By the same reasoning, if the angle had been 80° rather than 60°, the co-ordinates would have been (cos 80°, sin 80°), and at a general angle θ, ($\cos\theta$, $\sin\theta$).

This is a very important result as it allows you to extend the definition of $\sin\theta$ and $\cos\theta$ to angles greater than 90°.

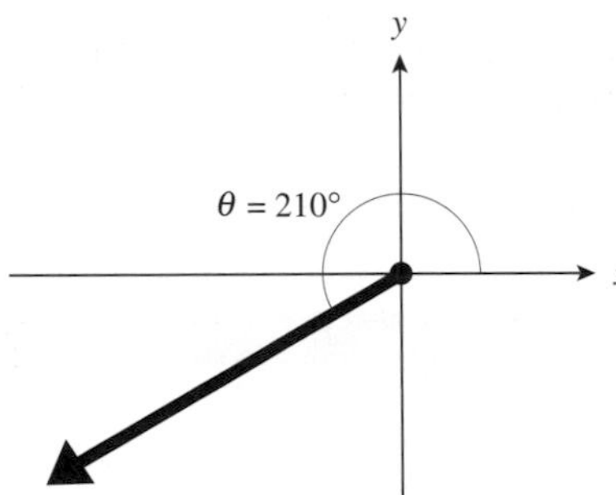

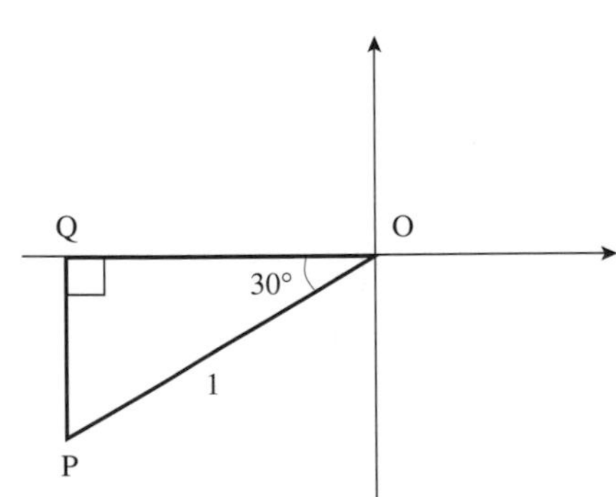

In this diagram the rotor arm has rotated through 210° and so the position of P is (cos 210°, sin 210°). However you can work out the co-ordinates of P by looking at triangle OPQ (left). In this triangle

$$\cos 30^\circ = \frac{OQ}{OP} \Rightarrow OQ = 1\cos 30^\circ = 0.866$$

and

$$\sin 30^\circ = \frac{QP}{OP} \Rightarrow PQ = 1\sin 30^\circ = 0.5.$$

So the co-ordinates of P are (–0.866, –0.5).

This tells you that cos 210° = –0.866 and sin 210° = –0.5.

You can use these values to find tan 210°.

$$\tan 210^\circ = \frac{\sin 210^\circ}{\cos 210^\circ} = \frac{-0.5}{-0.866} = 0.5774$$

Use your calculator to check these answers.

Activities

1. Take a sheet of graph paper with a 1cm grid, and paste it onto a sheet of stiff card. Mark the origin in the middle of the graph paper, and draw x and y axes using a scale of 10 cm to 1 unit for each. Now make a cardboard rotor arm 10 cm long and pin one end to the origin, so that it is free to rotate.

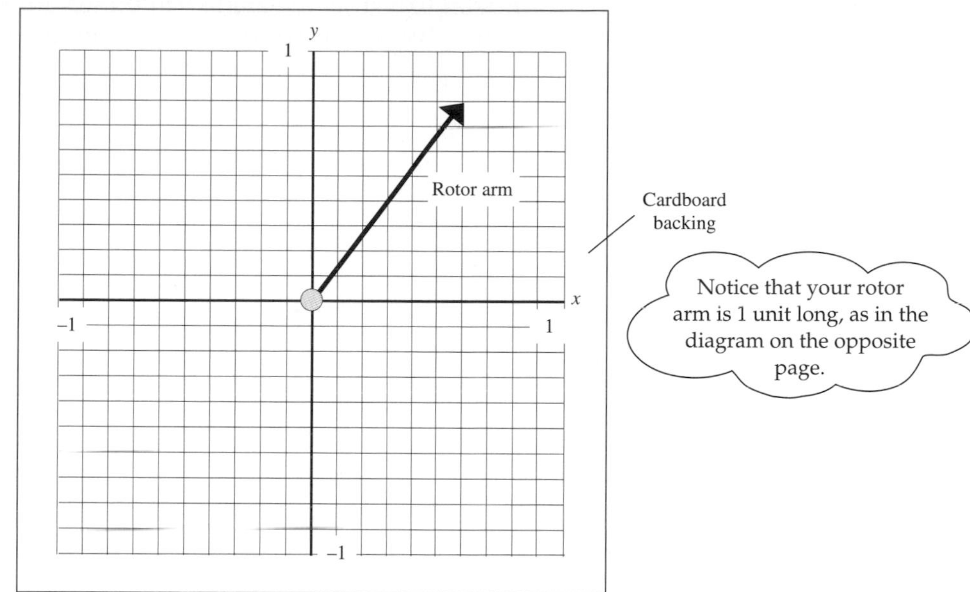

2. Before making any measurements, copy and complete the table on the right by entering + or – for each of the regions.

Angle θ	0° to 90°	90° to 180°	180° to 270°	270° to 360°
$\cos\theta$	+			
$\sin\theta$	+			

3. Now position the rotor arm at 30° and estimate the values of $\cos\theta$ and $\sin\theta$. Repeat this for 60°, 90° and so on up to 360°. Copy and complete this table.

Angle θ	0°	30°	60°	90°	120°	150°	180°	210°	240°	270°	300°	330°	360°
$\cos\theta$	1			0									
$\sin\theta$	0			1									

4. Now draw graphs of $\cos\theta$ and $\sin\theta$ for 0° to 360°, on axes like those below.

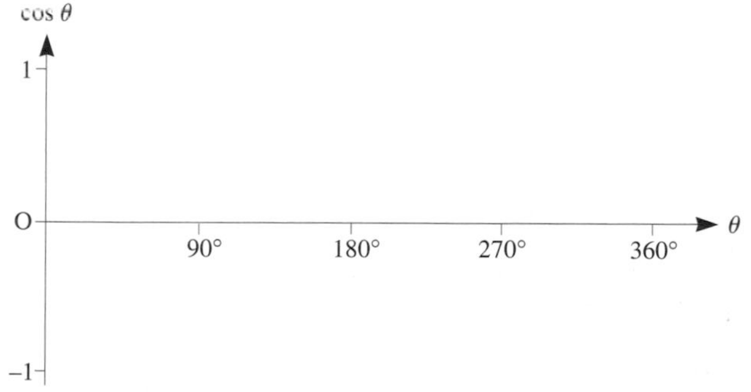

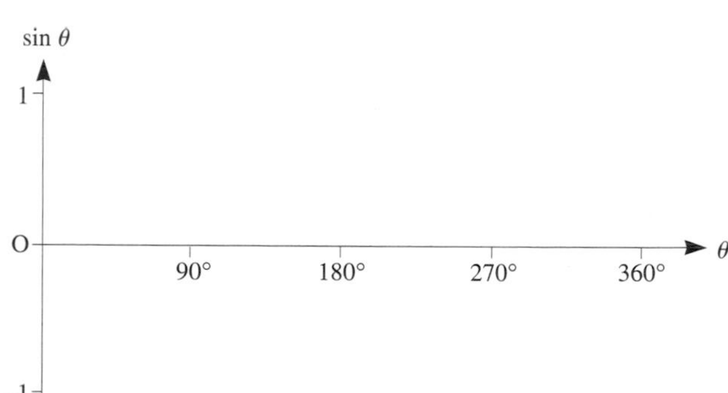

The graphs of cos θ and sin θ

In the activity on the previous page you drew the graphs of cos θ and sin θ against θ for values of θ between 0° and 360°. However there is no reason to restrict the angle θ to these values; the same patterns repeat themselves as the rotor arm goes round and round.

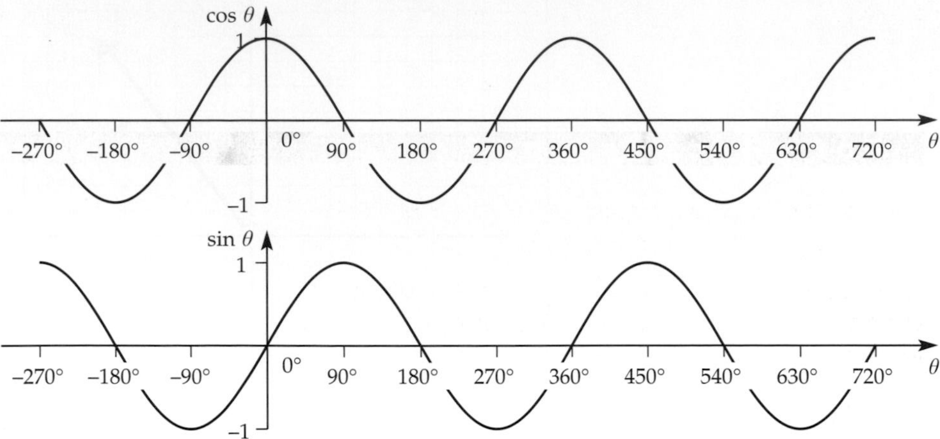

You will see that the two graphs are the same except that the one for sin θ is the cos θ graph shifted 90° to the right.

These graphs are very important because of their wave form. This is the form of most natural vibrations and of some that are man-made, like the mains voltage in our electricity supply.

A wave form might be of the type $y = 2\cos\theta$ or $y = 3\sin\theta$ and these graphs are shown below.

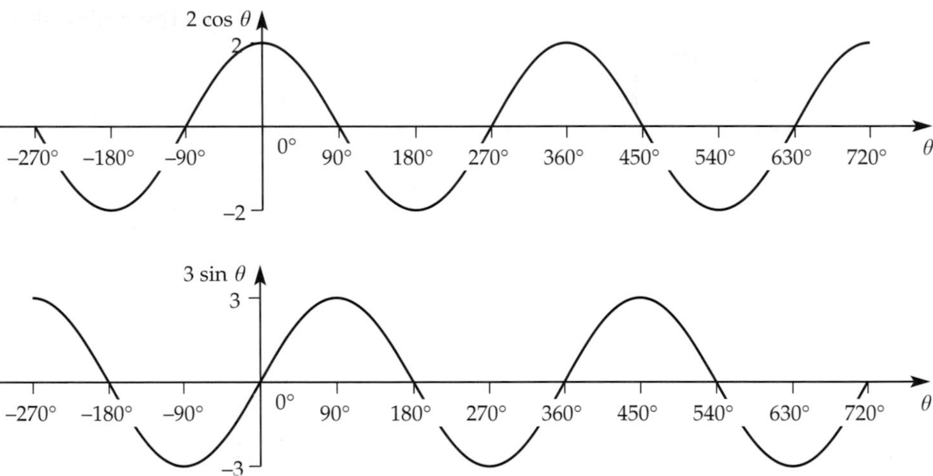

Compare these graphs with those of $y = \cos\theta$ and $y = \sin\theta$ at the top of the page. You can see that

- the graphs of $y = \cos\theta$ and $y = 2\cos\theta$ are the same except that the displacement in the y direction is double.
- the graphs of $y = \sin\theta$ and $y = 3\sin\theta$ are the same except that the displacement in the y direction is treble.

The graph of $y = \tan\theta$ is the focus of the investigation on page 167.

Exercise

Use your calculator to find, correct to 4 decimal places,

1. $\sin 210°$
2. $\cos 300°$
3. $\sin 225°$
4. $\tan 45°$
5. $\sin(-30°)$
6. $\cos(-45°)$
7. $\tan 60°$
8. $\cos 150°$
9. $\cos 210°$
10. $\sin 330°$

In questions 11 to 14, sketch the graphs of the trigonometrical functions in the given interval.

11. $y = 4\cos\theta$ for $\theta = 0°$ to $360°$.
12. $y = 2.5\sin\theta$ for $-180° \leq \theta \leq 180°$.
13. $y = 2\cos\theta + 1$ for $\theta = -180°$ to $180°$.
14. $y = 3\sin\theta - 4$ for $0° \leq \theta \leq 360°$.

For discussion

The graph of $\sin\theta$ has a shape which probably looks familiar to you. It is called a sine wave.

Here are four situations where similar shapes occur. Discuss whether these are true sine waves or even close to them.

(i) Water ripples

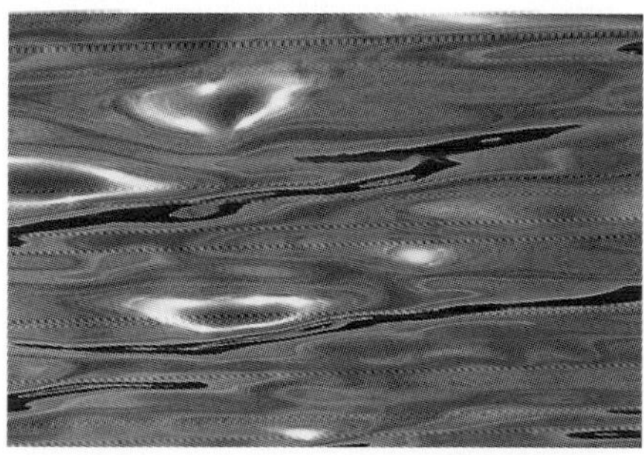

(ii) A heart beat

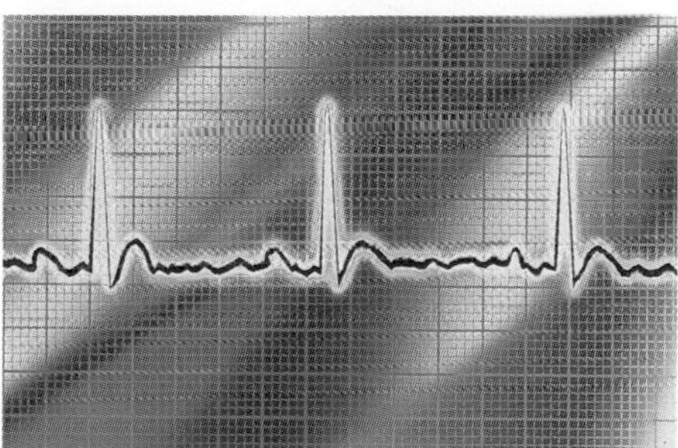

(iii) Vibrations on a guitar string

(iv) The mains electricity voltage

voltage

time

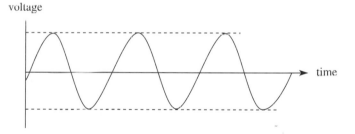

Investigation

Use your calculator to find the values of $\tan\theta$ for angles between $-360°$ and $540°$, and so draw the graph of $\tan\theta$. You will notice that problems occur when $\theta = 90°$. Take some values near to it (like $80°$, $85°$, $95°$ and $100°$) and then close in to see what happens. Are there any other values of θ which cause problems?

The sine rule

You can only use the sine, cosine and tangent ratios in right-angled triangles. When a triangle is not right-angled then you can use the sine rule (described below) and/or the cosine rule (see page 170).

Sine rule

In any triangle ABC

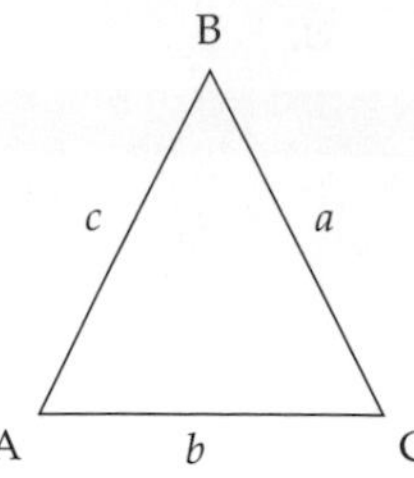

It is convenient to use an a to represent the side opposite angle A and so on.

$$\frac{a}{\sin A} = \frac{b}{\sin B} = \frac{c}{\sin C}$$

This is used to find a side.

This can be inverted to give

$$\frac{\sin A}{a} = \frac{\sin B}{b} = \frac{\sin C}{c}$$

This is used to find an angle.

Example

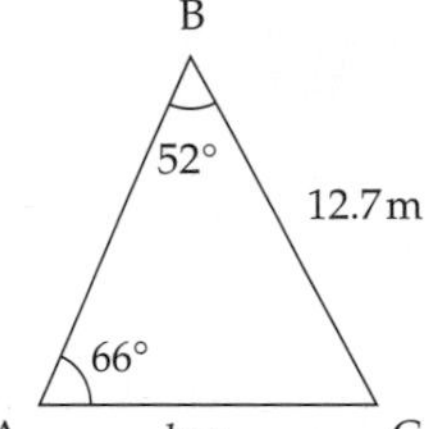

Find the length AC in the triangle ABC on the left.

Solution

Use the sine rule starting with the unknown side.

$$\frac{b}{\sin B} = \frac{a}{\sin A}$$

$$\frac{b}{\sin 52°} = \frac{12.7°}{\sin 66°} \quad \text{giving } b = \frac{12.7 \sin 52°}{\sin 66°} = 11.0 \text{ (to 1 d.p.)}$$

Example

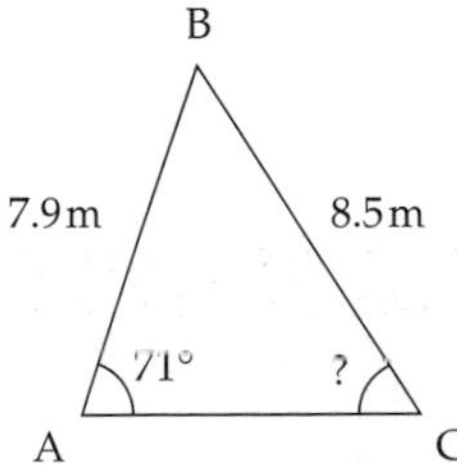

Find $\angle C$ in the triangle ABC on the left.

Solution

Use the sine rule inverting the formula and starting with the unknown angle.

$$\frac{\sin C}{c} = \frac{\sin A}{a}$$

$$\frac{\sin C}{7.9} = \frac{\sin 71°}{8.5} \quad \text{giving } \sin C = \frac{7.9 \sin 71°}{8.5} = 0.8787\ldots$$

so $\angle C = 61.5°$ (to 1 d.p.)

Exercise

In questions 1 to 4 find the unknown side x.

1.

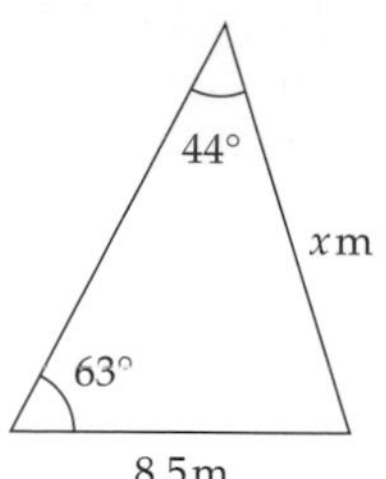

2.

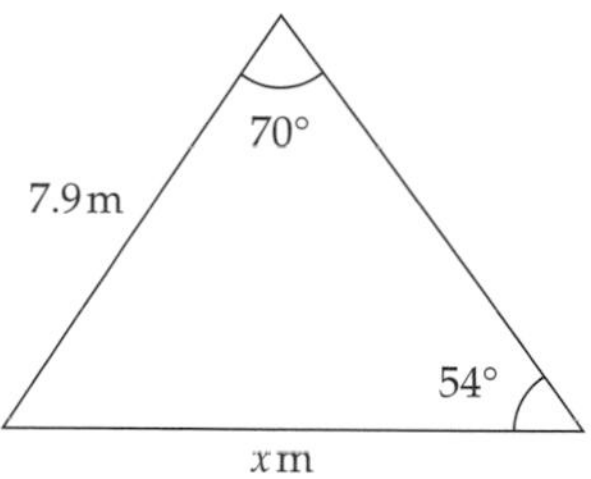

3.

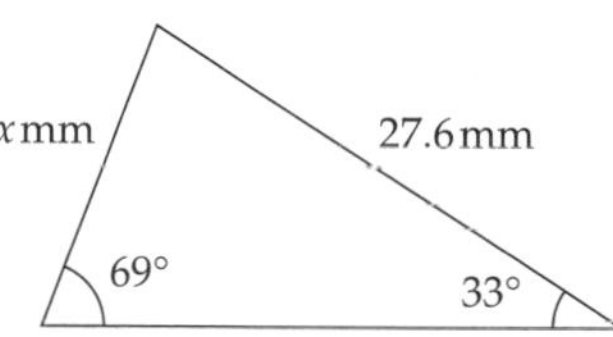

4.

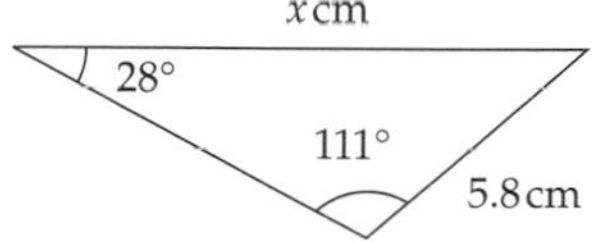

5. In triangle ABC, $\angle A = 48°$, $\angle B = 63°$ and $BC = 25.6$ cm. Calculate
   (i) AC (ii) $\angle C$ (iii) AB.
6. In triangle DEF, $\angle D = 78°$, $\angle E = 66°$ and $DE = 9.3$ m. Calculate the length of DF.

In questions 7 to 10 find the unknown angle θ.

7.

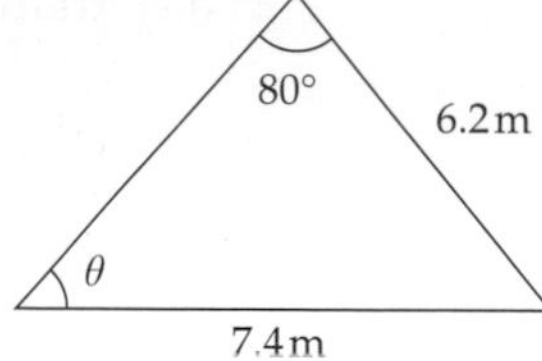

8.

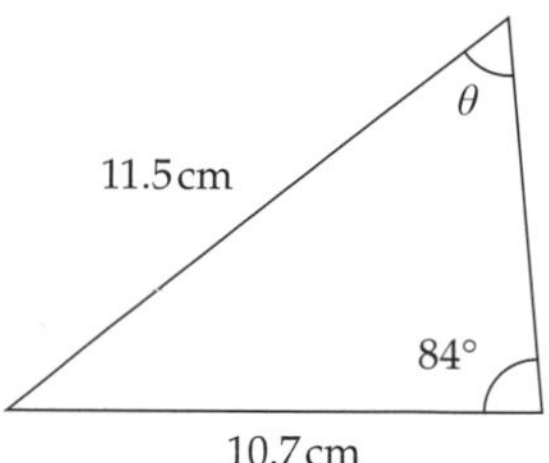

9.

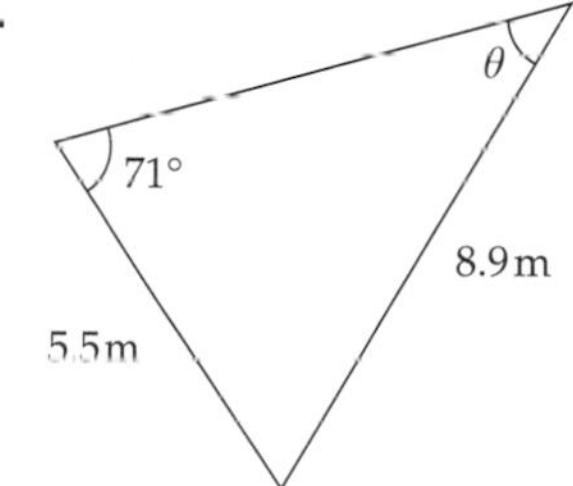

10.

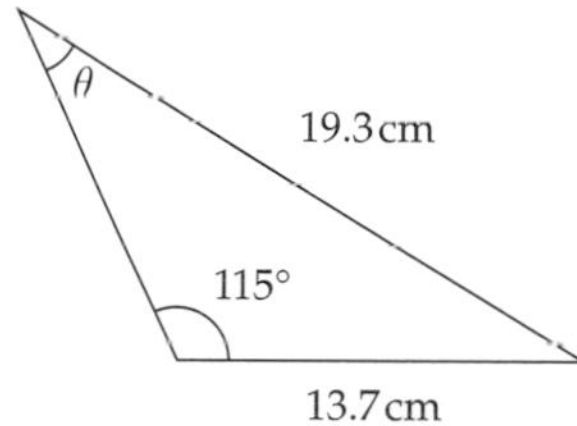

11. In triangle PQR, $\angle P = 109°$, $QR = 52.3$ cm and $PQ = 28.8$ cm.
    Calculate
    (i) $\angle R$ (ii) $\angle Q$ (iii) PR.
12. In triangle ABC, $AB = 8.3$ m, $AC = 11.2$ m and $\angle B = 74°$.
    Calculate $\angle A$.

Investigation

Look at the diagrams for each of questions 7 to 10 and you will see that the angles you are finding are opposite the *shorter* of the two given sides.

In triangle ABC, $AB = 5$ cm, $AC = 6$ cm and $\angle C = 50°$. $\angle A$ is opposite the *longer* side.

1. You have to find $\angle A$ by
   (i) using the sine rule, and
   (ii) by constructing triangle ABC.
2. What conclusions do you reach?

The cosine rule

In any triangle ABC

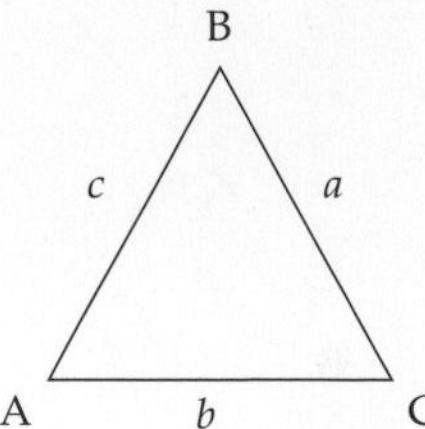

$a^2 = b^2 + c^2 - 2bc \cos A$

This is used to find a side.

This can be rearranged to give

$\cos A = \dfrac{b^2 + c^2 - a^2}{2bc}$

This is used to find an angle.

Example

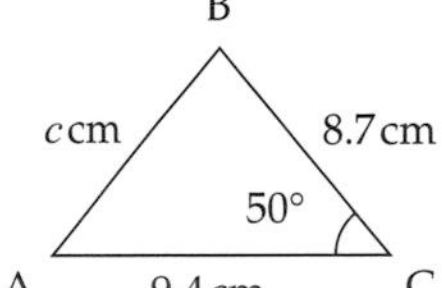

Find the length AB in the triangle ABC on the left.

Solution

Use the cosine rule in the form

$$c^2 = a^2 + b^2 - 2ab \cos C$$

This is the form given above with $a \rightarrow c$, $b \rightarrow a$ and $c \rightarrow b$.

$$\begin{aligned} c^2 &= 8.7^2 + 9.4^2 - 2(8.7)(9.4) \cos 50° \\ &= 75.69 + 88.36 - 105.13 \\ &= 58.92 \end{aligned}$$

So $c = \sqrt{58.92} = 7.68$ (to 2 d.p.).

Example

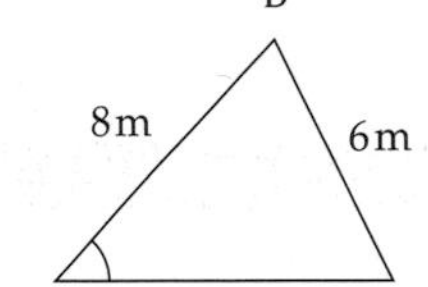

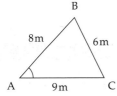

Find $\angle$A in the triangle ABC on the left.

Solution

Use the cosine rule in the form

$$\cos A = \frac{b^2 + c^2 - a^2}{2bc}$$

$$\cos A = \frac{9^2 + 8^2 - 6^2}{2(9)(8)} = \frac{81 + 64 - 36}{144} = \frac{109}{144}$$

So $\angle A = 40.8°$ (to 1 d.p.).

Exercise

In questions 1 to 4 find the unknown side x.

1.

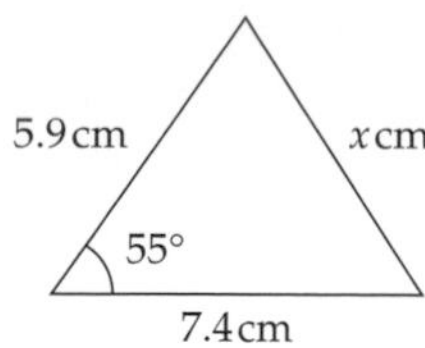

2.

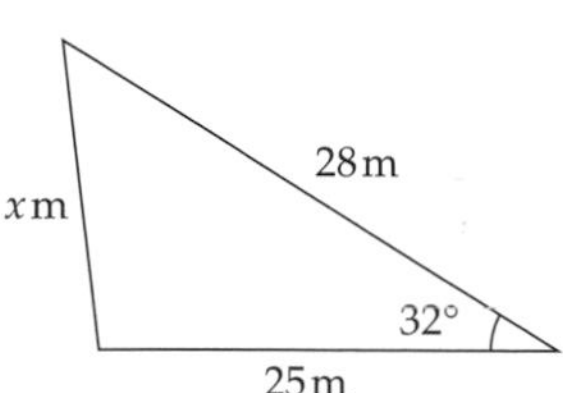

3.

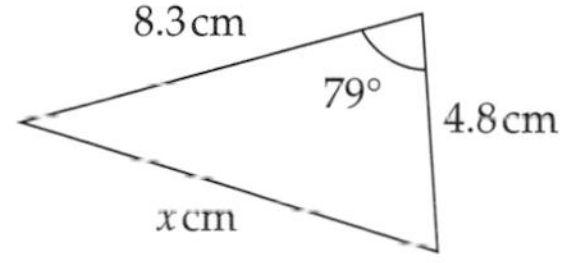

4.

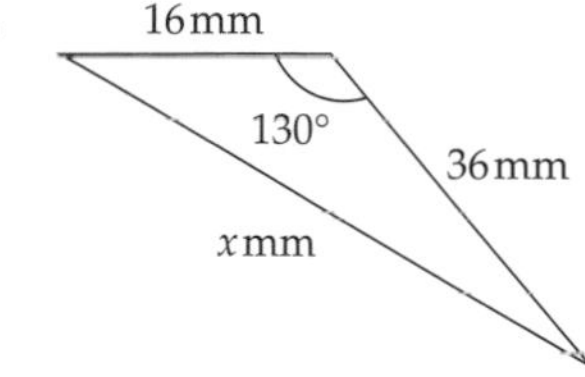

5. In triangle PQR, PQ = 20 m, QR = 24 m and $\angle Q = 43°$. Calculate the length PR.

6. In triangle XYZ, XY = YZ = 12 cm and $\angle Y = 40°$. Calculate the length XZ.

In questions 7 to 10 find the unknown angle θ.

7.

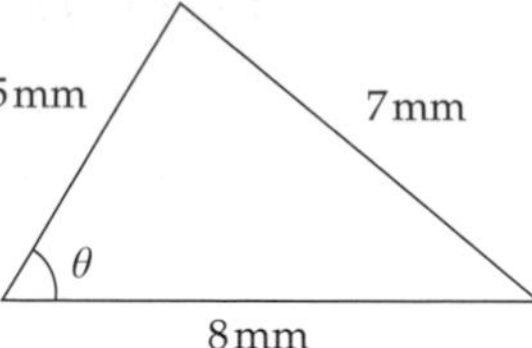

8.

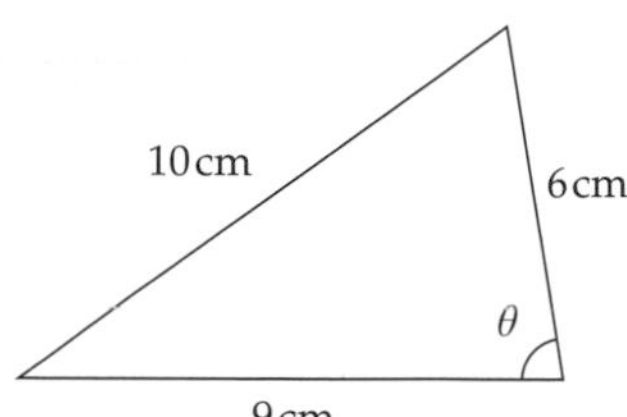

9.

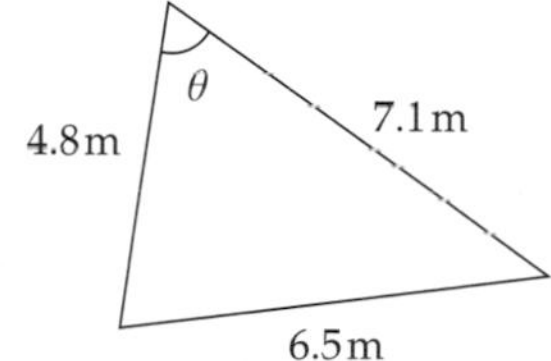

10.

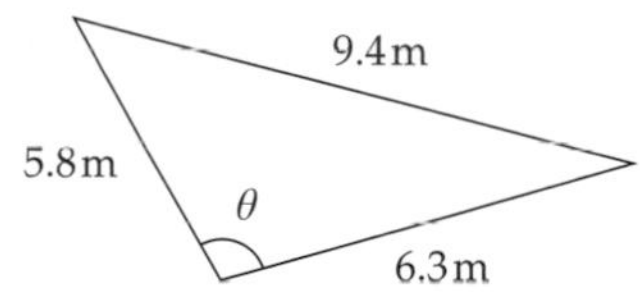

11. In triangle ABC, AB = 6 mm, BC = 7 mm and AC = 9 mm. Calculate
(i) $\angle A$ (ii) $\angle B$ (iii) $\angle C$.

12. In triangle ABC, AB = AC = 8 cm and BC = 5 cm. Calculate $\angle A$.

Investigation

In a triangle ABC, suppose the longest side is c cm and hence the largest angle is C.

The cosine rule gives

$$c^2 = a^2 + b^2 - 2ab \cos C$$

1. What happens to this formula when C = 90°?

2. When $c^2 < a^2 + b^2$ what can you deduce about angle C?

3. When $c^2 > a^2 + b^2$ what can you deduce about angle C?

Special cases of the sine and cosine rules

The sine rule and cosine rule enable you to find unknown sides and angles in any triangle. Sometimes a single application of one rule will solve a problem. In other cases more than one step is required to find the unknown side or angle.

How would you find the unknown side in the following cases?

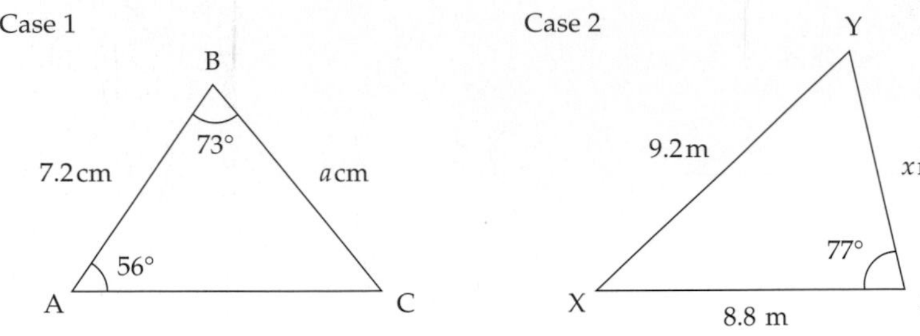

Remember: the sine rule involves two sides and two angles, the cosine rule involves three sides and one angle.

Case 1: This involves two sides and two angles but you cannot use the sine rule directly to find BC. However, you can work out ∠C to be 51° (since the three angles in a triangle add up to 180°) and then find a by using the sine rule as follows.

$$\frac{a}{\sin 56°} = \frac{7.2}{\sin 51°}$$

$$\text{giving } a = \frac{7.2 \sin 56°}{\sin 51°}$$

$$= 7.68 \text{ (to 2 d.p.).}$$

Case 2: This involves three sides and an angle. You can use the cosine rule but the unknown occurs more than once, leading to a quadratic equation. An alternative method of finding x, shown below, is to first find ∠Y using the sine rule, then find ∠X (the third angle of the triangle) and finally use the sine rule again to find x.

$$\frac{\sin Y}{8.8} = \frac{\sin 77°}{9.2}$$

$$\text{giving } \sin Y = \frac{8.8 \sin 77°}{9.2}$$

$$\text{and} \quad \angle Y = 68.7°.$$

Then $\angle X = 180° - (77° + 68.7°) = 34.3°$.

Finally, using the sine rule again

$$\frac{x}{\sin 34.3°} = \frac{9.2}{\sin 77°}$$

$$\text{giving } x = \frac{9.2 \sin 34.3°}{\sin 77°}$$

$$= 5.32 \text{ (to 2 d.p.).}$$

Exercise

In questions 1 to 4 find the unknown side or angle.

1.
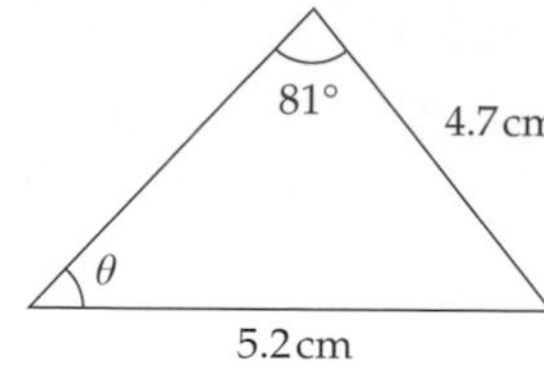

2.
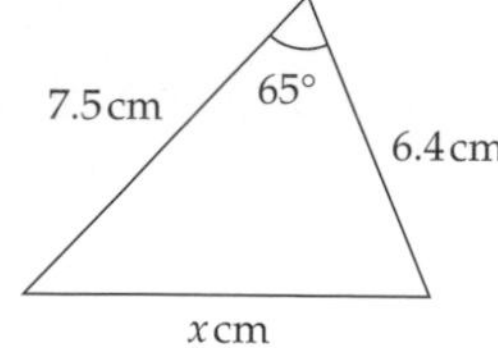

3.
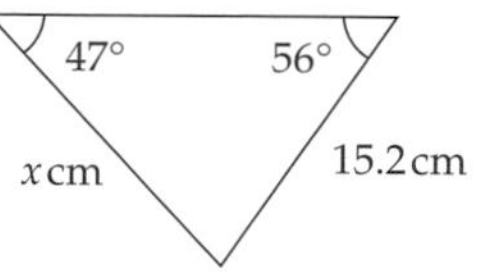

4.
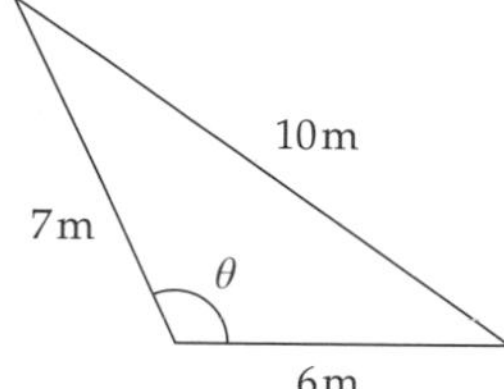

5. Craig sails 5 km south and then 6 km south-east. Work out how far he is from his starting position.

6. The diagram shows the side plan of a house.

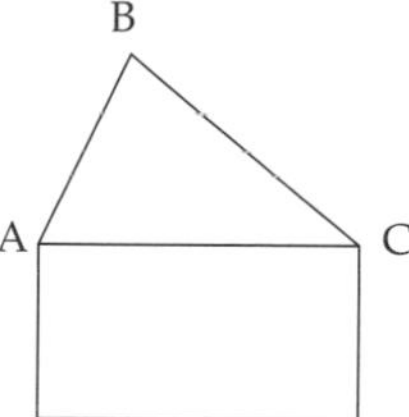

AC is horizontal, AB = 2.5 m, angle A = 80° and angle B = 45°.

Work out the length BC.

In questions 7 to 9 find the unknown side or angle.

7.
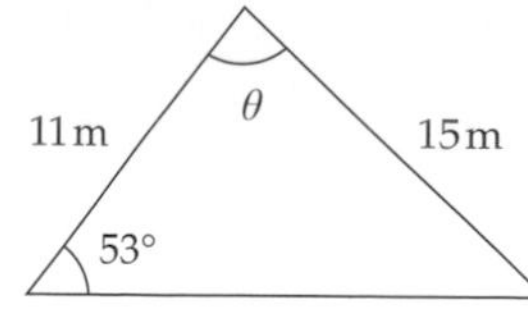

8.
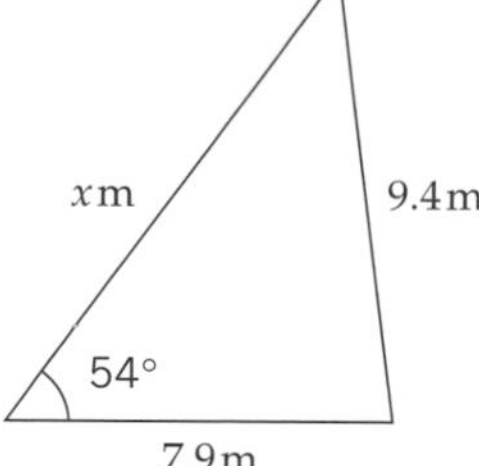

9.
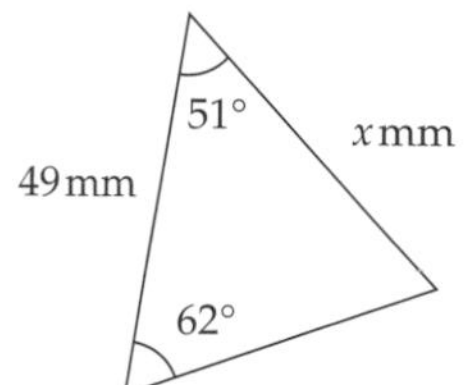

10. In triangle RST, angle R = 54°, RS = 76 mm and RT = 82 mm. Calculate angle S.

11. Point A is 4 km west of point B. Abdul walked from A on a bearing of 120°. Bethany walked from B on a bearing of 250°.

Their paths cross at C. Calculate the distances AC and BC. You may find it helpful to draw a diagram.

12. In triangle EFG, EF = 6 cm, FG = 7 cm and EG = 8 cm, Calculate the smallest angle in the triangle.

Activity: Proofs

Copy and complete these proofs using the diagram below.

1. Area of triangle $= \frac{1}{2}bc\sin A$

$\frac{h}{c} = \sin A$

$h = \ldots\ldots\ldots\ldots$

Area of triangle $= \frac{1}{2} \ldots \times h$

$\therefore$ Area $= \frac{1}{2}bc\sin A$

(Also $\frac{1}{2}ca\sin B$ or $\frac{1}{2}ab\sin C$)

2. Sine rule $\frac{\sin A}{a} = \frac{\sin B}{b}$

Area $= \frac{1}{2}bc\sin A = \frac{1}{2}ca \ldots\ldots\ldots$

(Divide by ......) $\frac{\sin A}{a} = \frac{\ldots\ldots\ldots}{b}$

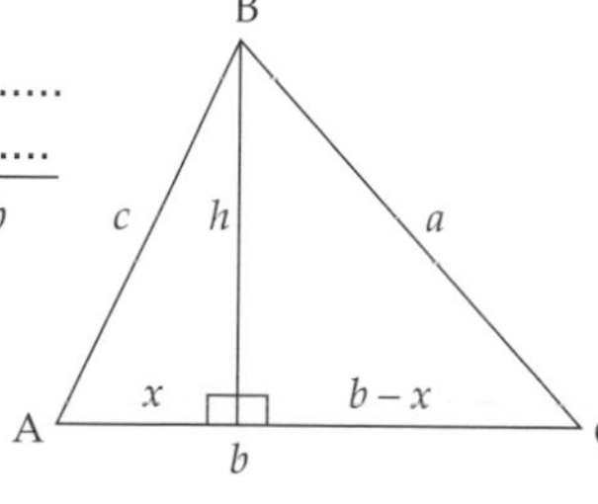

3. Cosine rule $\cos A = \frac{b^2 + c^2 - a^2}{2bc}$

$\frac{x}{c} = \cos A$

$x = \ldots\ldots$

$a^2 = (b - x)^2 + h^2$ (Pythagoras)

$a^2 = b^2 - 2bx + \ldots + \ldots$

and $c^2 = x^2 + h^2$ (Pythagoras)

Subtract $a^2 - c^2 = \ldots\ldots$

$2bc\ldots\ldots = b^2 + c^2 - a^2$

$2bc\cos A = \ldots\ldots$

(Divide by ...) $\cos A = \frac{b^2 + c^2 - a^2}{2bc}$

Answers

Page 3 *Exercise*

1. (i) 1, 2, 7, 14 (ii) 1, 2, 3, 4, 6, 8, 12, 24 (iii) 1, 2, 4, 5, 10, 20 (iv) 1, 2, 4, 7, 8, 14, 28, 56 (v) 1, 3, 9, 11, 33, 99
2. (i) 2, 4, 6, 8, 10 (ii) 5, 10, 15, 20, 25 (iii) 7, 14, 21, 28, 35 (iv) 11, 22, 33, 44, 55 (v) 17, 34, 51, 68, 85
3. (i) TRUE (ii) TRUE (iii) FALSE (iv) TRUE (v) FALSE (vi) TRUE
4. (i) 3 (ii) 10 (iii) 6 (iv) 18 (v) 14 (vi) 1 (vii) 6 (viii) 25
5. (i) 10 (ii) 12 (iii) 30 (iv) 54 (v) 120 (vi) 8 (vii) 12 (viii) 72
6. 23, 29
7. (i) 29, 31, 37 (ii) YES, 2 but it is the only one!

Investigations

1. The consecutive sum of three numbers is always a multiple of 3;
   the consecutive sum of five numbers is always a multiple of 5, etc.
   But for an even number of numbers:
   the consecutive sum of four numbers is always a multiple of 2;
   the consecutive sum of six numbers is always a multiple of 3, etc.
2. 6, 28
   496 has factors 1, 2, 4, 8, 16, 31, 62, 124, 248, 496.

Activity

Item	Number of units to be ordered
A4	2
Erasers	14
Print cartridges	2
HB Pencils	3
Box files	15
30 cm rulers	1
Staple guns	5

Page 5 *Exercise*

1. (i) $\frac{1}{2}=\frac{2}{4}=\frac{3}{6}=\frac{4}{8}=\frac{5}{10}=\frac{6}{12}$
   (ii) $\frac{1}{4}=\frac{2}{8}=\frac{3}{12}=\frac{4}{16}=\frac{5}{20}=\frac{6}{24}$
   (iii) $\frac{3}{5}=\frac{6}{10}=\frac{9}{15}=\frac{12}{20}=\frac{15}{25}=\frac{18}{30}$
   (iv) $\frac{5}{12}=\frac{10}{24}=\frac{15}{36}=\frac{20}{48}=\frac{25}{60}=\frac{30}{72}$
2. (i) $\frac{1}{2}$ (ii) $\frac{1}{4}$ (iii) $\frac{1}{4}$ (iv) $\frac{1}{4}$ (v) $\frac{1}{2}$ (vi) $\frac{3}{4}$ (vii) $\frac{2}{3}$ (viii) $\frac{2}{3}$ (ix) $\frac{4}{9}$ (x) $\frac{2}{3}$
3. $\frac{2}{3}=\frac{10}{15}$, $\frac{3}{5}=\frac{9}{15}$, so $\frac{2}{3}$ is the larger.
4. $\frac{5500}{55000}=\frac{1}{10}$; $\frac{2000}{20000}=\frac{1}{10}$. They are the same.
5. (i) $\frac{1}{2}$ (ii) $\frac{1}{2}$ (iii) $\frac{17}{24}$ (iv) $\frac{5}{8}$ (v) $\frac{7}{24}$
6. (i) $\frac{1}{4}$, $\frac{1}{3}$, $\frac{1}{2}$ (ii) $\frac{1}{4}$, $\frac{3}{10}$, $\frac{1}{2}$
7. 'Post-consumer' pulp.
8. (i) $1\frac{1}{3}$ (ii) $2\frac{1}{2}$ (iii) $1\frac{3}{7}$ (iv) $2\frac{1}{3}$ (v) $2\frac{1}{2}$ (vi) $3\frac{1}{3}$ (vii) $3\frac{1}{4}$ (viii) $18\frac{2}{3}$
9. (i) $\frac{9}{4}$ (ii) $\frac{23}{6}$ (iii) $\frac{37}{4}$ (iv) $\frac{101}{10}$ (v) $\frac{12}{11}$
10. (i) $\frac{10}{60}$ (ii) $\frac{1}{6}$ (iii) No: $\frac{1}{6}$ of 20 litres is $3\frac{1}{3}$ litres so she needs $3\frac{1}{3}$ litres of blue paint.
11. (i) $\frac{1}{4}$ (ii) $\frac{1}{5}$ (iii) $\frac{1}{10}$

Page 7 *Exercise*

1. (i) $\frac{7}{8}$ (ii) $\frac{1}{2}$ (iii) $\frac{5}{8}$ (iv) $\frac{23}{30}$ (v) $\frac{29}{24}=1\frac{5}{24}$ (vi) $\frac{23}{40}$ (vii) $3\frac{5}{8}$ (viii) $8\frac{1}{8}$
2. $\frac{1}{6}$, $\frac{5}{6}$
3. (i) $\frac{3}{8}$ (ii) $\frac{2}{5}$ (iii) $\frac{1}{16}$ (iv) $\frac{5}{24}$ (v) $1\frac{1}{4}$ (vi) $1\frac{3}{4}$ (vii) $\frac{3}{5}$ (viii) $1\frac{5}{12}$
4. Yes, with $\frac{13}{20}$ m to spare.
5. (i) $\frac{1}{4}$ (ii) $\frac{1}{8}$ (iii) $\frac{3}{16}$ (iv) $\frac{1}{100}$ (v) $\frac{3}{4}$ (vi) $\frac{6}{7}$ (vii) 8 (viii) $16\frac{2}{3}$
6. $21\frac{1}{8}$ m^2
7. (i) $1\frac{1}{2}$ (ii) $1\frac{1}{9}$ (iii) $17\frac{1}{2}$ (iv) 12 (v) $4\frac{2}{3}$ (vi) $\frac{1}{8}$ (vii) $1\frac{2}{3}$ (viii) $\frac{5}{8}$
8. $\frac{1}{10}$ **9.** (i) $\frac{2}{3}$ (ii) $\frac{4}{11}$ **10.** 400 **11.** £3000
12. (i) 128 cm (ii) 122 cm (iii) 4 **13.** B

Page 9 *Exercise*

1. (i) 248.65 (ii) 60.66 (iii) 6.87 (iv) 14.66
2. £5.73 **3.** (i) 1.05 m (ii) 0.95 m
4. (i) 71.6 (ii) 186.68 (iii) 5.04 (iv) 3.68 (v) 0.20355
5. £202.45 **6.** (i) 6.9 (ii) 0.12 (iii) 140
7. 86.5 m
8. (i) 0.26 m (or 26 cm) (ii) 0.54 m (or 54 cm) (iii) 1.37 m (iv) Mike's car: 0.7875 m; Debra's car: 0.8975 m (v) 1.685 m
9. (i) Company A (ii) Company A (iii) 2.25 km

Page 11 *Exercise*

1.

Fraction	Decimal	Percentage
$\frac{1}{4}$	0.25	25
$\frac{1}{5}$	0.2	20
$\frac{2}{5}$	0.4	40
$\frac{5}{8}$	0.625	62.5
$\frac{3}{40}$	0.075	7.5
$\frac{1}{10}$	0.1	10
$\frac{6}{25}$	0.24	24
$\frac{1}{3}$	$0.\dot{3}$	$33\frac{1}{3}$
$\frac{3}{5}$	0.6	60
$2\frac{1}{2}$	2.5	250
$\frac{9}{20}$	0.45	45
$\frac{3}{25}$	0.12	12

2. $\frac{1}{8} = 0.125 = 12.5\%$, so the bonus represents the larger amount. However, in the longer term the 10% salary increase may be worth more as it is permanent.

3. | | | |
|---|---|---|
| Light bulb 60 W | Infra-red lamp 250 W | Electric kettle 750 W |
| Vacuum cleaner 750 W | Storage heater 2500 W | Cooker 10 500 W |

4. B
5. Wednesday

Page 13 *Exercise*

1. (i) £480 (ii) £60 (iii) £177.50
2. (i) £32.90 (ii) £368.95 (iii) £102.39 (iv) £42.44 (v) £52.22
4. (i) $15.1\dot{5}\%$ increase (ii) 3.02% increase (iii) 12% decrease (iv) 12.5% increase (v) 38.24% decrease
5. (i) £810.75 (ii) £74.32 (iii) £608.06 (iv) £148.64
6. (i) £64.80 (ii) £15.61 (iii) £48.37 (iv) £12.33
7. (iii) £55.00 (iv) £21.00

Page 15 *Exercise*

1. (i) 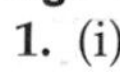

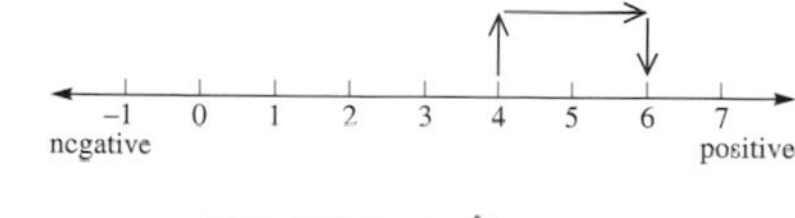

(ii)

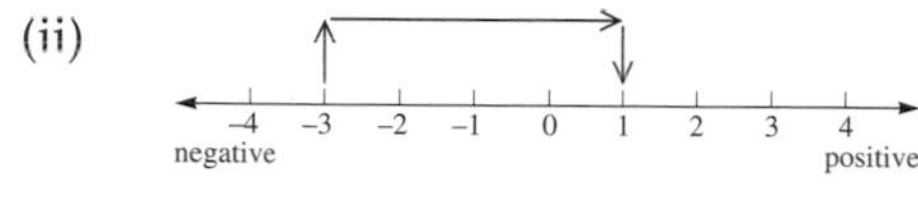

(iii)

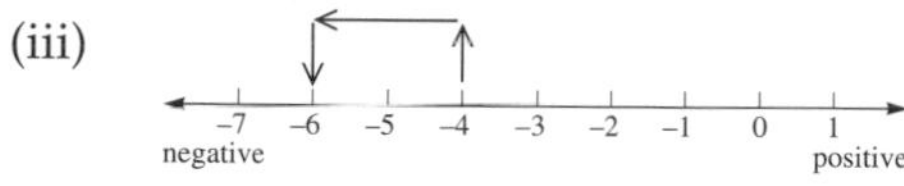

2. (i) Black £29 (ii) Black 50p (iii) Red £7.50 (iv) Red £1.43 (v) Black £8.29 (vi) Red £9.11
3. (i) Fall 12°C (ii) Rise 1°C (iii) Rise 19°C (iv) Fall 9°C
4. (i) +13 (ii) +3 (iii) –16.8 (iv) +8.2 (v) –10 (vi) +4 (vii) +8.5 (viii) –10
5. (i) –6 (ii) 4 (iii) –3 (iv) +4.2 (v) –4 (vi) –4 (vii) –50 (viii) +0.75
6. (i) 14 (ii) 11 (iii) 11 (iv) 11 (v) 8 (vi) 18 (vii) 2 (viii) 7
7. (i) –£1000 (ii) (–500) × (+2) = (–1000)
8. (i) 419 m (ii) 14 m below sea level
9. (i) 6.5 litres per day (ii) 650 litres

Page 17 *Exercise*

1.

2^5	2 × 2 × 2 × 2 × 2	32
2^4	2 × 2 × 2 × 2	16
2^3	2 × 2 × 2	8
2^2	2 × 2	4
2^1	2	2
2^0	1	1
2^{-1}	$\frac{1}{2}$	$\frac{1}{2}$
2^{-2}	$\frac{1}{2\times2}$	$\frac{1}{4}$
2^{-3}	$\frac{1}{2\times2\times2}$	$\frac{1}{8}$
2^{-4}	$\frac{1}{2\times2\times2\times2}$	$\frac{1}{16}$

As you move row by row down the last column, the numbers are divided by 2 each time.

2. (i) 9 (ii) 64 (iii) 81 (iv) 125 (v) 10 000
3. (i) $\frac{1}{4}$ (ii) $\frac{1}{81}$ (iii) $\frac{1}{25}$ (iv) $\frac{1}{1000}$ (v) $\frac{1}{4}$
4. (i) 2^4 (ii) 3^3 (iii) 10^2 (iv) 10^{-2} (v) 2^{-4} (vi) 7^{-2} (vii) 4^2 (viii) 5^0
5. (i) 10^{10} (ii) 10^3 (iii) 10^{19} (iv) 10^9 (v) 10^8 (vi) 10^{12} (vii) 10^4 (viii) 10^4 (ix) 10^{-6} (x) 10^6 (xi) 10^{-9} (xii) 10^{-36}
6. (ii) (a) 40 (b) 126 (c) 8800 (d) 225 (e) 1 000 000
7. (i) $2^2 \times 3 \times 5$ (ii) $2^2 \times 7^2$ (iii) $2^4 \times 3 \times 5^2$ (iv) $2^5 \times 5^3$ (v) 5^7
8. (ii) (a) $\frac{3}{4}$ (b) $\frac{2}{9}$ (c) $\frac{5}{12}$ (d) $\frac{9}{10}$ (e) $\frac{1}{1000}$
9. (i) $2^{-1} \times 3 \times 5^{-1}$ (ii) $2^2 \times 3^{-2}$ (iii) $3^2 \times 11^{-1}$ (iv) $5^2 \times 3^{-2} \times 2^{-2}$ (v) $2^3 \times 3^{-3}$
10. (i) £15 (ii) £31 (iv) Jan 1st 2010

Page 19 *Exercise*

1. (i) 16 cm² (ii) 81 cm² (iii) 100 cm² (iv) 6.25 cm²
2. 540 cm by 540 cm (5.4 m by 5.4 m)
3. 243.36 cm²
4. (i) 4.36 cm (ii) 3.61 cm (iii) 7.07 cm (iv) 9.15 cm
5. (i) 216 cm³ (ii) 512 cm³ (iii) 1000 cm³ (iv) 3.375 cm³
6. (i) 3.11 cm (ii) 5.19 cm (iii) 4.03 cm (iv) 5.95 cm
7.

Side (cm)	1	2	3	4	5	6	7
Area (cm²)	1	4	9	16	25	36	49

(i) About 6.5 cm (ii) 6.56 cm (to 3 sig. figs)

Activity

(i) $a = \sqrt{2}$, $b = \sqrt{3}$, $c = \sqrt{4}$ and so on.

Investigations

1. (i) 30 cm (ii) 10 cm (iii) 9 : 1 (iv) 3 : 1
2. (i) 8 cm (ii) 16 cm (iii) 1 : 8 (iv) 1 : 2

Page 21 *Exercise*

1. (i) 3.57 m (ii) 345 cm (iii) 1.328 g (iv) 0.59 kg (v) 3400 ml (vi) 1.734 m
2.

Top panes:	366 mm × 300 mm
Middle panes:	376 mm × 300 mm
Lower panes:	383 mm × 300 mm
Overall height:	1246 mm
Overall width:	672 mm
Glazing bar and frame:	25.4 mm

3. (i) cm/minute (ii) km/h (iii) cm² (iv) l/min (v) cm²/second (vi) ml/minute (vii) °C/min OR suitable alternatives
4. (i) 58.29 mph (ii) $13.\dot{3}1$/day (iii) 9 g/cm³
5. (i) 3.62 inches (ii) 8.75 miles (iii) 541 (iv) 77.50 euros (v) 2 gallons (vi) 15 miles

Activities

2. (iii) 14.2 km

Page 23 *Exercise*

1. (i) $3000 = 3 \times 10 \times 10 \times 10 = 3 \times 10^3$
   (ii) $343 = 3.43 \times 100 = 3.43 \times 10^2$
2. (i) 4×10^2 (ii) 6×10^4 (iii) 3.2×10^3 (iv) 4.32×10^6 (v) 2.376×10^2 (vi) 9.91×10^{14}
3. China: 1.131×10^9, UK: 5.76×10^7
4. (i) 2000 (ii) 400 000 (iii) 62 000 (iv) 763 (v) 797
5. (i) London to Sydney (ii) 5535 km, 17 005 km
6. (i) $0.003 = \frac{3}{1000} = \frac{3}{10\times10\times10} = \frac{3}{10^3} = 3 \times 10^{-3}$
   (ii) $0.000\,214 = \frac{2.14}{10000} = \frac{2.14}{10^4} = 2.14 \times 10^{-4}$

7. (i) 3×10^{-3} (ii) 1.7×10^{-4} (iii) 1.59×10^{-3} (iv) 1.726×10^{-5} (v) 1.2×10^{-1}
8. (i) 0.004 (ii) 0.000 56 (iii) 0.0489 (iv) 0.999 89
9. (i) 1.1×10^{-2} mm (ii) 0.011 mm
10. 0.000 3, 3×10^{-2}, 3, 30, 3×10^{2}
11. (i) 3.7×10^{6} (ii) 3.2×10^{-4} (iii) 7.3×10^{10} (iv) 2.4×10^{23} (v) 1.23×10^{-9} (vi) 6.48×10^{-1} (vii) 5×10^{11} (viii) 2.5×10^{23} (ix) 4×10^{3} (x) 2.5×10^{2}

Activities

1. (i) 1×10^{16} tonnes (ii) 1.275×10^{4} km (iii) 8.85×10^{3} m (iv) 1.1×10^{12} km^3 (v) 3×10^{8} ms^{-1} (vi) 9.109×10^{-28} g (vii) 3.3×10^{-9} seconds per metre

Page 25 *Exercise*

1. (i) 5:3 (ii) 3:2 (iii) 3:1 (iv) 4:1 (v) 1:3 (vi) 1:20 (vii) 40:9 (viii) 3:1
2. 40 men, 50 women
3. 4000
4. Megan: £32 000; Greg: £64 000
5. (i) (a) 20 m (b) 76 m (ii) (a) 10 cm (b) 260 cm
6. 1 tonne cement, $2\frac{1}{2}$ tonnes sand, $3\frac{1}{2}$ tonnes aggregate
7. 4 hours
8. Lead: 49 kg; zinc: 14 kg; tin: 7 kg
9. Leo: £16 000; Mel: £12 000; Nathan: £4000
10. (i) (a) 5.50 m by 2.50 m (b) 4.50 m by 2.50 m (ii) (a) 15.75 m^2 (b) 12.25 m^2 (iii) 4 cm and 3 cm (iv) 15 cm^2

Page 27 *Exercise*

1. (i) 204.75 (ii) 30.48 (iii) 33.15 (iv) 48.27 (v) 39.25
2. (i) (a) 10 oranges at £1.20 (b) 16 oz at 76p (c) '10 pack' at £10.10
3. $33.\dot{3}$ kg
4. 10 days
5. 96p
6. $2\frac{1}{2}$ minutes
7. 2 hours 24 minutes
8. 55 bottles (rounded up since part bottles are not possible)
9. One. He actually needs $6\frac{2}{3}$ kg, so he has some spare.
10. 7

Activities

2. (i) (a) 41 078 (to nearest whole number) (b) 42 373 (c) 334 864
   (ii) Con: 276; Lab: 227; Lib.Dem: 118
3. (i) About 13 km/l
   (ii) 12.7 km/l

Page 29 *Exercise*

1. (i) (a) 7 (b) 6 (c) 11 (d) 67 (ii) (a) 50 (b) 90 (c) 200 (d) 170 (e) 1840
   (iii) (a) 100 (b) 900 (c) 800 (d) 200 (e) 200
2. (i) True (ii) True (iii) False 4750 (iv) False 255
4. (i) (a) 200 (b) 2000 (c) 500 (d) 0.7 (e) 10.00
   (ii) (a) 1860 (b) 144 (c) 76.3 (d) 5290 (e) 16.4
   (iii) (a) 0.58 (b) 0.45 (c) 32.63 (d) 1.78 (e) 10.07
5. Simon £1.12 Sharon £3.11 Min £5.46 Chris £4.16
6. £86
7. £17 each would allow a tip of £4, i.e. 8.5%
   £17.50 each would allow a tip of £5.50, i.e. 11.7%

Page 31 *Exercise*

2. (i) $7 \times 2 = 14$ (ii) $25 \times 3 = 75$ (iii) $400 \div 10 = 40$ (iv) $70 - 10 = 60$ (v) $2 \times 800 = 1600$ (vi) $(20 \times 20) \div 50 = 8$ (vii) $600 \times 1 = 600$ (viii) $69 + 1 = 70$ (ix) $1500 \div 50 = 30$ (x) $100 \div 0.5 = 200$
3. Daily cost (1996 rates)
   $= 1100 \times 20 + 600 \times 26$ (pence)
   $\simeq 10 \times 20 + 6 \times 25$ (pounds)
   $= £(200 + 150) = £350$
   The working week is 5 days and there are 52 weeks in a year, so 1 yr $\simeq$ 250 days. Multiply the daily cost by 1000 and divide by 4:
   approx. annual cost = £35 000 $\simeq$ £9000
4. (ii) Correct answer: £128.96
6. About 1600
7. Three species: one species laid 1 or 2 eggs; one species laid 4, 5 or 6 eggs, the third laid a large number
8. Spirits 3; fortified wine 1; beer 4; shandy 3
9. (ii), (iii), (v)

Page 33 *Exercise*

1.

	Max	Min		Max	Min
(i)	15.5	14.5	(ii)	405.5	404.5
(iii)	4 min 25 s	4 min 15 s	(iv)	7.55	7.45
(v)	17.35	17.25	(vi)	10.165	10.155

2. 729.5 – 11.5 = 718 cm (ii) 0.139%
3. Smallest: 12.0075 m² Largest: 13.0175 m² 4.06%
4. (i) £239.58 (ii) £233.75 (iii) £233.16 **6.** Yes
7. $0.769\text{m/s} \le v \le 0.910\text{m/s}$ (to 3 sig. figs.)

Page 35 *Exercise*

1. (i) 24 cm (ii) 36 cm^2
2. (i) 26 m (ii) 40 m^2
3. (i)12.5 cm^2 (ii) 12 cm^2 (iii) 6 cm^2 (iv) 66 cm^2 (v) 24 cm^2 (vi) 35 cm^2
4. (i) 8.5 cm^2 (ii) 9.5 cm^2
5. 184 cm^2
6. 60 mm^2
7. (i) 41.75 m^2 (ii) 889 (assuming perfect cutting)

Page 37 *Exercise*

1. (i) Circumference = 18.85 cm; area = 28.27 cm^2
   (ii) Circumference = 62.83 cm; area = 314.16 cm^2
2. (i) Perimeter = 12.57 cm; area = 12.57 cm^2
   (ii) Perimeter = 7.71 cm; area = 3.53 cm^2
   (iii) Perimeter = 5.71 cm; area = 2.01 cm^2
   (iv) Perimeter = 28.19 cm; area = 41.56 cm^2
3. (i) Area = 12 379.58 m^2; perimeter = 429.50 m
   (ii) Area = 28.27 cm^2; perimeter = 24.85 cm
4. 585.84 cm^2 **5.** 549.78 mm^2

Page 39 *Exercise*

1. (i) 400 cm³ (ii) 367.5 cm^3
2. (i) 75 000 cm^3 (ii) 210 000 cm^3
3. (i) 12 000 cm^3 (ii) 500 cm^3
4. 2.4 m^3
5. 157.5 m^3
6. (i) (a) 300 cm × 200 cm × 100 cm
   (b) 3000 mm × 2000 mm × 1000 mm
   (ii) (a) 6 m^3 (b) 6 000 000 cm^3 (i.e. 6×10^6 cm^3)
   (c) 6 000 000 000 mm^3 (i.e. 6×10^9 mm^3)
   (iii) 1 000 000 (1 million)

Activities

2. False (i) $\sqrt[3]{2}$ (ii) $\sqrt[3]{3}$ (iii) $\sqrt[3]{27} = 3$

Page 41 *Exercise*

1. (i) 37.70 cm^3 (ii) 236.55 cm^3 (iii) 7.07 cm^3
2. (i) 1562 cm^3 (ii) 1.562 litres
3. (i) 7.85 cm^3 (ii) 31.42 cm^3
4. (i) Volume: 4.19 mm^3; surface area: 12.57 mm^2
   (ii) Volume: 8181 cm^3; surface area: 1963 cm^2
5. 6445

Activity

Volume of the Earth $\simeq 1.1 \times 10^{12}\text{ km}^3$
Surface area $\simeq 5.1 \times 10^8\text{ km}^2$
Average mass of 1 m^3 of the Earth $\simeq 5436$ kg
Core of Earth is very dense compared with soil.

Page 43 *Exercise*

1. (i) 240 cm^3 (ii) 120 cm^3
2. 140 cm^2
3. 20 mm by 20 mm
4. (i) 209 cm^3 (ii) 4712 cm^3
5. (i) 427 cm^2 (ii) 628 cm^2
6. 33.3 cm
7. 6.0 cm
8. 322 g
9. 6.51 cm

Investigations

1. (i) 1:2 (ii) 1:2
2. (i) 1:4 (ii) 1:8

Page 45 *Exercise*

1. (i) $N + 2$ (ii) $\frac{N}{2}$ (iii) $4N$ (iv) $\frac{2N}{3}$ (v) $N - 3$ (vi) $N + 6$
   (vii) $3N$ (viii) $N - 4$ (ix) $2N + 2$ (x) $N - 1$ (xi) $\frac{12}{N}$
   (xii) $\frac{N}{2} + 2$ (xiii) $4N - 2$
2. (i) £4.24 (ii) $100d - (25x + 72y + 132)$ (in pence)
3. (i) £255 (ii) £$(m + 6p)$
4. (i) £304 (ii) £$(40x + 8z)$
5. (i) £38 (ii) £$(36x - c)$
6. (i) 5 hours (ii) $(\frac{x}{4} + \frac{x}{6})$ hours ($= \frac{5x}{12}$ hours)
7. (i) 440 square feet (ii) $2h(a + b)$
8. (i) $y = 2x$ (ii) $x = 2y$ (or $y = \frac{1}{2}x$) (iii) $x + y = 100$
   (iv) $x = y - 20$
   (v) $y = x + 15$ (vi) $x - 10 = y + 10$ (or $x = y + 20$)
   (vii) $2(y - 20) = x + 20$ (viii) $x = 0$ (ix) $y = -10$
9. (i) $p + q = 40$ (ii) $p = 3q$ (iii) $2(p + 10) = q - 10$
   (iv) $2p = q - 5$ (or $p = \frac{q-5}{2}$) (v) $p = 2q$ (or $q = \frac{p}{2}$)
   (vi) $q = 0$
10. (i) Receipts (in £) for Thursday
    (ii) Receipts (in £) for Friday
    (iii) Receipts (in £) for Saturday
    (iv) Receipts (in £) for Friday and Saturday
    (v) Total number of tickets sold
    (vi) Mean number of tickets sold per performance
    (vii) The number by which Saturday's ticket sales exceed Friday's sales
    (viii) The number by which Friday's ticket sales exceed Thursday's sales
11. (i) The number of pieces of fruit that Jangir buys
    (ii) The cost of apples in pence
    (iii) The cost of bananas in pence
    (iv) The cost of the fruit in pence
    (v) The cost of apples in pounds
    (vi) The cost of bananas in pounds
    (vii) The cost of the fruit in pounds

Page 47 *Exercise*

1. (i) 45°C (ii) 50°C (iii) 5°C (iv) –25°C
2. (i) 72 m (ii) 52 m
3. £3349.385. (The bank would round this down to £3349.38)
4. (i) (a) £21.20 (b) £147.20 (ii) 15.8 pence
5. (i) 10 (ii) 8
6. (i) 1460.16 cm^3 (ii) 1950.75 cm^3
7. £250
8. (i) 12 m^2 (ii) (a) 100 m^2 (b) 27 m^2
   (iii) (a) 1000 m^3 (b) 135 m^3
9. (i) 0 (ii) –15 (iii) 33
10. (i) 9.52 cm (ii) 9.90 cm

Page 49 *Exercise*

1. (i) $6a$ (ii) $-2a$ (iii) $7c + d$ (iv) $-3e + 4f$
   (v) $9a + 5x$ (vi) $3k - 2n$ (vii) $3r - 4t$ (viii) $-3x + 5y$
   (ix) $-2q - 3r$ (x) $4a + 3c - 3f$ (xi) $2a - 3t$ (xii) $-9d$
2. (i) $6xy$ (ii) $-28xy$ (iii) $2x$ (iv) -5 (v) $30k^2$ (vi) $-24tuv$
   (vii) $-2d$ (viii) $\frac{-1}{3f}$
3. (i) x^7 (ii) y^3 (iii) z^8 (iv) a^3 (v) b^5 (vi) c^4 (vii) $14a^2$ (viii) $6c^2$
   (ix) a^5b^2 (x) a^5b^2 (xi) a^4 (xii) 3
4. (i) $6a - 8$ (ii) $-4n + 2$ (iii) $2p^2 + 2pq$ (iv) $-12x + 18y$
   (v) $24b + 4c$ (vi) $7c^2 - 7cd$ (vii) $6r - 3s - 3t$
   (viii) $-15x + 35y$

Investigation

1. Surface areas: $6p^2$; $24p^2$; $54p^2$ Volumes: p^3; $8p^3$; $27p^3$
2. (i) 8 (ii) 27
3. 1; 19
4. (i) 3 (ii) 3

Page 51 *Exercise*

1. (i) $x = 9$ (ii) $y = 12$ (iii) $d = 4$ (iv) $n = 32$ (v) $c = 5$ (vi) $x = 6$
   (vii) $x = 2$ (viii) $n = 5$ (ix) $d = 2.5$ (x) $x = -4$ (xi) $y = 7$
   (xii) $x = 4$ (xiii) $n = 10.5$ (xiv) $d = 43$ (xv) $x = 4$ (xvi) $x = 3$
   (xvii) $x = -0.6$ (xviii) $x = 15$ (xix) $y = 13$ (xx) $n = -0.8$
2. 46 000
3. (i) £15 (ii) $15x = 30\,000$; 2000
4. (i) $6x$ (ii) $240x$ (iii) $120x$ (iv) $360x = 25\,200$; 70p
5. $3x + 1.6 = 10.3$. 2.9 kg; 3.5 kg; 3.9 kg
6. (i) $80x$ (ii) $x + 25$ (iii) $50(x + 25)$
   (iv) $80x + 50(x + 25) = 2550$; 10p
7. (i) $9x$ (ii) $5(20 - x)$ (iii) $100 + 4x$ (iv) $100 + 4x = 148$
   (v) 12
8. (i) (a) $x + 25$ (b) $2x + 50$
   (ii) $4x + 75 = 135$; $x = 15$; red 80.
9. (i) (a) $x + 25$ (b) $12x$ (c) $10(x + 25)$
   (ii) $12x + 10(x + 25) = 2010$; $x = 80$ and a flannel costs £1.05

Page 53 *Exercise*

1. (i) Seven is greater than five. (ii) Six is less than eight.
   (iii) Three is less than four which is less than six.
   (iv) Ten is greater than eight which is greater than seven.
   (v) Two to the power ten is greater than a thousand.
   (vi) N is greater than three. (vii) A is less than eight.
   (viii) x and y both equal three. (ix) a is not equal to two.
   (x) x is not equal to zero. (xi) The value of π squared is approximately ten. (xii) a is not equal to b.

2. (i) $32 > 3$ (ii) $0.5^2 < 0.5$ (iii) $N = 8$ (iv) $A > 6$ (v) $z \neq 0$ (vi) $p < 3.5$ (vii) $A = B$ (viii) $x > y$ (ix) $a = b = 7$ (x) $10 < y < 20$
3. (i) 3, 4, 5, 6 (ii) 3, 2, 1, 0, –1, –2 (iii) –1, 0, 1 (iv) 49, 48, 47 (v) 4, 5, 6, 7 (vi) –1, 0 (vii) 3, 4, 5 (viii) 3, 4 (ix) 5, 6, 7, 8 (x) 3, 4, 5
4. (i) $R < 28$ (ii) $t < 9$ (iii) $x < 15$ (iv) $x > 5$ (v) $a < 3$ (vi) $b < 3$ (vii) $p > 10$ (viii) $h > 3$ (ix) $x < 4$ (x) $x > 3$ (xi) $b < -1$ (xii) $c < -2$ (xiii) $x > 5$ (xiv) $x \leq 3$ (xv) $x > 5$ (xvi) $x < 3$ (xvii) $x \geq 3$ (xviii) $x \geq 0.5$
5. (i) (a) $6x \geq 720$ (b) $x \geq 120$ (ii) 30 seconds
6. (i) $1200 \leq 200x \leq 1600$ (ii) 6, 7 or 8
7. (i) $1 \leq c \leq 4$ (ii) (a) $7 \leq c \leq 28$ (b) $14 \leq c \leq 56$
8. (i) $1.25x + 24$ (ii) $1.25x + 24 \leq 35$; 8
9. (i) $15b \leq 3000$ (ii) £200
10. (i) $652 + 62x$ (ii) $652 + 62x \leq 1000$; $x = 1, 2, 3, 4$, or 5

Page 55 *Exercise*

1. (i) 12; $3n$ (ii) 13; $1 + 3n$ (iii) 15; $3 + 3n$ (iv) 23; $-1 + 6n$ (v) 1.8; $1 + 0.2n$ (vi) 6; $22 - 4n$
2. $75n + 45$ seconds
3. $150 - 10t$; midday
4. £$(420 - 5n)$
5. (i) £30 (ii) £75 (iii) £$(n + 25)$
6. (i) (a) 5.75 hours (b) 7 hours (ii) $\frac{5c+3}{4}$ hours
7. (i) 83 (ii) 68 (iii) $98 - 3n$
8. (i) 4 h 15 min (ii) 4 h 2.5 min (iii) 4 h + $2.5n$ min
9. (i) 200 (ii) 300 (iii) $100 + 20n$

Page 57 *Exercise*

1. (i) Because the amount of icing depends on the area to be covered, and the area of a circle is πr^2. (ii) $2n^2$
2. (ii) $n(n - 1)$
3. (i) 16, 20, 24, 28; $4n + 12$ (ii) $(n + 2)$ and $(n + 4)$ (iii) $(n + 2)(n + 4)$
4. (i) £15 000 × 1.05 = £15 750 (ii) £15 000 × 1.05^2 = £16 537.50 (iii) £15 000 × 1.05^{10} = £24 433.42 (iv) £15 000 × 1.05^n
5. (i) $10 \times 0.5 = 5$ (ii) $10 \times 0.5^2 = 2.5$ (iii) $10 \times 0.5^3 = 1.25$ (iv) 10×0.5^n
6. Alpha: (i) £67 500 (ii) £50 625 (iii) £28 476.56 (iv) £90 000 × 0.75^n
   Epsilon: (i) £34 000 (ii) £28 900 (iii) £20 880.25 (iv) £40 000 × 0.85^n
   Omega: (i) £13 125 (ii) £11 484.38 (iii) £8792.72 (iv) £15 000 × 0.875^n
7. (i) 2^{n-1} (ii) 2^{63} pence (iii) 3^{n-1} pence (iv) 3^{63} pence

Investigation

(i) n (ii) $n + 1$ (iii) $\frac{1}{2}n(n + 1)$ (v) 210; 10
(vi) Double it and see if it has two consecutive numbers as factors.

Page 59 *Exercise*

1. (i) $x = 16, y = 4$ (ii) $x = 7, y = 9$ (iii) $x = 3, y = 2$ (iv) $x = 6, y = 1$ (v) $x = -1, y = 6$ (vi) $x = 4, y = 2$ (vii) $x = 5, y = 5$ (viii) $x = 2, y = -3$ (ix) $x = 7, y = 1$ (x) $x = -1, y = -2$ (xi) $x = 5, y = 0$ (xii) $x = 10, y = -2$
2. (i) 5 minutes (ii) 2 hours 45 minutes
3. $p = 10, q = 5$
4. $x = 2, y = 3$
5. P = £100 million, L = £80 million
6. 15 ml; 5 ml
7. 10 trays of sandwiches, 5 pizzas
8. £8; £10
9. £1.20; 50p
10. 60 Alpha units; 40 Beta units
11. Arborough: 800; Essminster: 700

Page 61 *Exercise*

1. (i) $x^2 + x - 12$ (ii) $x^2 + x - 2$ (iii) $x^2 - 8x + 15$ (iv) $x^2 + 5x + 4$ (v) $x^2 + 5x + 6$ (vi) $x^2 - 4x - 21$ (vii) $x^2 - x - 2$ (viii) $x^2 - 8x + 12$ (ix) $x^2 - 6x + 9$ (x) $x^2 - 1$ (xi) $x^2 + 5x - 6$ (xii) $x^2 - 16$ (xiii) $x^2 + 4x + 4$ (xiv) $x^2 - 2x + 1$ (xv) $9x^2 + 6x + 1$ (xvi) $6x^2 + 5x + 1$ (xvii) $10x^2 + 28x + 16$ (xviii) $14x^2 - 5x - 1$ (xix) $18x^2 - 21x - 4$ (xx) $10x^2 - 3x - 1$ (xxi) $2x^2 + 5x - 12$ (xxii) $6x^2 - 19x - 7$ (xxiii) $24x^2 - 25x + 6$ (xxiv) $30x^2 - 11x + 1$ (xxv) $15x^2 - 37x + 20$
2. (i) (a) $x + 4$ (b) $x + 2$ (c) $(x + 2)(x + 4)$ (ii) $x^2 + 6x + 8$
3. (i) $12 - 2x$ (ii) $10 - 2x$ (iii) $(12 - 2x)(10 - 2x)$ (iv) $120 - 44x + 4x^2$
4. (i) (a) πr^2 (b) $r + x$ (c) $\pi(r + x)^2$ (ii) $\pi x(2r + x)$
5. (i) (a) ab (b) $e(a - c - d)$ (c) $ab - ae + ce + de$ (ii) $a(b - e) + ce + de$, or $bc + (b - e)(a - c - d) + bd$

Page 63 *Exercise*

1. $5(x + 2)$
2. $3(x - 3)$
3. $5(x + 5)$
4. $7(x - 2)$
5. $3(2x + 1)$
6. $5(3x + y)$
7. $4(5x - 4y)$
8. $6(4x - 3y)$
9. $a(b + 2)$
10. $a(b + 6c)$
11. $a(c + d)$
12. $b(c + d)$
13. $3c(2a + b)$
14. $7x(2x + 3y)$
15. $a(b + c - d)$
16. $3a(c + 2d - f)$
17. $a(p + q + r)$
18. $e(a - d - f)$
19. $a(a - b - e)$
20. $4a(3a - 2b + 5c)$
21. $(a + 3)(x + y)$
22. $(a + b)(a + c)$
23. $(x + y)(a + b)$
24. $(a - d)(e + b)$
25. $(a - b)(a + e)$
26. $(p + q)(a + b)$
27. $(a + c)(a - 5)$
28. $(a - 2)(b - 1)$
29. $(x - y)(3x + 7)$
30. $(x + y + z)(a - b)$
31. $(a + c)(b - d)$
32. $(m - 2n)(a - b)$
33. $(y - z)(y + 2)$
34. $(b + c)(2a + 1)$
35. $(a - b)(a + 2)$
36. $(m + n + p)(a + b)$
37. $(3a - 1)(a - 2)$
38. $(a + 2)(5a - 1)$
39. $(2x + 1)(x + 7)$
40. $(5x - 3)(x - 1)$
41. $(x - 2)(3x + 4)$
42. $(3x + 5)(2x - 1)$
43. $(3y + 5)(y + 1)$
44. $(x + 2)(7x + 2)$
45. $(1 - 3t)(1 - 8t)$
46. $(2d - 1)(5d + 1)$
47. $(4y - 1)(3y - 1)$
48. $(t - 3)(t - 4)$
49. $(2x - 1)(3x + 4)$
50. $2(a - 1)(4a + 3)$
51. $2(x + 2)(2x + 3)$
52. $3(2x - 3)(3x + 2)$
53. $3(x + 1)(8x - 5)$
54. $2(x + 2)(3x + 5)$
55. $5(4x + 1)(2x + 9)$
56. $(2x - 1)(6x - 5)$
57. $(a + 2)(14a + 3)$
58. $(2x - 5)(3x + 2)$
59. $(x - 3)(3x - 11)$
60. $(a - 2b)(4a - 5b)$

Activities

2. (i) $a^2 + c^2 = b^2$ (ii) $\frac{\pi b^2}{2}$, $\frac{\pi a^2}{2}$, $\frac{\pi c^2}{2}$ (iii) Yes
3. $\pi(15 - t)^2(19 - x - t)$ (ii) 4 mm

Page 65 *Exercise*

1. $(x + 7)(x - 3)$
2. $(x + 1)(x + 2)$
3. $(a + 9)(a - 8)$
4. $(a - 8)(a - 3)$
5. $(a - 7)(a + 3)$
6. $(x + 9)(x + 2)$

7. $(a-7)(a-1)$ 8. $(b-4)(b-1)$
9. $(x-1)(x+4)$ 10. $(y+4)(y-3)$
11. $(x-10)(x-1)$ 12. $(x+7)(x-2)$
13. $(x+20)(x+3)$ 14. $(x+15)(x-2)$
15. $(x-6)(x-6)$ 16. $(2x+1)(x+1)$
17. $(4y+1)(y+1)$ 18. $(p+2)(3p-1)$
19. $(n+2)(2n-3)$ 20. $4(t+6)(t-3)$
21. $(3a-1)(a-2)$ 22. $(a+2)(5a-1)$
23. $(x+7)(2x+1)$ 24. $(x-1)(5x-3)$
25. $(3x+4)(x-2)$ 26. $(3x+5)(2x-1)$
27. $(3y+5)(y+1)$ 28. $(3c+2)(4c-3)$
29. $(6x+1)(2x-3)$ 30. $(8t-1)(3t-1)$

Investigation

1. (i) $(x+9)(x-9)$(ii) $(x+10)(x-10)$ (iii) $(x+8)(x-8)$ (iv) $(x+5)(x-5)$ (v) $(x+11)(x-11)$ (vi) $(x+12)(x-12)$ (vii) $(2a+5)(2a-5)$ (viii) $(4a+7)(4a-7)$ (ix) $9(3+a)(3-a)$ (x) $4(3b+4)(3b-4)$ (xi) $(x+y)(x-y)$ (xii) $(2x+3y)(2x-3y)$

Page 67 *Exercise*

1. $x = 0$ or 3 2. $x = 4$ or 0
3. $x = \frac{3}{2}$ or 5 4. $x = -\frac{1}{5}$ or $\frac{2}{5}$
5. $x = -1$ or 2 6. $x = -4$ or 4
7. $x = -4$ or $\frac{1}{4}$ 8. $x = -3$ or $\frac{5}{2}$
9. $t = 2$ or 3 10. $t = -2$ or -3
11. $x = 1$ or 2 12. $t = -2$ or 3
13. $x = 3$ or 5 14. $x = -3$ or -5
15. $x = -7$ or -4 16. $x = 7$ or 4
17. $x = 4$ or 5 18. $x = -3$ or 4
19. $y = \frac{1}{2}$ or 2 20. $x = -1$ or $\frac{2}{3}$
21. $x = -\frac{1}{3}$ or $\frac{3}{2}$ 22. $x = -3$ or $-\frac{1}{4}$
23. $x = \frac{3}{2}$ or 2 24. $x = \frac{5}{3}$ (repeated)
25. (i) $(16-2x)$ by $(10-2x)$ (ii) $x = 2$; reject $x = 11$. Lawn 12 m × 6 m
26. $t = 2$; reject $t = 0$ 27. $x = 1$ or 2

Page 69 *Exercise*

1. $x = -2$ or 7 2. $x = 1$ or 11
3. $x = 5.32$ or -1.32 4. $x = -0.29$ or 10.29
5. $x = 0.76$ or 5.24 6. $x = -5.41$ or -2.59
7. $x = -3.37$ or -1.63 8. $x = -2.66$ or 0.66
9. $x = -0.06$ or 6.06 10. $x = 3.37$ or 1.63
11. $x = 0.82$ or -2.42 12. $x = -1.55$ or 0.80
13. $x = -2.78$ or -0.72 14. $x = -2.38$ or 1.05
15. $x = 1.08$ or -3.41 16. $x = -0.26$ or 1.93
17. $x = 0.51$ or 2.82 18. $x = 2.87$ or -0.87
19. $x = -0.44$ or 3.44 20. $x = 6.89$ or 0.11
21. $x = -1.24$ or 3.24
22. $x = 2$, $R = C = 196$ or $x = 50$, $R = C = 2500$
23. 4, reject -25 24. 80 metres or 120 metres 25. 0.5

Investigation

(i) You need to find the square root of a negative number
(ii) The curve does not cross the x axis
(iii) (a) two roots (b) one (repeated) root (c) no roots

Page 71 *Exercise*

1. $\frac{5x}{6}$ 2. $\frac{9x}{10}$
3. $\frac{x}{2}$ 4. $\frac{x}{12}$
5. $\frac{8x}{15}$ 6. $\frac{2x}{15}$
7. $-\frac{x}{15}$ 8. $\frac{31x}{20}$
9. $\frac{x}{28}$ 10. x
11. $\frac{a}{5}$ 12. $\frac{25b}{36}$
13. $\frac{x+3}{3}$ 14. $\frac{3x+14}{6}$
15. $\frac{39-4x}{35}$ 16. $\frac{x-5}{6}$
17. $\frac{x+7}{12}$ 18. $\frac{x+16}{10}$
19. $x = 6$ 20. $x = 10$
21. $x = 8$ 22. $x = 24$
23. $x = 30$ 24. $x = 15$
25. $x = -15$ 26. $x = -40$
27. $x = -7$ 28. $x = \frac{3}{4}$
29. $a = 5$ 30. $b = 36$
31. $x = 4$ 32. $x = 1$
33. $x = 1$ 34. $x = -1$
35. $x = -8$ 36. $x = -\frac{2}{5}$
37. (i) $\frac{x}{8}$ hours (ii) $\frac{x}{12}$ hours (iii) $\frac{5x}{24}$ hours (iv) 6 miles.
38. (i) $\frac{x}{3}$ hours (ii) $\frac{x}{5}$ hours (iii) $\frac{8x}{15}$ hours (iv) 1.5 miles.

Page 73 *Exercise*

1. $\omega = \frac{2\pi}{T}$
2. (i) $b = \frac{2A}{h}$ (ii) $h = \frac{2A}{b}$
3. (i) $a = \frac{v^2-u^2}{2s}$ (ii) $s = \frac{v^2-u^2}{2a}$
4. (i) $L = \frac{p}{2} - B$ (ii) $B = \frac{p}{2} - L$
5. $V = \sqrt{\frac{5hR}{12d}}$ 6. (i) $h = \frac{V}{\pi r^2}$ (ii) $r = \sqrt{\frac{V}{\pi h}}$
7. (i) $u = v - at$ (ii) $a = \frac{v-u}{t}$
8. (i) $l = \frac{\lambda x}{T}$ (ii) $x = \frac{lT}{\lambda}$
9. $H = \frac{v^2}{64}$ 10. $l = \frac{gT^2}{4\pi^2}$
11. (i) 25.5 mm (ii) $T = \frac{2}{3}(L - C)$ (iii) 2 mm
12. (i) 320 (ii) $t = \frac{320m-c}{8}$ (iii) 120 minutes
13. (i) 48 (ii) $y = \frac{N-4-2x}{2}$ (iii) 30 m (iv) $N = 4x + 4$ (v) $x = \frac{N-4}{4}$ (vi) 40 m square
14. (i) 3.06 (ii) $P = \frac{RC}{100}$ (iii) £1575
15. (i) 4550 (ii) $H = \frac{C}{S}$ (iii) 81.25

Page 75 *Exercise*

1.

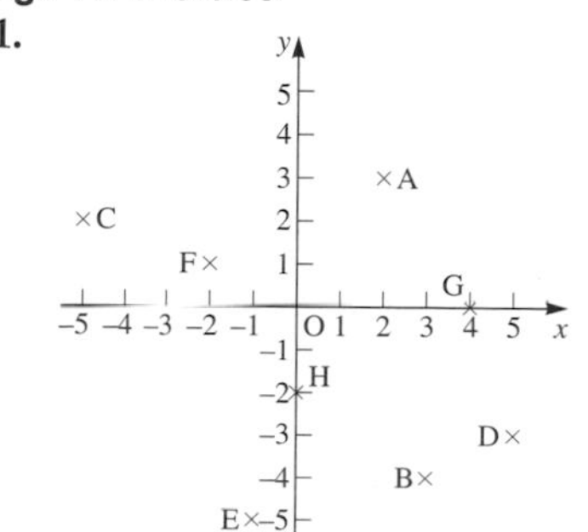

2. A(3,2); B(4,3); C(1,–2); D(–2, 1); E(–3,–3); F(4,–1); G(0,3); H(2,0) I(–4,0); J(0,–4)

3.

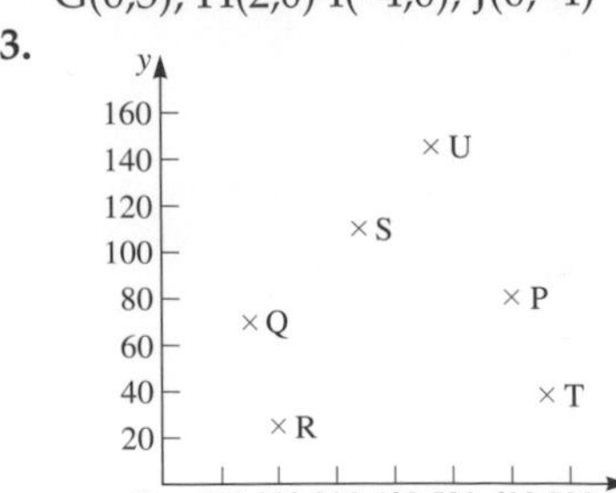

4. (a) 5.4 (b) 13 (c) 35.5 (d) 70 (e) –5.2 (f) 3 (g) –0.44 (h) 0.087

Activity

O (0,0,0); B (8,9,0); C (0,9,0); D (0,0,6); E (8,0,6); F (8,9,6); G (0,9,6)

Page 77 *Exercise*

1. (i) £60 (ii) £88 (iii) 175 miles
2. (i) 96 cm (ii) 127 cm (iii) 12 $\frac{3}{4}$ years
3. (i) 42 mpg (ii) 50 mph (iii) 28–76 mph
4. (i) £40 000 (ii) £26 000 (iii) 15 years

Page 79 *Exercise*

1. (i)

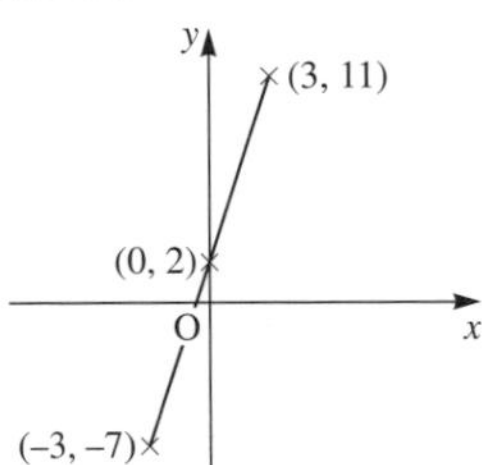

(ii)

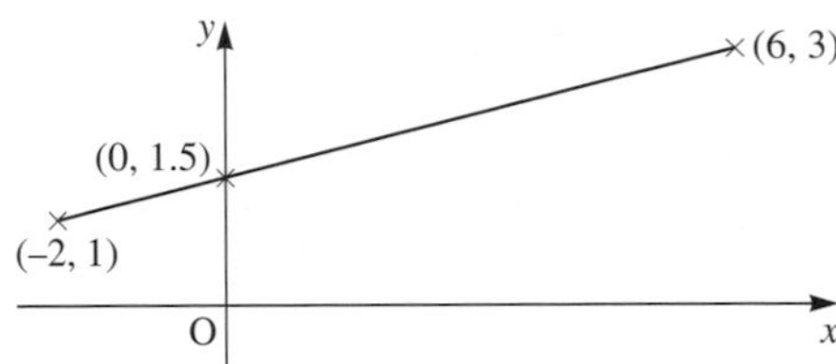

(iii)

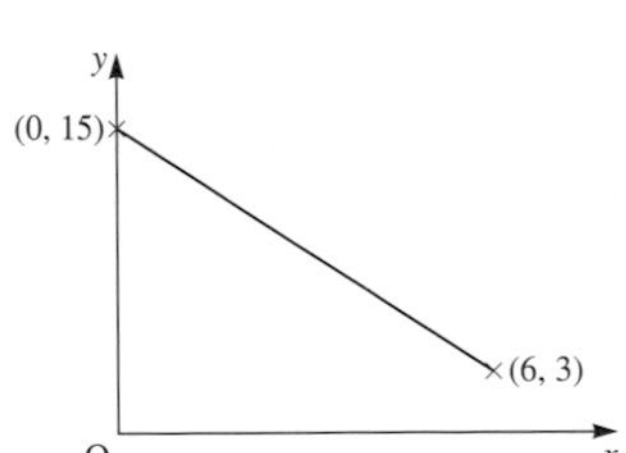

(iv)

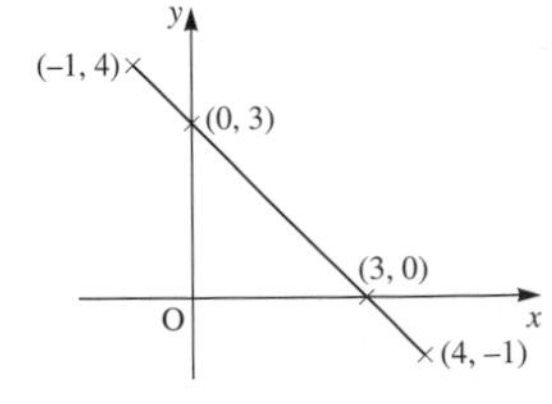

(v)

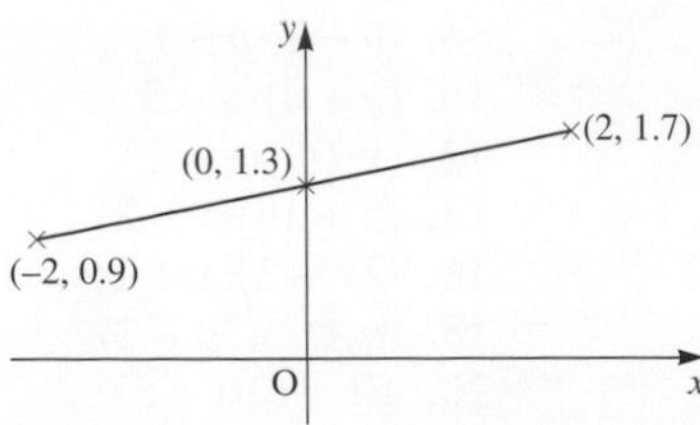

(vi)

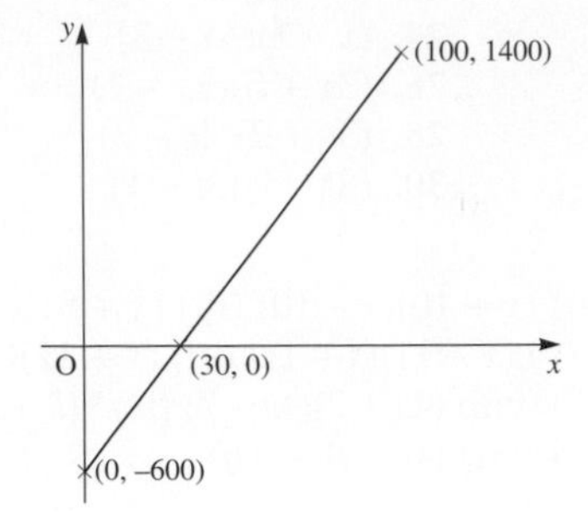

(vii)

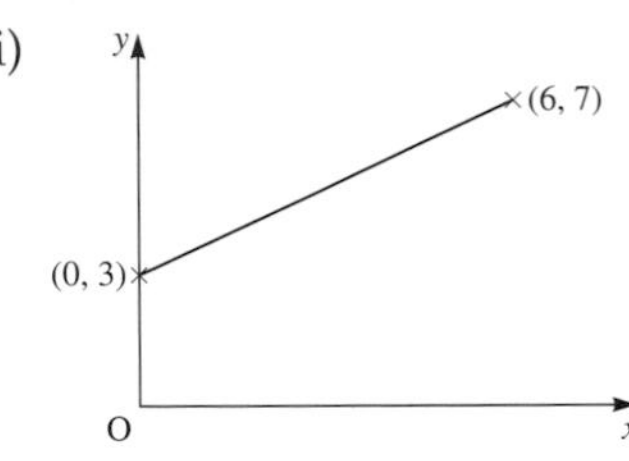

(viii)

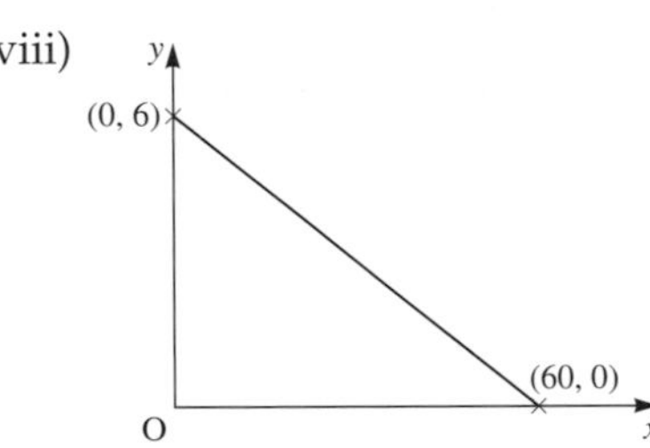

2.

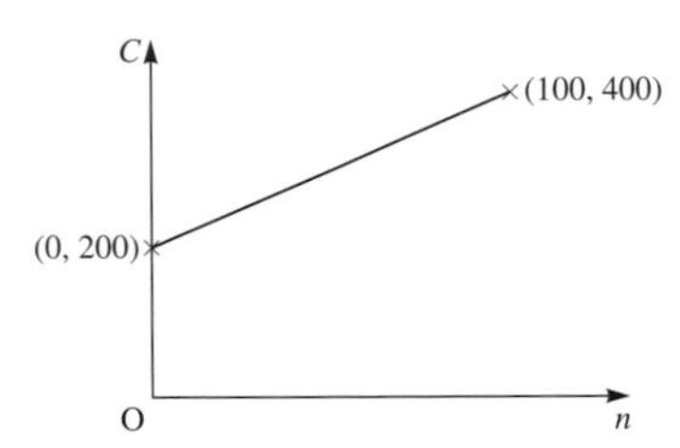

(i) £270 (ii) 68 items

3.

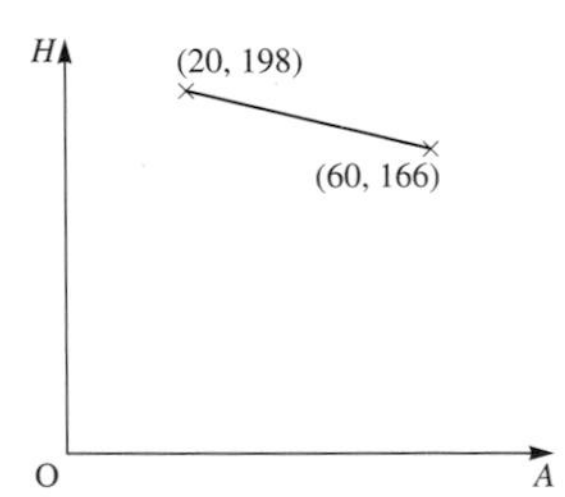

(i) 179/180 bpm (ii) 31 years

4. 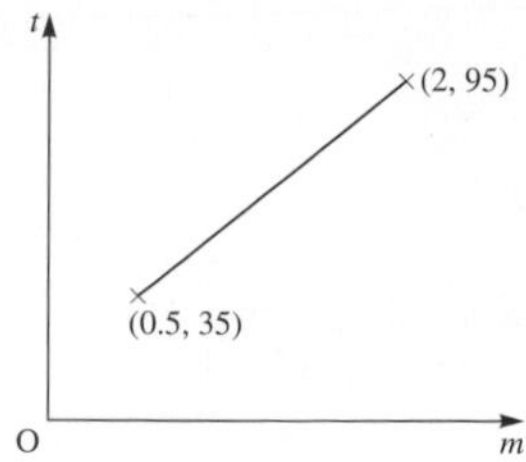

(i) 47 minutes (ii) 1.5 kg

5. 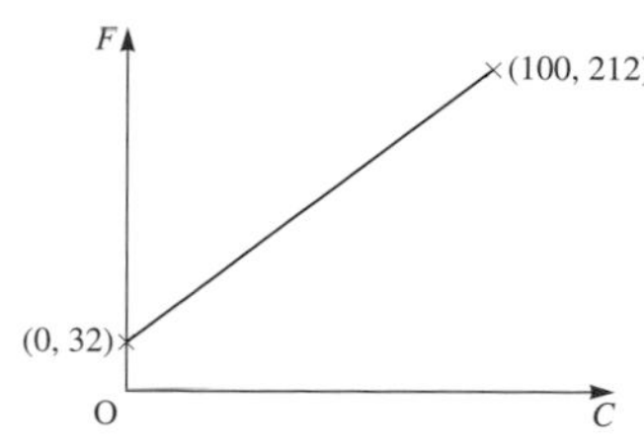

(i) 68°F (ii) 36.6/36.7°C

6. (i) 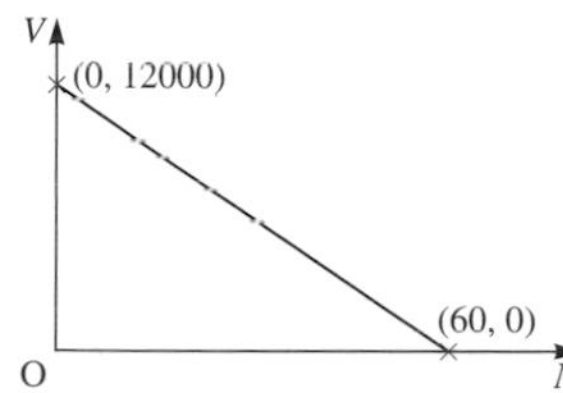

(ii) 20 months (iii) 45 months (iv) £4200

Activities

2. (i) About 81 cm (ii) Length about 11.7 inches, width about 8.3 inches
(iii) About 71 cm (iv) Width about 24 inches, length about 47 inches

Page 81 *Exercise*

1. (i) 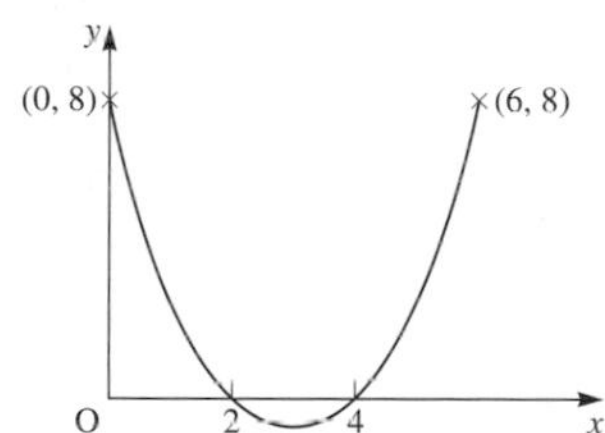

(ii) 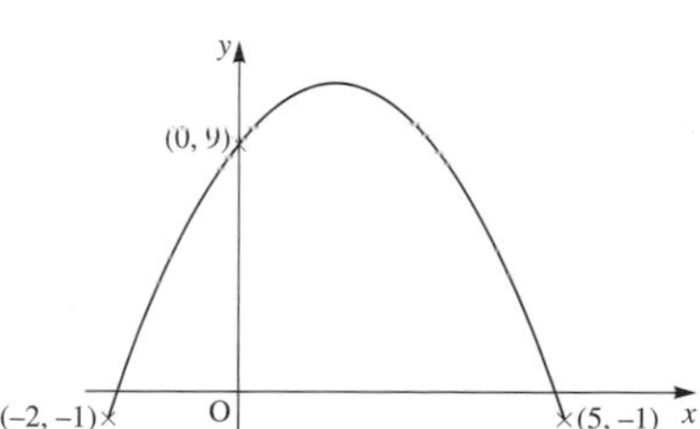

(iii) 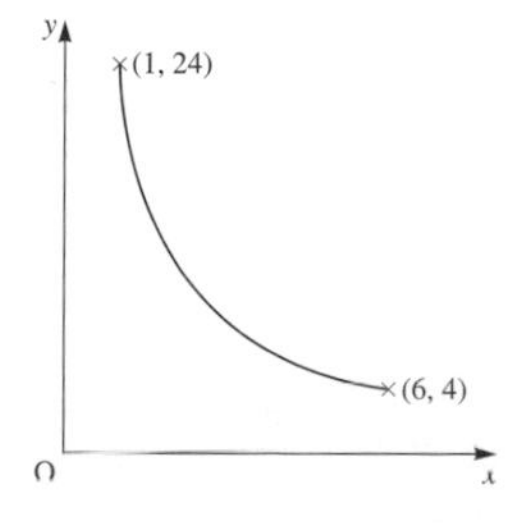

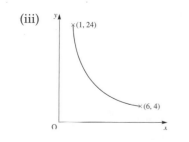

(iv) 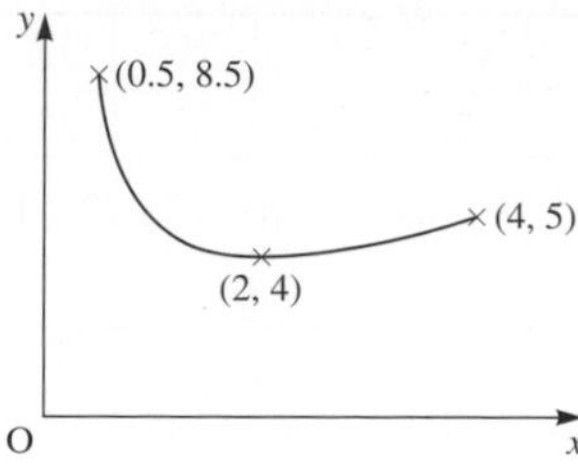

(v) 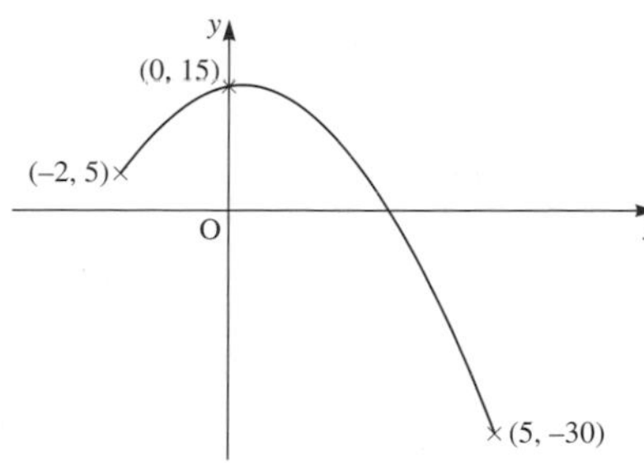

(vi) 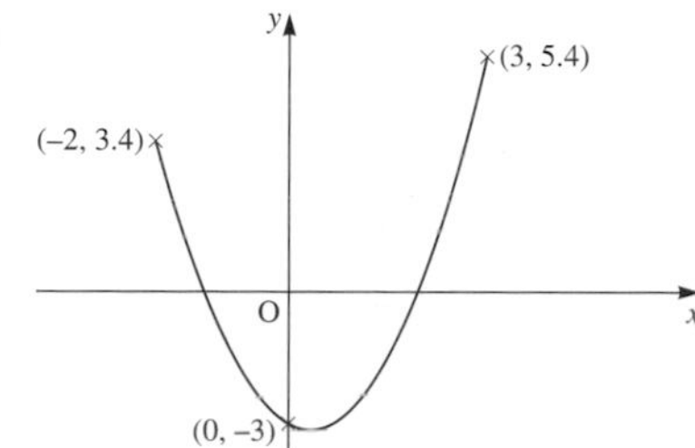

(vii) 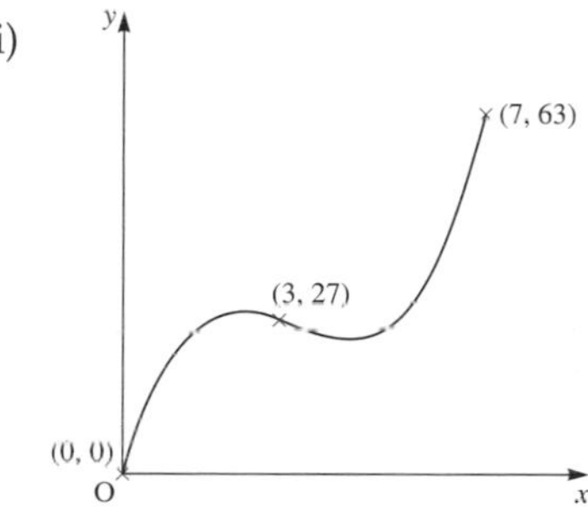

(viii) 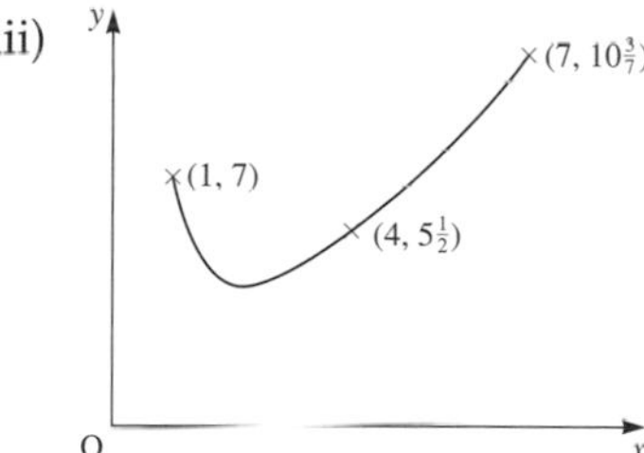

2. (i) 40 – 41 cm² (ii) 1.7 – 1.8 cm

3. 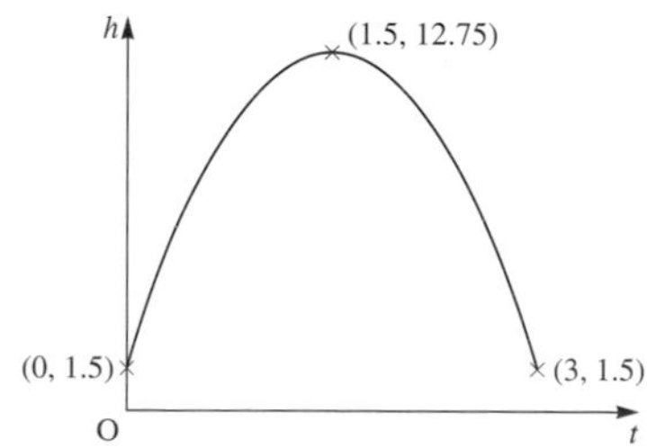

(i) 10.3 m (ii) $t = 0.57$ or 2.43 (iii) 1.5 (iv) 12.75 m

4.

Q	2	4	6	8	10	12	14
AUC	45	27	22.3	21	21	21.7	22.7

(i) £25.30 (ii) £20.90, 890 items (iii) $6.3 \leq Q \leq 12.7$

5. (ii)

x	0	1	2	3	4	5	6
V	0	180	256	252	192	100	0

(iv) 263 cm^3 when $x = 2.43$ (v) $0.71 \leq x \leq 4.59$

Page 83 *Exercise*

1. (i) (a) 2 (b) 2 (ii) (a) $-\frac{1}{2}$ (b) $-\frac{1}{2}$
   (iii) It does not matter which two points are used.
2. (a) (1,2), (3,4), (6,7); gradient = 1; y intercept = 1
   (b) (–1,7), (0,5), (4,–3); gradient = –2; y intercept = 5
   (c) (–4,8), (–2,6), (3,1); gradient = –1; y intercept = 4
   (d) (–1,4), (2,4), (4,4); gradient = 0; y intercept = 4
   (e) (0,–3), (4,0), (8,3); gradient = $\frac{3}{4}$; y intercept = –3
   (f) (–3,–1), (0,0), (3,1); gradient = $\frac{1}{3}$; y intercept = 0
3. (i) gradient = $\frac{1}{2}$; y intercept = 1.5
   (ii) gradient = –1; y intercept = 6
   (iii) gradient $-\frac{1}{3}$; y intercept = 4
   (iv) gradient = $\frac{2}{5}$; y intercept = –1.4
   (v) gradient = 2; y intercept = –4
   (vi) gradient = $-\frac{3}{4}$; y intercept = 6
4. (a) 5 (b) –0.04

Page 85 *Exercise*

1. (i) gradient = 4; y intercept = –1
   (ii) gradient = –3; y intercept = 7
   (iii) gradient = –1; y intercept = 8
   (iv) gradient = $\frac{3}{2}$; y intercept = –6
   (v) gradient = 3; y intercept = –5
   (vi) gradient = $\frac{5}{2}$; y intercept $\frac{1}{4}$
   (vii) gradient = $\frac{7}{4}$; y intercept = 0
   (viii) gradient = $-\frac{2}{5}$; y intercept = 4
2. (i) gradient = –2; y intercept = 5; $y = -2x + 5$
   (ii) gradient = $\frac{1}{20}$; y intercept = 0 $y = \frac{1}{20}x$ or $y = 0.05x$
   (iii) gradient = $\frac{1}{5}$;
   y intercept = –2 $y = \frac{1}{5}x - 2$ or $y = 0.2x - 2$
3. (i) –1 (ii) 30 (iii) $y = -x + 30$

Page 87 *Exercise*

1. $m = 0.153$; $c = 4.80$
2. $E = 1.50$; $r = 0.82$
3. $m = 0.039$; $c = 20.04$
4. $a = 0.295$; $b = 28.0$

Activity

(iii) $a \approx 1.93$; $b \approx 6$

Page 89 *Exercise*

1. (i) 0.65 or 3.85 (ii) –1.62 or 0.62 (iii) –3.73 or –0.27
   (iv) ± 2.45 (v) 0.44 or 4.56 (vi) 1.13 or 3.54
   (vii) 0.47, 1.65 or 3.88 (viii) 4.15

Page 91 *Exercise*

1. (i) (2.67, 4.33) (ii) (2.80, 3.20)
   (iii) (1.38, 2.38); (3.62, 4.62)
   (iv) (–2.81, 1.09); (2.31, 3.66)
   (v) (1.38, 3.62); (3.62, 1.38)
   (vi) (–1.49, –2.98); (1.66, 3.32); (2.83, 5.66)
2. (i) Process A; £200 (ii) 233; cost = £167
3. 1.40 s, 4.20 m
4. (ii) $x = 36.2$, $y = 27.6$ or $x = 13.8$, $y = 72.4$
5. (i) During the 48th month (ii) Diana's

Page 93 *Exercise*

1. (i) $\frac{2}{3}$ ms^{-2} (ii) 0 (iii) –1 ms^{-2}
2. (i) 48.5 million (ii) 2166 (iii) 900 000 per year
3. (ii) Approximately £28 per item (iii) –£10 per item
4. (i) Approximately 60 m/min^{-1}
   (ii) Approximately 160 m/min^{-1} (iii) 5.6 (iv) 7.2

Activity

(i) 0815 (approx); 0.85 (ii) $12\frac{1}{2}$ hours
(iii) 1130 (iv) – 0.9 m/h^{-1}

Page 95 *Exercise*

1. (i) 75 m (ii) 50 m (iii) 275 m
2. (i) 3
3. (ii) Approximately 2200 microcoulomb
4. (ii) Approximately 114 dm^3

Activity

(ii) 2.1, 3.7, 4.5, 4.5, 3.7, 2.1; Area ≃ 41.2
(iii) Overestimate: convex curve, so $S > H$ each time.

Page 103 *Exercise*

1.

No. of defective items	0	1	2	3	4
No. of batches	7	11	7	4	1

2.

No. of spelling errors	No. of students
1–5	7
6–10	7
11–15	3
16–20	3
21–25	1

3. One possibility is

Attendance	No. of games
0 – 4999	16
5000 – 9999	8
10 000 – 14 999	4
15 000 – 19 999	1
20 000 – 24 999	1
25 000 – 29 999	3
30 000 – 34 999	1
35 000 – 39 999	3
40 000 – 44 999	1

4. (ii)

Time parked (nearest minute)	1	2	3	4	5	6
Number of drivers	1	3	9	2	3	2

5. One possibility is

Weekly wage	No. of students
£35.00 – £39.99	9
£40.00 – £44.99	6
£45.00 – £49.99	4
£50.00 – £54.99	3
£55.00 – £59.99	2

Activities

2. (i) 8 (ii) 70%

Page 105 *Exercise*

1. (i) 150 (ii) Flyway, 225 (iii) X is Wingside, Y is Airland
2. (ii) 1988 (iii) 2 (iv) 1996
3. (i) (a) 125 (b) 212 (ii) X is 1987, Y is 1983 (iii) 1992
4. (i) (c)

Page 107 *Exercise*

2. (i) 9.9 (ii) 11.3 (iii) 3:1 (iv) 1568 (2.8 × 560)
4. (i) 126 000 (ii) Hertsmere
   (iii) Stevenage and Three Rivers (iv) 40 000

Page 109 *Exercise*

2. (ii) (a) fat 31.25% (b) starch 30%
   (c) sugar 26.25% (d) protein 12.5%
5. (i) 15 (ii) 10 (iii) 5 (iv) 20 (v) D

Page 111 *Exercise*

1. Music 136°, Dogs 104°, Domestic activities 44°, Voices 24°, DIY 16°, Others 36°
2. (i) 50% (ii) $\frac{1}{9}$ (iii) 14% (iv) 1:6
4. 4.47 cm
5. (i) (a) 10% (b) 2 000 000 (ii) Social security
   (iii) Fell by 6% but increased in number
   (iv) 1.2 times greater

Page 113 *Exercise*

1. (ii) $2000 \leq c < 2500$
2. (i) 9–11 (ii) 12–14 (iii) 11 minutes 30 seconds (v) 12–14
3. (i) (a) 24.5 and 29.5 (b) 29.5 and 34.5
   (iv) One possibility is $x = 29.5$, $y = 49.5$
4. (ii) 12 (iii) 2

Page 115 *Exercise*

1. (i)

Insurance premium (£)	Cumulative frequency
100	0
200	7
300	24
400	48
500	81
600	103
700	113
800	118
900	120

   (ii) £330 (iii) 14 (iv) £555 (v) 11
2. (i) The number of births in 2000 was down by 82 000 (11%) compared to 1993. In 1993 the majority of births were to women in their twenties but by 2000 the proportion of births to women in their thirties was showing a significant increase.
   (ii) 330 000
   (iii) 31 000
   (iv) 31.5 years
   (v) Approximately 14%

Activities

1. (i) $a = 70$ (ii) $b = 41$ (iii) $c = 81$ (iv) $d = 55$
   (v) $e = 90$ (vi) $f = 32$ (vii) $g = 12$

Page 117 *Exercise*

1. 9.62
2. (i) (a) £400 (b) £175 (c) £100
3. (i) There are only 6 courts
   (ii) (a) 4.46 (b) 6
4. (i) (a) 200.75 (b) 183.25
5. (i) (a) 110 (b) 210
6. (i) (a) 4.5 (b) 4 (c) 7 (iii) 36 000 (if the mean is used)
7. 1 260 000 **8.** 5 min 00 s

Page 119 *Exercise*

1. 1.65
2. (i) 2.39
   (ii) 2800
3. (i) (a) 10.75 (b) 10.75
4. 64.8
5. (i) Class frequencies are 3, 5, 4, 2, 2, 1, 0, 0, 1
   (ii) £7000, £9000, ...
   (iii) £11 444.44 ie £11 400 (3 s.f.)
   (v) £11 277.78 ie £11 300 (3 s.f.)

Page 121 *Exercise*

1. (i) Hastings 75.8°F, Casablanca 74.7°F
   (ii) Hastings 16°F, Casablanca 4°F
2. (i) Machine 1 (a) 252.5 (b) 9
   Machine 2 (a) 249.25 (b) 66
   Machine 3 (a) 260.625 (b) 23
3. (ii) (a) £265 (b) £220 (c) £310 (iii) £90
4. (ii) (a) 211 mg dl^{-1} (b) 191 mg dl^{-1} (c) 232 mg dl^{-1}
   (iii) 41 mg dl^{-1}
5. (i) 1997, £223.60; 1998, £233.70; 1999, £242.80; 2000, £248.90
   (ii) 2000 (iii) (a) £368.20 (b) £382.15 (c) £395.90
   (d) £408.35

Page 123 *Exercise*

1. (ii) Positive linear correlation
2. (ii) No correlation
3. (ii) Negative linear correlation
   (v) Just over 50 000
4. (ii) Very slight positive correlation
   (iv) Strong positive correlation

Page 127 *Exercise*

1. (iii) 61.25, 60.0, 57.5, 58.75, 57.5, 55.0, 53.75
2. (ii) 15.0, 18.0, 18.5, 19.67, 18.67, 17.17, 16.0
3. (i) (a) Le Jardin 17 000, Le Château 26 000
   (b) Le Jardin 2000, Le Château 56 000
4. (iii) 45.5, 47.0, 49.0 50.75, 51.25, 52.25, 55.5, 54.5, 56.5

Page 129 *Exercise*

1. (i) $\frac{1}{6}$ (ii) $\frac{1}{3}$ (iii) $\frac{5}{6}$
2. (i) $\frac{1}{13}$ (ii) $\frac{1}{4}$ (iii) $\frac{1}{26}$ (iv) $\frac{3}{4}$
3. (i) $\frac{27}{50}$ (ii) $\frac{3}{25}$ (iii) $\frac{39}{50}$
4. (i) $\frac{1}{4}$ (ii) $\frac{1}{16}$
5. (i) $\frac{19}{100}$ (ii) (a) unchanged (b) increased to $\frac{1}{5}$
6. (i) Agree (ii) Disagree (iii) Disagree

Page 131 *Exercise*

1. (i) $\frac{1}{2}$ (ii) $\frac{1}{5}$ (iii) $\frac{3}{10}$ **2.** $\frac{29}{30}$
3. (i) 50 (ii) (a) $\frac{17}{25}$ (b) $\frac{3}{5}$
4. (i) 0.135 **5.** No

Page 133 *Exercise*

1. (i) $\frac{2}{13}$ (ii) $\frac{7}{13}$ (iii) $\frac{11}{26}$
2. (i) $\frac{2}{3}$ (ii) $\frac{1}{2}$ (iii) $\frac{1}{3}$ (iv) $\frac{5}{6}$
3. (i) $\frac{3}{16}$ (ii) $\frac{3}{4}$ (iii) $\frac{1}{4}$ (iv) $\frac{9}{16}$ (v) $\frac{15}{16}$
4. $\frac{1}{4}$
5. (i) $\frac{3}{4}$ (ii) $\frac{7}{20}$ (iii) $\frac{9}{10}$

Page 135 *Exercise*

1. (i) $\frac{361}{400}$ (ii) $\frac{19}{200}$
2. (i) $\frac{49}{100}$ (ii) $\frac{21}{50}$ (iii) $\frac{9}{100}$
3. (i) $\frac{1}{2}$ (ii) (a) $\frac{1}{4}$ (b) $\frac{1}{2}$
4. (i) $\frac{1}{36}$ (ii) $\frac{1}{18}$ (iii) $\frac{1}{6}$
5. (i) $\frac{2}{5}$ (ii) $\frac{1}{2}$ (iii) $\frac{2}{25}$
6. (i) $\frac{4}{25}$ (ii) $\frac{48}{125}$ (iii) $\frac{529}{625}$
7. $\frac{1}{200}$

Page 137 *Exercise*

1. 0.58
2. (i) 0.72 (ii) 0.26 (iii) 0.02
3. (i) 0.02 (ii) 0.26
4. (i) $\frac{1}{495}$ (ii) $\frac{19}{198}$
5. (i) $\frac{1}{12}$ (ii) $\frac{23}{144}$
6. (i) $\frac{6}{25}$ (ii) $\frac{8}{25}$
7. (i) $\frac{9}{25}$ (ii) $\frac{12}{25}$ (iii) $\frac{24}{125}$

Page 141 *Exercise*

1. (i) $a = 42°$ (ii) $b = c = 65°$ (iii) $d = 56°$ (iv) $e = 68°$, $f = 44°$ (v) $g = h = i = 60°$ (vi) $j = 104°$
2. (i) $a = 7$ (ii) $b = 12$ (iii) $c = 5, d = 24$ (iv) $e = 13.5, f = 24$ (v) $g = 20, h = 7$ (vi) $i = 13, j = 14$

Page 143 *Exercise*

1. 5 cm
2. 10 cm
3. 5 cm
4. 15 cm
5. 29 cm
6. 20 cm
7. 9 cm
8. 1 cm
9. 7 cm
10. (i) Yes (ii) No (iii) Yes (iv) Yes

Page 145 *Exercise*

1. 192 yards
2. 24 m
3. 6 m
4. 0.65 cm
5. (i) 15 cm (ii) 17 cm
6. (i) 5 (ii) 13
7. 1.8 km
8. (i) 80.6 m (i) 250 m

Page 147 *Exercise*

1. & 2. (i) 1.28 (ii) 0.71 (iii) 0.50 (iv) 1.00 (v) 0.95
3. (i) 30° (ii) 45.0° (iii) 32.0° (iv) 20.7° (v) 53.1° (vi) 45° (vii) 68.2° (viii) 30° (ix) 39.2° (x) 0°

Page 149 *Exercise*

1. $a = 26.6°$ $b = 63.4°$
2. $e = 78.5°$ $f = 11.5°$
3. $i = 21.5°$ $j = 68.5°$
4. $c = 53.1°$ $d = 36.9°$
5. $g = 21.8°$ $h = 68.2°$
6. $k = 62.8°$ $l = 27.2°$

Activities

1. (i) 26.6° (ii) 45° (iii) 63.4°
2. (i) 0.2° (ii) 5.7° (iii) 2.9°
3. 0.056°
4. (a) Horizontal
   (b) sloping from top left to bottom right
5. Approximately $\frac{1}{12}$

Page 151 *Exercise*

1. 6 cm
2. 5.20 cm
3. 12.14 miles
4. 5.07 mm
5. 9.11 cm
6. 9.06 m
7. 5.60 m
8. 8.39 cm
9. 5.41 cm
10. 18.12 mm
11. 33.7°
12. 10.89 m
13. 3.72 km
14. 7.63 m
15. (clockwise from O) (0, 14.14), (0, 34.14), (14.14, 48.28), (34.14, 48.28), (48.28, 34.14), (48.28, 14.14), (34.14, 0) (14.14, 0)
    A (14.14, 14.14), B (14.14, 34.14) C (34.14, 34.14), D (34.14, 14.14)

Page 153 *Exercise*

1. (i) 17.5° (ii) 1.43 m (iii) 0.9 m
2. (i) 11.47 km (ii) 9.40 km (iii) 33.8 minutes
3. (ii) 40°, 70° (iii) 89.7 cm (iv) 2242 cm^2
4. (i) 4.33 m (ii) 57.2° (iii) 10.23 m
5. (i) 26.7 m (ii) 34.7 m (iii) 32.7°

Page 155 *Exercise*

1. (i) 2.83 km (ii) 2.83 km (iii) 5.83 km (iv) 6.48 km (v) 25.9° (vi) 115.9°
2. (i) 43.3 N (ii) 70.7 N (iii) 114.0 N (iv) 25 N (v) 70.7 N (vi) 45.7 N (vii) 122.8 N (viii) 21.8°

Investigation

All six walkers end up at the Inn, 8.33 km from X on a bearing of 041.3°.

Page 157 *Exercise*

1. (i) 5, 53.1° (ii) $\sqrt{50}$, 135°
   (iii) $\sqrt{10}$, 198.4° (iv) $\sqrt{20}$, 296.6°
   (v) 3, 90°
2. (i) $\sqrt{40}$, 71.6° (ii) $\sqrt{26}$, 168.7°
   (iii) 10, 306.9° (iv) $\sqrt{5}$, 206.6°
3. $\overrightarrow{OP} = \begin{pmatrix} 6.93 \\ 4.00 \end{pmatrix}$ $\overrightarrow{OQ} = \begin{pmatrix} -4.50 \\ 7.79 \end{pmatrix}$
   $\overrightarrow{OR} = \begin{pmatrix} -4.02 \\ -5.73 \end{pmatrix}$ $\overrightarrow{OS} = \begin{pmatrix} 7.43 \\ -6.69 \end{pmatrix}$
4. ±12
5. ±7

Page 159 *Exercise*

1. (i) $8\mathbf{i} + 5\mathbf{j}$ (ii) $-\mathbf{i} + 6\mathbf{j}$
   (iii) $9\mathbf{i} - \mathbf{j}$ (iv) $9\mathbf{i} + 21\mathbf{j}$
   (v) $\mathbf{i} - 6\mathbf{j}$ (vi) $\mathbf{i} - 3\mathbf{j}$
   (vii) $10\mathbf{i} - 4\mathbf{j}$ (viii) $12\mathbf{i} + 6\mathbf{j}$
2. (i) $\sqrt{82}$, 353.7° (ii) $\sqrt{10}$, 288.4°

3. (i) $\begin{pmatrix}1\\10\end{pmatrix}$ (ii) $\begin{pmatrix}-6\\5\end{pmatrix}$
   (iii) $\begin{pmatrix}8\\28\end{pmatrix}$ (iv) $\begin{pmatrix}3\\-9\end{pmatrix}$
   (v) $\begin{pmatrix}7\\5\end{pmatrix}$ (vi) $\begin{pmatrix}2\\-6\end{pmatrix}$
   (vii) $\begin{pmatrix}2.5\\-1\end{pmatrix}$ (viii) $\begin{pmatrix}8\\2\end{pmatrix}$
4. (i) $a = 5, b = 1$ (ii) $k = 2, n = -5$
5. (i) $a = -3, b = -3$ (ii) $a = 5, b = 1$
6. (i) 10**i** + **j** (ii) 11**i** – 8**j** (iii) 2**i** + 21**j** (iv) –14**i** + 22**j**
   (v) 4**i** – 10**j** (vi) 12**i** + 9**j** (vii) 41**i** – 5**j** (viii) 5**i** + 7**j**
7. (i) $\begin{pmatrix}7\\11\end{pmatrix}$ (ii) $\begin{pmatrix}8\\-4\end{pmatrix}$ (iii) $\begin{pmatrix}5\\31\end{pmatrix}$ (iv) $\begin{pmatrix}-17\\0\end{pmatrix}$ (v) $\begin{pmatrix}-6\\1\end{pmatrix}$
   (vi) $\begin{pmatrix}-23\\-32\end{pmatrix}$ (vii) $\begin{pmatrix}0.5\\5\end{pmatrix}$ (viii) $\begin{pmatrix}10\\19\end{pmatrix}$
8. (i) $\sqrt{193}$, 239.7° (ii) $\sqrt{493}$, 82.2°
9. (i) 0 (= 0**i** + 0**j**)
   (iii) When **a** + **b** + **c** = 0 then **a**, **b** and **c** form a triangle.
10. (i) $t = 2, u = -7$ (ii) $t = 23, u = 5$

Investigation
(ii) 21.8° (iii) 2 min 24 s (iv) 160 m

Page 163 *Exercise*
1. (i) (b) 26.9 m, 21.8°
   (ii) (b) 27 m, 25.1 m, 21.7°
   (iii) (a) 14.5 m (b) 12.0°
2. (ii) 4.62 m (iii) 24.8° (iv) 42.6°
3. (i) 90°, 20.6°, 13.2°, 16.7° (ii) 49.5°
4. (i) 3.69 m (ii) 2.83 m (iii) 2.37 m (iv) 48.1 °

Page 167 *Exercise*
1. –0.5000
2. 0.5000
3. –0.7071
4. 1.0000
5. –0.5000
6. 0.7071
7. 1.7321
8. –0.8660
9. –0.8660
10. –0.5000

Page 169 *Exercise*
1. 10.9
2. 9.18
3. 16.1
4. 11.5
5. (i) 30.7 cm (ii) 69° (iii) 32.2 cm
6. 14.5 m
7. 55,6°
8. 67.7°
9. 35.8°
10. 40.0°
11. (i) 31.4° (ii) 39.6° (iii) 35.3 cm
12. 60.6°

Page 171 *Exercise*
1. 6.28
2. 14.9
3. 8.76
4. 47.9
5. 16.5 m
6. 8.21 cm
7. 60°
8. 80.9°
9. 62.8°
10. 101.9°
11. (i) 51.0° (ii) 87.3° (iii) 41.8°
12. 36.4°

Page 173 *Exercise*
1. 63.2°
2. 7.53
3. 17.2
4. 100.3°
5. 10.2 km
6. 3.01 m
7. 91.1°
8. 11.5
9. 47.0
10. 67.3°
11. AC = 1.79 km, BC = 2.61 km
12. 46.6°

Index